Analytical Uses of Immobilized Biological Compounds for Detection, Medical and Industrial Uses

NATO ASI Series

Advanced Science Institutes Series

A series presenting the results of activities sponsored by the NATO Science Committee, which aims at the dissemination of advanced scientific and technological knowledge, with a view to strengthening links between scientific communities.

The series is published by an international board of publishers in conjunction with the NATO Scientific Affairs Division

A Life Sciences B Physics	Plenum Publishing Corporation London and New York
C Mathematical and Physical Sciences	D. Reidel Publishing Company Dordrecht, Boston, Lancaster and Tokyo
D Behavioural and Social Sciences E Applied Sciences	Martinus Nijhoff Publishers Dordrecht, Boston, Lancaster
F Computer and Systems Sciences G Ecological Sciences H Cell Biology	Springer Verlag Berlin, Heidelberg, New York, London, Paris, and Tokyo

Series C: Mathematical and Physical Sciences Vol. 226

Analytical Uses of Immobilized Biological Compounds for Detection, Medical and Industrial Uses

edited by

George G. Guilbault

Department of Chemistry,
University of New Orleans, Louisiana, U.S.A.

and

Marco Mascini

Department of Chemistry,
University of Florence, Italy

D. Reidel Publishing Company

Dordrecht / Boston / Lancaster / Tokyo

Published in cooperation with NATO Scientific Affairs Division

Proceedings of the NATO Advanced Research Workshop on
Analytical Uses of Immobilized Biological Compounds
for Detection, Medical and Industrial Uses
Florence, Italy
May 4-8, 1987

Library of Congress Cataloging in Publication Data

NATO Advanced Workshop on Analytical Uses of Immobilized Biological Compounds for Detection, Medical, and Industrial Uses (1987: Florence, Italy)
Analytical uses of immobilized biological compounds for detection, medical, and industrial uses / edited by George G. Guilbault and Marco Mascini.
p. cm. — (NATO ASI series. Series C, Mathematical and physical sciences; vol. 226)
"Proceedings of the NATO Advanced Research Workshop on Analytical Uses of Immobilized Biological Compounds for Detection, Medical, and Industrial Uses, Florence, Italy, May 4–8, 1987"—T.p. verso.
"Published in cooperation with NATO Scientific Affairs Division."
Includes index.

ISBN-13: 978-94-010-7804-7 e-ISBN-13: 978-94-009-2895-4
DOI:10.1007/ 978-94-009-2895-4

1. Biochemistry—Technique—Congresses. 2. Immobilized enzymes—Congresses. 3. Immobilized proteins—Congresses. 4. Biosensors—Congresses. I. Guilbault, George G. II. Mascini, Marco. III. North Atlantic Treaty Organization. Scientific Affairs Division. IV. Title. V. Series: NATO ASI series. Series C, Mathematical and physical sciences; no. 226.
QP519.7.N38 1987
660′.63—dc 19 87–32076
CIP

Published by D. Reidel Publishing Company
P.O. Box 17, 3300 AA Dordrecht, Holland

Sold and distributed in the U.S.A. and Canada
by Kluwer Academic Publishers,
101 Philip Drive, Norwell, MA 02061, U.S.A.

In all other countries, sold and distributed
by Kluwer Academic Publishers Group,
P.O. Box 322, 3300 AH Dordrecht, Holland

D. Reidel Publishing Company is a member of the Kluwer Academic Publishers Group

Softcover reprint of the hardcover 1st edition 1988

DEDICATION

To Susan and Mareza

Table of Contents

PREFACE

On May 4-8, 1987, a NATO Advanced Research Workshop on the Analytical Uses of Immobilized Biological Compounds was held in Florence, Italy. The Director of the Workshop was Professor George G. Guilbault of the University of New Orleans, and the Co-Director was Professor Marco Mascini of the University of Florence

It was the purpose of this meeting to assemble scientists from all NATO Countries with an interest in immobilized biological compounds, to discuss

- methods of immobilization
- properties of immobilized compounds
- enzyme electrodes and biosensors
- optical devices utilizing immobilized enzymes
- microbial sensors and clinical uses of immobilized enzymes
- flow injection analysis using enzymes
- immobilized biological compounds in chemical defense detection
- pharmaceutical analysis
- uses in industrial analysis
- enzyme reactors
- air pollution detectors
- immunosensors
- medical uses and applications
- solid state and FET sensors

Goals to be achieved by the conference were

- to permit an exchange of views and experience in all these areas
- to review and critically assess the state-of-the-art in these fields
- to set guidelines for future research and establish collaborative projects between scientists in NATO laboratories in the above areas.

Thirty-seven lectures were given by 36 speakers in all of the above areas. Sessions were devoted to (1) methods and properties of immobilized enzymes (2) clinical and pharmaceutical analysis (3) enzyme, bio- and microbial sensors, (4) defense applications, (5) solid state/FET devices, (6) optrodes and spectroscopic applications, (7) immunosensors, (8) flow injection analysis, and (9) industrial and analytical applications. Finally, two hours were devoted to an open discussion of future status, new directions and joint projects.

This book is a publication of most of the lectures given at this workshop.

We wish to thank for their financial support of this conference: Cassa di Risparmio di Firenze, Eli Lilly (Indianapolis), Esacontrol SpA (Genova), Instrumentation Laboratory (Milano), Universal Sensors (New Orleans), Universita degli Studi di Firenza and especially NATO for the ARW grant that made the Conference possible. We also thank our organizing committee (Drs. Coulet, Patriarche, Campanella and Palleschi) for the organization and local support, and D. Moscone, R. Pillotan and S. Salleri for providing the secretarial services that made the Conference run smoothly. Finally, the typing support of Mrs. Gayle Barlow is gratefully acknowledged.

CONCLUSIONS

In the last afternoon, a round table discussion of selected speakers (see program), the Directors and audience centered on the status of the field of Analytical Uses of Immobilized Biological Compounds, cooperation between labs and the future.

Several participants expressed appreciation to the Conference for the possibility to meet and develop contacts with scientists of other NATO countries doing similar research. Collaborative projects were established between (1) the Universities of Rome, Florence and New Orleans on NAD/NADH dependent dehydrogenases, modified electrodes and immobilization techniques for substrates of clinical interest, (2) the University of Porto in Portugal and Barcelona in Spain in new electrodes for whole blood flow analysis, (3) U. S. Defense Labs in Edgewood, Md. and the University of Rouen in France on Receptor Electrodes, (4) University of Brussels, Belgium, and New Orleans on New Methods for Artificial Sweeterners and Carbohydrates, (5) University of Cincinnati and University of Lund on Modified Electrodes, and (6) Technical University of Denmark and University of New Orleans on Flow Injection Analysis of Artificial Sweeterners.

Discussion on new areas was devoted to (1) Receptor Electrodes and Methods for use in Defense and Analysis (2) Modified Electrodes as better analytical methods (3) Miniaturized solid state (FET type electrodes) for the sensors of the future (4) Immobilized Enzymes in Medical Shunts (5) Enzyme Immunoassay and Immuno Probes and (6) Optrodes - now finally getting established.

Because of the highly productive nature of the Conference, it was decided to have a second NATO Conference in 1990.

LIST OF CONTRIBUTORS

Salvador Alegret - Department of Chemistry, Universitat Autonoma de Barcelona, E08193 Bellaterra, Spain

M. A. Angellotti - Institute of Analytical Chemistry, University of Bologna, Italy

Piet Bergveld - Department of Electrical Engineering, Twente University, P. O. Box 217, 7500 AE Enschede, The Netherlands

L. J. Blum - Laboratoire de Genie Enzymatique - Atelier de Biotechnologie CNRS - Universite Claude Bernard Lyon 1-43, Bd du 11 Novembre 1918, 69622 Villeurbanne Cedex, France

L. Campanella - Department of Chemistry, University of Rome I, "La Sapienza", Piazzale A. Moro, 5, Rome, Italy

G. Carrea - Institute of Chemistry of Hormones, CNR Milan, Italy

Pierre R. Coulet - Laboratoire de Genie Enzymatique - Atelier de Biotechnologie, CNRS- Universite Claude Bernard, Lyon 1-43, Bd du 11 Novembre 1918, 69622 Villeurbanne Cedex, France

M. Alves da Silva - Department of Chemistry, University of Coimbra, 3000 Coimbra, Portugal

Christopher C. Feistel - Liston Scientific Corp., Newport Beach, CA

Larry Gayleard - Liston Scientific Corp., Newport Beach, CA

M. H. Gil - Department of Chemistry, University of Coimbra, 3000 Coimbra, Portugal

S. Girotti - Institute of Chemical Sciences, University of Bologna, Italy

Masao Gotoh - NOK Corporation, Fujisawa 251, Japan

G. G. Guilbault - Department of Chemistry, University of New Orleans, New Orleans, LA

Elizabeth Hall - The Biotechnology Centre, University of Cambridge, Downing Street, Cambridge, CB2 3EF, United Kingdom

H. B. Halsall - Department of Chemistry, University of Cincinnati, Cincinnati, Ohio 45221-0172

Elo H. Hansen - Chemistry Department A, Technical University of Denmark, Building 207, DK-2800 Lyngby, Denmark

W. R. Heineman - Department of Chemistry, University of Cincinnati, Cincinnati, Ohio 45221-0172

A. Atilla Hincal - Hacettepe University, Faculty of Pharmacy, Pharmaceutical Technology Department, 06 100 Ankara, Turkey

Isao Karube - Research Laboratory of Resources Utilization, Tokyo University of Technology, 4259 Nagatsuta-cho, Midori-ku, Yokohama, 227 Japan

H. Suheyla Kas - Hacettepe University, Faculty of Pharmacy, Pharmaceutical Technology Department, 06 100 Ankara, Turkey

J.-M. Kauffmann - Free University of Brussels, Institute of Pharmacy, Campus Plaine 205/6, Bd du Triomphe, Brussels, Belgium

Izumi Kubo - Research Laboratory of Resources Utilization, Tokyo University of Technology, Nagatsuta-cho, Midori-ku, Yokohama, 227 Japan

Glenn J. Lubrano - Universal Sensors, 5258 Veterans Blvd., Suite D, Metairie, LA 70006

C. D. Luther - Liston Scientific Corp., Newport Beach, CA

Adelio A. S. C. Machado - Chemistry Department, Faculty of Science, P4000 Porto, Portugal

Angelo Manzoni - Research and Development Department, Instrumentation Laboratory S.P.A., Viale Monza 338, 20128 Milan, Italy

Marco Mascini - Istituto Chemica Analitica, Universita di Firenza, Via Gino Capponi 9, 50121 Firenze, Italy

J. Ngeh-Ngwainbi - Universal Sensors, 5258 Veterans Blvd., Suite D, Metairie, LA 70006

Valerie M. Owen - Biotechnology Centre, Cranfield Institute of Technology, Cranfield, Beds MK43 OAL, United Kingdom

G. Palleschi - Department of Chemical Science and Technology, 2nd University of Rome, Via O. Raimondo 8, 00173 Rome, Italy

Gaston J. Patriarche - Free University of Brussels (U.L.B.), Institute of Pharmacy, Campus Plaine 205/6, Bd du Triomphe Brussels, Belgium

Ayse Kevser Piskin - Hacettepe University, Faculty of Medicine, Medical Biology Department, Ankara, Turkey

Erhan Piskin - Hacettepe University, Chemical Engineering Department, Ankara, Turkey

R. Rizzoli - Institute of Chemical Sciences, University of Bologna, Italy

A. Roda - Institute of Analytical Chemistry, University of Bologna, Italy

P. Simoni - Institute of Analytical Chemistry, University of Bologna, Italy

Koji Sode - Research Laboratory of Resources Utilization, Tokyo Institute of Technology, 4259 Nagatsuta-cho, Midori-ku, Yokohama, 227 Japan

Charles F. Stemburg - Liston Scientific Corp., Newport Beach, CA

J. D. R. Thomas - Applied Chemistry Department, Redwood Building, UWIST, P. O. Box 13, Cardiff CF1 3XF, Wales, United Kingdom

M. Tomassetti - Department of Chemistry, University of Rome, "La Sapienza", Piazzale A. Moro, 5, Rome, Italy

Anthony P. F. Turner - Bioelectronics Division, Biotechnology Centre, Cranfield Institute of Technology, Cranfield, Bedfordshire, MK43 OAL, United Kingdom

P. Vagdama - Department of Clinical Biochemistry, The Medical School, Newcastle University, Newcastle Upon Tyne, NE2 4HH, United Kingdom

Bart H. Van der Schoot - Department of Electrical Engineering, Twente University, P. O. Box 217, 7500 AE Enschede, The Netherlands

B. J. Vinicke' - Free University of Brussels (U.L.B.), Institute of Pharmacy, Campus Plaine 205/6, Bd du Triomphe, Brussels, Belgium

J.-C. Vire - Free University of Brussels (U.L.B.), Institute of Pharmacy, Campus Plaine 205/6, Bd du Triomphe, Brussels, Belgium

Howard H. Weetall - Cornig Glass Works, Ciba-Corning Diagnostics Corp., 63 North Street, Medfield, MA 02052

Otto S. Wolfbeis - Analytical Division, Institute of Organic Chemistry, KF-University, 8010 Graz, Austria

ENZYMES IMMOBILIZED ON INORGANIC SUPPORTS

Howard H. Weetall
Corning Glass Works
Ciba-Corning Diagnostics Corp.
63 North Street
Medfield, MA 02052

ABSTRACT. Enzymes immobilized on inorganic supports by covalent attachment show unique characteristics with regard to pH optimum, thermal profile and kinetics. Covalent attachment of enzymes requires that the support first be treated with a silane coupling agent whereby organic functional groups can be covalently attached to the support. Following this the enzyme may be covalently coupled to the now available organic functional groups on the support.

1. INTRODUCTION

Immobilized enzymes have, over the past twenty years, moved from the laboratory to the industrial plant as a method for the commerical production of food stufs, pharmaceuticals, and fine chemicals. The advantages of enzymes, and in particular immobilized enzymes, for processing, are many. The major advantages are:

1) Specificity of reactions is high.
2) Gentleness of conditions for the chemical reactions are generally better than for most chemical reactions.
3) Temperature ranges are generally between 4°-60°C.
4) Immobilized enzymes are reuseable.
5) Immobilized enzymes allow closer control of the reaction.
6) Immobilized enzyme reactions can be less expensive than some chemical reactions when overall processing costs are considered.

Although this paper will deal mostly with inorganic supports, the characteristics of these supports are, in fact, similar to enzymes attached to organic matrices.

The choice of an inorganic system as a support material was based upon several perceived advantages.

1) Inorganic materials can be prepared in most pore diameters and particle sizes ranging from 30A^{o} - 2000A^{o} pore diameters and 400 U.S. mesh size to anything as large as marbles. This allows one to optimize a process in regards to diffusion limitation, pressure drop

G. G. Guilbault and M. Mascini (eds.),
Analytical Uses of Immobilized Biological Compounds for Detection, Medical and Industrial Uses, 1–16.

and other similar parameters.

2) Inorganic materials are impervious to biological attack by enzymes produced by contaminating and even non-contaminating bacterial systems, as compared to some of the more common carriers such as celluloses, polydextrans, and polyamines.

3) Inorganic materials do not change morphology under different solvent and pH conditions as do many organic polymers.

4) Inorganics can be doped with activator ions which can then be eliminated from the feed. Together these advantages increase the lase and the speed by which scale up is possible, as well as decrease many of the usual scale up and operational problems.

When choosing a support for a specific application it is always wise to know something of the new properties of the support material and the conditions that material will encounter. In the case of inorganic supports we have carried out exhaustive studies on the physical and chemical characteristics of the carriers with regard to the presence of acids, bases, chelates, organic solvents, increasing and decreasing pressure drops; and even sterilizing agents.

With this information available we can choose the carrier vest suited for operation in a higher or lower ionic strength environment of some specific pH, temperature and range of flow-rates.

2. PHYSICAL AND CHEMICAL CHAACTERISTICS

The carriers described here are all porous ceramics. Many of the carriers are similar to those developed by R.A. Messing (1). These materials can be prepared in a variety of pore morphologies and particle sizes. Table 1 gives the physical characteristics of some of these ceramic support materials.

Table 1

Physical characteristics of several ceramic support materials

Composition	Size (US mesh)	Pore diameter range (A°)	Pore diameter average (A°)	Pore volume (ml/g)
TiO_2 98%, MgO 2%	30/45	205-500	410	0.53
SiO_2 75%, Al_2O_3 25%	30/60	205-575	435	0.89
SiO_2 89.3%, ZrO_2 10.7%	30/60	110-575	235	1.30
SiO_2 100%	30/60	185-700	435	0.76
SiO_2 100%*	30/60	310-655	550	2.2
SiO_2 90%, ZrO_2 10%	30/60	185-700	435	0.76
SiO_2 75%, TiO_2 25%	30/45	875-205	465	0.76
Controlled Pore Glass	30/60	450-600	550	0.50

*Modified binder used.

The method used for determining these data is mercury intrusion. The pressure required to cause the mercury to fill the pores of a particle is proportioned to the pore diameter. The volume required determines the pore volume (Figure 1).

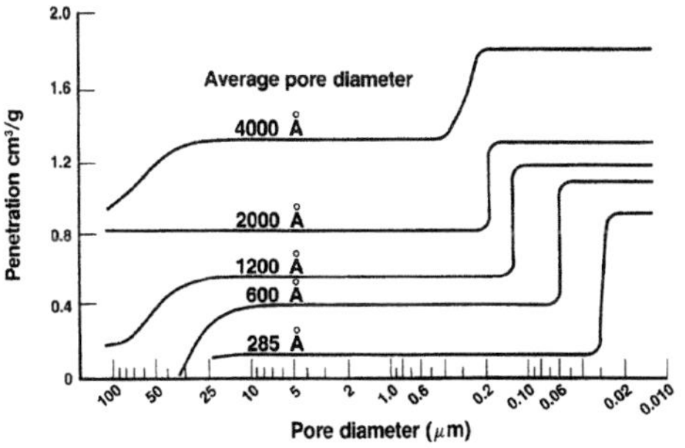

Figure 1. Mercury intrusion data for several controlled pore glasses indicating average pore diameter, pore ranges and pore volumes (penetration, volume per gram of CPG) for each material. (1.0   m = 10000A.)

Chemical durability is determined by carrying out a series of static and dynamic tests utilizing strong acids, alkalies and buffers (2). This type of study aids one in choosing a carrier for a particualr application.

2.1. Covalent Attachment

Inorganic supports can be covalently coupled ot organic materials by a variety of methods. Two of the presently used approaches involve activation with $TiCl_4$ or adsorption of some amino carrying polymer such as poly-L-lysine followed by reaction with the protein of interest.

We have developed and optimized the process which involves the use of intermediate silane coupling agents. The advantages offered by the silane is mainly stability. Since the covalent linkage is directly to the inorganic support, one find that enzyme leakage is non-existent over a pH range of about pH5-8. In addition the cost of silane is far less than poly-L-lysine, the most common aminoopolymer used and far less obnoxious than $TiCl_4$

The most common silane coupling agent used is gamma-amino propyltriethoxysilane. Silane such as this one will ract with any available oxide or hydroxide groups on the surface of glass, ceramics or metal oxides.

The general scheme is given in Figure 2.

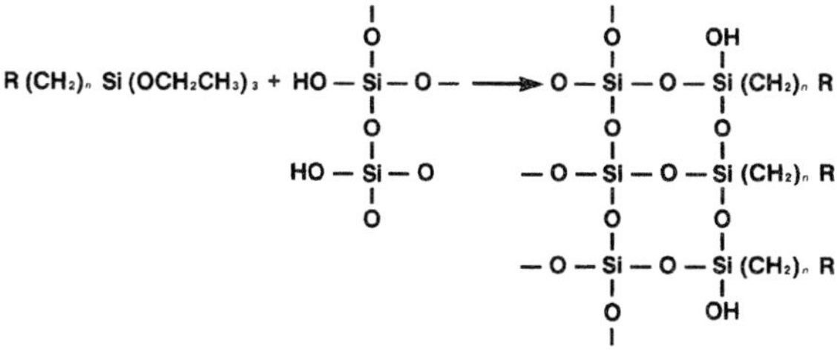

Figure 2. Reaction between gamma-aminopropyl-triethoxysilane and an inorganic support. The ethoxy group reacts with the available silanol groups forming a siloxane linkage. In addition, the silane appears to polymerize over the support surface.

Once an alkylamine group has been attached to the inorganic support it takes only standard organic coupling reactions to covalently bond the protein of interest (3). Figures 3-8 show a few of the most common methods.

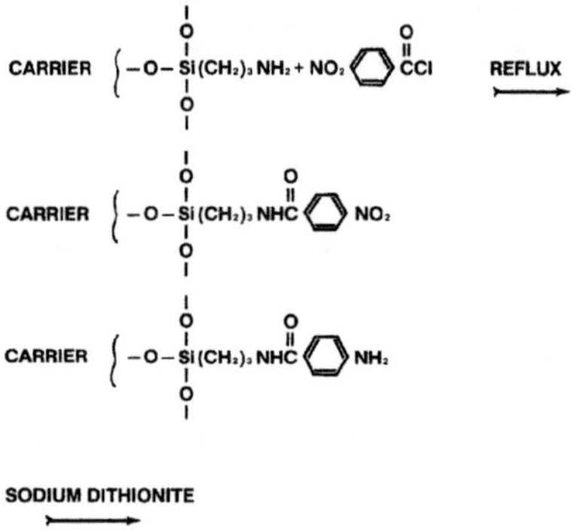

Figure 3. Conversions of the alkylamine carrier to an arylamine can be accomplished via reaction with p-nitrobenzylchloride followed by reduction.

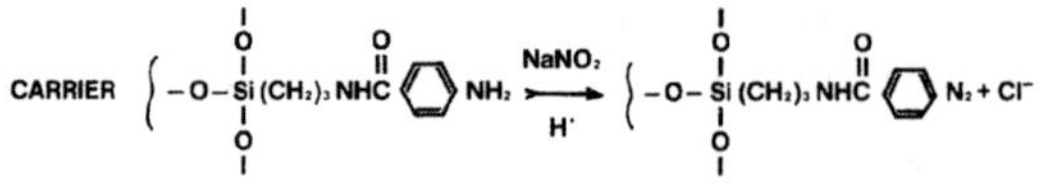

Figure 4. The arylamine carrier can be characterized and coupled to any protein via available tyrosine residues.

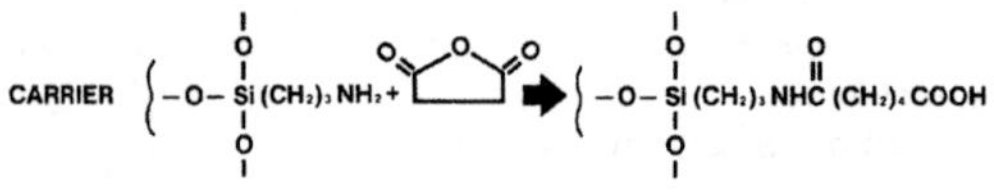

Figure 5. An alkylamine carrier can easily be conveted to a carboxylated carrier by reaction with succinic anhydride.

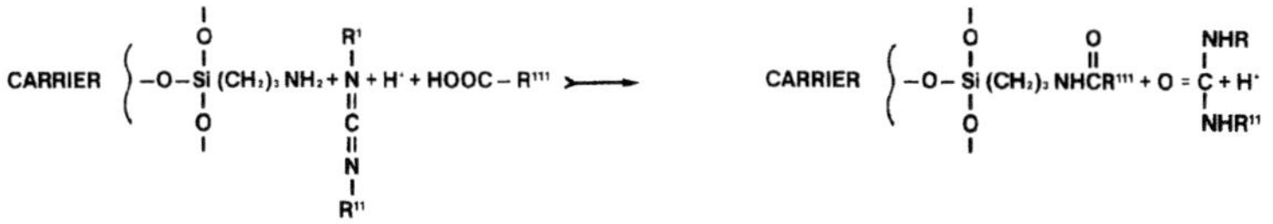

Figure 6. Either the alkylamine or the carboxylated amine in the presence of carbodiimide will couple to protein. However best results are achieved with the carbodiimide activated carboxylated derivative (not shown).

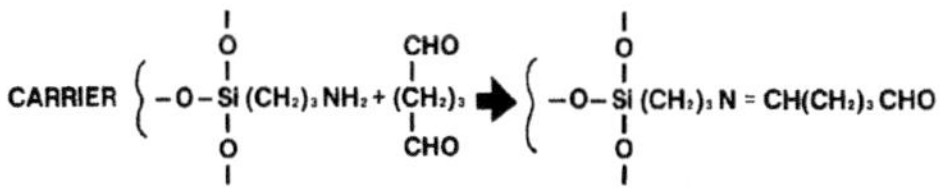

Figure 7. One of the most gentle coupling methods uses glutaraldehyde. The reaction shown here, in a simplified form, will combine the alkylamine support directly to an amine on the enzyme through this bifunctional coupling agent.

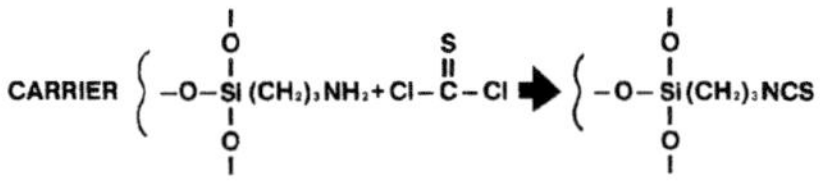

Figure 8. The use of thiophosgene to activate the alkylamine is another way of combining two amine groups. However, because of the toxic nature of the coupling agent it is not highly recommended.

The covalent attachment of enzyme to an inorganic support is not only dependent on the method of attachment for high efficiency and activity recovery. The efficiency of coupling is also dependent on the quantity of enzyme offered. In the case of lactase covalently coupled to three different inorganic supports, it was found that the highest efficiency was obtained at the lowest enzyme levels offered. (Figure 9).

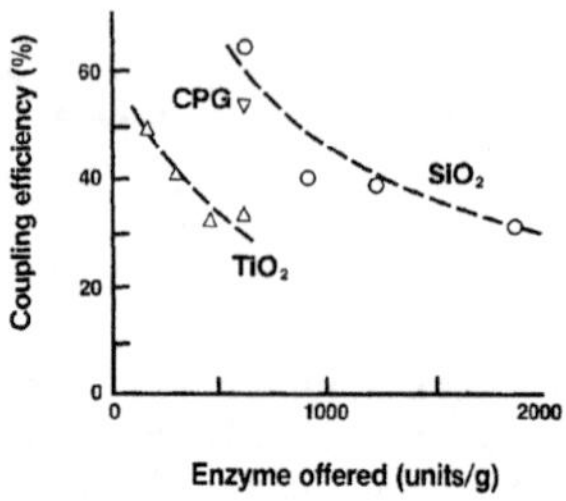

Figure 9. Coupling efficiency of B-galactosidase to three inorganic supports. The figure shows that highest coupling efficiences are achieved at the lowest concentration of offered enzyme, even though the total bound activity may be higher at the highest concentrations of offered enzyme.

The phenonmenon is most likely the result of decreased diffusion control of substrate and product. If the enzyme concentration is high enough that the rate of substrate diffusion to the enzyme and product diffusion away from the enzyme may act as the controlling steps on the overall reaction. Decreasing enzyme loading decreases the chances of substrate limitation controlling the reaction.

Once a protein is covalently coupled to the support it is necessary to compare its physio-chemical characteristics to that in the native state. In the case of immobilized enzymes one must investigate such parameters as pH profile, thermal profile, kinetics and even substrate specificity. With many immobilized enzyme systems several of the parameters will appear changed from the native state.

2.2 pH Profile

The immobilization of an enzyme on any charged support will generally cause a shift in the apparent pH optimum. The mechanism for this observed pH shift has been elegantly explained by Goldstein et al (4,5). Using highly charged enzyme supports they were able to show the microenvironmental effects of these supports on the enzymes' pH characteristics. Highly negatively charged carriers created a lower pH at the boundary layer between the carrier and the bulk solution. Thus the enzyme finds itself in a more acidic environment as compared to the bulk of the solution. The opposite result occurs with positively charged supports. Even in the cases of uncharged supports, the overall charges on the enzyme may be sufficient to cause changes in the apparent pH optimum. Generally, however, the greater the charge on the support, the greater the effect, particularly with charged substrates. Figure 10 shows the shift int he apparent pH optimum for a yeast B-galactosidase covalently coupled to a porous ceramic support.

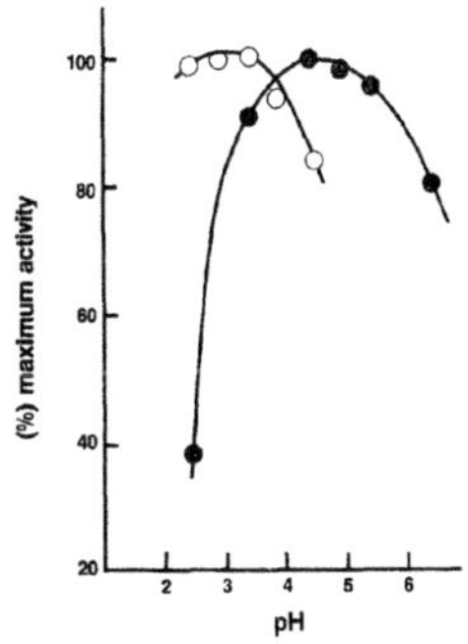

Figure 10. pH optimum of soluble lactase (filled circles) and the enzyme immobilized on porous SiO_2 (open circles).

2.3 Enzyme Stability

Upon immobilization enzymes can change their thermal characteristics. These changes generally cause apparent improvement in the thermal stability of the enzyme.

Enzyme denaturation generally occurs as a result of changes in tertiary structure, oxidation of some lieable groups at high temperature or some other physical modification of the protein. In some cases immobilization can increase thermal stability by increasing the molecule's rigidity (6) thus preventing changes in the molecules tertiary structure upon heating.

In addition to thermal effects, large changes in the apparent kinetics can occur upon immobilization. Generally the apparent K_M of an immobilized enzyme is increased over that of a soluble form. This increase is usually related to a variety of parameters including charge of substrate and carrier, diffusion effects, and even the mode of operation. Table 2 gives an example of K_M values for several immobilized enzymes. The mechanisms involved in causing the apparent changes in kinetics are complex, but generally can be described via mathematical terms involving the usual engineering paramters covered in any good text on heterogeneous catalysis.

Table 2

Some K_m values for soluble and immobilized enzymes on inorganic supports

			K_m, x 10^3M		
Enzyme	Substrate	Support	Soluble	Immobilized	Ref
Lactate	Lactate	CPG	8.0	5.9	7
dehydrogenase	Pyruvate	CPG	0.125	0.39	7
Invertase	Sucrose	CPG	0.49	0.49	8
Glucose oxidase	Glucose	SiO_2	7.7	6.8	8
Alkaline phosphatase	PNPP	SiO_2	0.10	2.90	8
L-amino acid oxidase	L-leucine	SiO_2	1.00	4.00	8
Arylsulfatase	CPG	PNPS	1.85	1.57	9
Aminopeptidase	LpNA	CPG	0.28	1.07	10
Trypsin	TAME	CPG	0.12	1.53	11

CPG, controlled pore glass; PNPS, p-nitrophenyl sulfate; LpNA, L-leucine-p-nitroanilide; TAME, α-N-tosyl-L-arginine methyl ester; PNPP, p-nitrophenyl phosphate.

Enzymes which show excellent thermal stability and even more favorable kinetics do not necessarily show excellent operational stability. Operational stability is not only a function of thermal stability but includes such factors as carrier and feed composition, dimensional and chemical stability, presence of organic and inorganic inhibitors, activator concentrations, and operating pH values. Figure 11 gives an excellent example of stability of glucoamylase vs operating temperature.

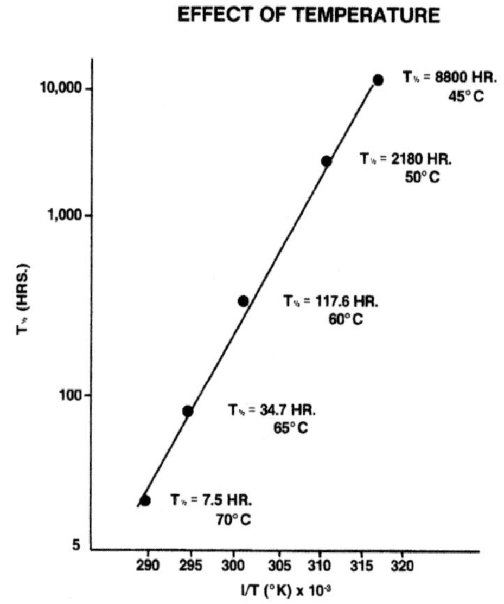

Figure 11. Half life values vs 1/T (K^{o}) for glucoamylase immobilized on porous SiO_2 with 30% w/w starch solutions as substrate. The columns were operated on a continuous basis 24 h/day for over 100 days. Half-lives at 45 and 50°C were extrapolated by regression analysis from collected data at those temperatures.

In the case of feed composition, one example which demostrate the effect can be found in the case of immobilized lactose. The operational half-life varies from a low of 5-7 days at 23°C using whole cheese whey to 100 days at 23°C using 5% lactose in buffer. The difference can be attributed to the presence of contaminants which can inhibit or bind to the enzyme in some fashion.

As previously mentioned, significant decreases in enzyme activity can be caused by denaturation processes. However, in addition the elution of enzyme from the support vs time can also contribute significantly to operational half-life. Even covalently coupled enzymes can be eluted due to bond scisson (12).

In our own laboratories we have run comparison studies on several enzymes using our inorganic carriers vs several organic supports including resins, celluloses and cross-linked polymers. In all cases some enzyme "leakage" occurs. The extent of the leakage varies as one would expect depending on the support and the operating conditions. There are no hard and fast rules.

In addition to our studies, several studies have been carried out on matrix stability and effects of coupling methodology on inorganic supports (13,14).

Significant differences in enzyme stability were observed for lactate dehydrogenase immobilized on controlled pore glass via azo linkage and Shiff base linkage with the Shiff bases being the more stable (14).

Similarly immobilized enzymes in many cases show increased stability to proteolyses. This type of increased stability shown in Table 3 has been used as the basis for a patented cleaning procedure for the Nutrosearch, Incorp. industrial immobilized lactase columns to remove accumulated absorbed proteinous material from the column surface.

Immobilization can also modify the characteristics of an enzyme such that its properties became uniquely changed.

Table 3

Stability of immobilized enzymes covalently coupled to controlled pore glass to protease hydrolysis[18]

		Activity	
Enzyme	Protease	Native state	Immobilized
Lactate dehydrogenase	Subtilisin	55	80
Isocitrate dehydrogenase	Trypsin	0	90
Aminoacylase	Pronase	68	88

In 1970 Neurath and I (15) found that DNAase when immbolized on porous SiO_2 lost its requirement for the addition of a divalent metal for activation. Analyses of all buffers, substrates and the support revealed no divalent metals capable ofactivating this enzyme.

Obivously some modification of the enzymes tertiary structure occured upon immobilization. Royer et al elegantly showed that major changes in tertiary structure can indeed occur upon immobiization to an inorganic support (16). Using trypsin immoblized in the presence or absence or benzylarginine ethyl ester (BAEE) they were able to cause a major change in the enzyme's kinetic constants and stability. He explained this phenomenon by hypothesizing that the enzyme, when coupled in the presence of the substrate, binds and is frozen in a more active configuration.

With this study in mind we examined trypsin immobilized on an inorganic support in the presence and absence of BAEE and observed an increase in stability in high concentrations of organic solvents (17). The results of the experiments showed large improvement in the enzyme kinetics in high concentrations of DMSO, methanol, acetone, propanol and ethanol. Solvent concentrations were as high as 95%. Also the enzyme showed high reaction rates for peptide hydrolysis in solvent concentration exceeding 90%, in many cases.

Recently we have used tannase immobilized on a porous ceramin to produce gallic acid esters of methanol to octanol as well as esters of several diols, by using the alcohol or diol as solvent and substrate (18). In these studies high conversion rates were obtained at organic solvent concentrations which were for all practical purposes greater than 99%.

3. COMMERCIAL APPLICATIONS

Inorganic supports have been successfully used for scale up of at least three enzyme systems: Glucoamylase, glucose isomerase and lactase.

3.1 Glucoamylase

Immobilized glucoamylase was scaled up at Iowa State University over an 18 month period. During that time optimum operating conditions were determined and a plant designed for the hydrolysis of 10 million pounds of starch per year (19). However, due to the belief that the process would add overall only minor value to this system it was never commercialized.

In recent years British Charcoal and other companies have commercialized glucoamylase for finishing in dextrose crystallization. The immobilized enzyme has also found use in high fructose corn syrup processing (HFBS). Tables 4a and 4b give an overall description of these processes.

Table 4a

DEXTROSE CRYSTALLIZATION:

THE LIQUID REMAINING AFTER INITIAL CRYSTALLIZATION CAN BE PASSED THROUGH AN IME TO INCREASE DEXTROSE BY 2 - 3%. THIS ALLOWS FOR HIGHER CRYSTALLIZATION LEVELS ON NEXT CRYSTALLIZATION.

Table 4b

HFCS RAFFINATE:

THE FRACTION RECOVERED AFTER CHROMATOGRAPHIC RECOVERY OF FRUCTOSE FROM A 42% HFCS SUBSTRATE.

IME TREATMENT OF THE RAFFINATE INCREASES DEXTROSE BY 2 - 3% (FROM SMALL OLIGOSACCHRIDES). THIS CAN THEN BE BLENDED BACK INTO THE MAINSTREAM FOR ISOMERIZATION.

PRODUCTION OF 360 TONS/HR OF DRY SUSTANCE REQUIRES 7200m³ TANK VOLUME. IME REQUIRES 6m³ FOR SAME PRODUCTIVITY LEVEL.

3.2 Glucose Isomerase

At present there are several potential HFCS processes (Table 5). The conversion of glucose to fructose is the alrgest immobilized enzyme process used today. The process flow is relatively similar for all the HFCS commerical processes.

Table 5

GLUCOSE ISOMERASE SYSTEMS

Company	Source	Immobilization Techinque	Reactor
Car Mi U S 3 625 828 3 654 081 3 779 869	Streptomyces olivaceus	Granular particles containing whole cells cross linked with glutaraldehyde	Column reactor
Clinton Corn Processing U S 3 788 945 3 623 953	Streptomyces wedmorensis	Enzyme immobilized on DEAE cellulose via ion exchange	Shallow bed reactors
Corning Glass Works U S 3 847 740 3 850,751 3 868,304 3 982,997 3 992,329	CPC patented strains of Streptomyces olivochromogenes	Enzyme adsorbed on controlled pore alumina carrier (0 46 mm D APD 140 220 A) (carrier can be regenerated)	Column reactors
Gist Brocades U S 3 834 848 3 838,007	Actinoplanes missouriensis (licensed from Anheuser Busch)	Gel entrapment of crude mycelium fixed enzyme followed by cross linking with glutaraldehyde	Various configurations (columns up to 6 m high bed depth)
ICI U S 3 645 848 3 821 086 3 935 068	Arthrobacter (licensed from R J Reynolds Tobacco Co)	Flocculating agent used to immobilize within cells solid paste then extruded and dried into cylindrical pellets	Column reactors (up to 5 m bed depths)
Novo B R 1 381 387	Bacillus coagulans	Enzyme mixed with inorganic diluent and formed into solid	Column reactors (up to 4 5 m depths)

* BREWERS DIGEST — September 1986

Table 6 shows a typical HFCS flow.

Table 6

HFCS FLOW

RAW SYRUP (PRETREATED WITH ACTIVATED CARBON & ION EXCHANGE)
↓
CONCENTRATE TO 45% d.s.
↓
ION EXCHANGE TO REMOVE Ca^{+2}
↓
ISOMERIZATION pH 8.3 - 8.5, 62 - 65°C.
↓
pH TO 4 - 5
↓
CHARCOAL FILTERED
↓
ION EXCHANGED
↓
CONCENTRATED (72.5% d.s.)
↓
STORAGE

Characteristics of the two most popular processes are shown in Tables 7 and 8. Table 7 shows the Novo process. Table 8 decribes the isomerization conditions in this porcess.

Table 7

SWEETZYME - NOVO

MIN. ACT.	150 IU/g
MIN. LIFE	1650 HRS. @ 61°C
DENSITY	1.4 cc/g
BULK DENSITY (WET)	270 - 310 Kg/m^3
NOMINAL PART. SIZE	300 - 1000μ
PRODUCTIVITY	> 2000 Kg d.s./Kg IME

Table 8

ISOMERIZATION CONDITIONS

d.s. CONTENT	35 - 45%
GLUCOSE CONTENT	92%
pH (IN)	8.3 - 8.5
TEMP (IN)	62 - 65°C
ACTIVATOR	4 x 10^{-4} Mg^{+2}
RESIDENCE TIME	0.5 - 3 HRS.
COLUMN:	DOWNFLOW, FIXED BED
TEMPERATURE:	58 - 60°C
pH INLET:	7.5 (AT 25°C)
pH OUTLET:	6.8 MINIMUM (AT 25°C)
Mg:	3mM
Ca:	<3 ppm
d.s.:	40 - 45%
DEXTROSE CONTENT:	93 - 96%
COLUMN SIZE:	5m HIGH (16 FT) 0.6m WIDE (2FT)
PRODUCTIVITY:	$T_{1/2}$ IS 55 DAYS AT 60°C. 4500 Kg FRUCTOSE/Kg DRY ENZYME.
BED VOLUMES/HR:	1 - 1.3 AT DAY 0 0.7 AT DAY 100
FINAL SYRUP COMPOSITION:	42% FRUCTOSE 54% GLUCOSE 2.5% DISACCHRIDES 1.5% TRISACCHRIDES

COST PER Kg FRUCTOSE SYRUP: 15 - 45¢

Table 9

GLUCOSE ISOMERASE GIST-BROCADES

ORGANISM:	ACTINOPLANES MISSOURIENSIS
IMMOBILIZATION:	GELATIN ENTRAPPMENT WITH GLUTARALDEHYDE CROSS-LINKING
SIZE:	1.5 MM PARTICLES, 750 Kg/m^3 BULK DENSITY
ACTIVITY:	1000 UNITS GI/Kg (1 UNIT ISOMERIZES 1g GLUCOSE TO FRUCTOSE/HR AT 45%) d.s., 60°C, pH 7.5.

3.3 Other Processes

There are a variety of other immobilized enzyme processes in use around the world. Some of the processes are summarized in Table 10. Of those listed only a few utilize an excess of $1,000,000 worth of enzyme (Table 10).

Table 10

PROCESSES

	STATE	WHERE
ETHANOL PRODUCTION	COMMERCIAL	KYOWA HAKKO
L - AMINO ACID RESOLUTION	COMMERCIAL	TANABE SEIYAKU
L - METHIONINE	COMMERCIAL	TANABE SEIYAKU
L - VALINE	COMMERCIAL	TANABE SEIYAKU
L - PHENYLALANINE	COMMERCIAL	TANABE SEIYAKU
L - ASPARTIC ACID (CELLS)	COMMERCIAL	TANABE SEIYAKU
FUMARIC ACID ⇌ (ASPARTASE)	L - ASPARTIC ACID	TANABE SEIYAKU
L - MALIC ACID	COMMERCIAL	
FUMARIC ACID ⇌ (FUMERASE)	L - MALIC ACID	TANABE SEIYAKU
L - ALANINE	COMMERCIAL	
L - ASPARTIC ACID → (L - ASPARTATE β - DECARBOXYLASE)	L - ALANINE	TANABE SEIYAKU
AMINO ACYLASE		AMANO
HYDANTOINASE		AMANO
LACTASE		CGW
NITRILASE		NITTO
PENICILLIN G ACYLASE		BAYER
PENICILLIN V ACYLASE		NOVO

Table 11

CURRENT 1 MM BIOPROCESSES

SUBSTRATE	PRODUCT	BIOCATALYST
STARCH	GLUCOSE/FRUCTOSE	α-AMYLASE GA, GI
STARCH	CYCLODEXTRINS	ALK-AMYLASE
LACTOSE	GLUCOSE/GALACTOSE	LACTASE
BENZPENNICILLIN	6-APA	PENICILLINASE
FUMARIC ACID	L-ASPARTIC	ASPARTASE
FUMARIC ACID	MALIC ACID	FUMARASE
L-D-ACYL AA	L-AA	AMINO ACYLASE
D, L-AMINO-CAPROLACTAM	L-LYSINE	L-AMINOLACTAM HYDROLASE
D, L-5-(P-HYDROZY-PHENYL)-HYDANTOIN	D-N-CARBAMOYL-P-HYDROXYPHENYL-GLYCINE	D-HYDRO-DANTOINASE

The promise of immobilized technology, although not yet completely met, is still potentially large. Many processes are still in development (Table 12).

Table 12

IMMOBILIZED SYSTEM
NOT YET COMMERCIALIZED

L - GLUTAMIC ACID	GLUTATHIONE
L - CITRULLINE	UROCANIC ACID
L - ISOLEUCINE	LACTIC ACID
L - TRYPTOPHAN	CITRIC ACID
L - LYSINE	GLUCONIC ACID
L - PHENYLALANINE	2-KETOGLUCONIC ACID
L - TYROSINE	

ISOPROPANOL
N-BUTANOL
2, 3-BUTANEDIOL

CYANIDE DEGRADATION
INULIN HYDROLYSIS

The additional capabilities presented by the use of genetically engineered enzymes should initiate a new round of enzyme process developments. Inorganic supports are presently being used in several other processes which are either still in research or in the pilot plant. The use of inorganic supports still offers the researcher many advantages over most organic materials due to its physical characteristics. New methods in preparing carriers of SiO_2, ZrO_2 and TiO_2 have reduced carrier costs to levels comparable with organic supports.

This investigator expects to see new and varied uses for inorganic materials with increasing frequency, not only for enzyme immobilization but for immunoadsorption, affinity chromatography, trace metal recovery and solid phase catalysis.

4. REFERENCES

1. Messing, R.A. US Patent 3, 892, 580.
2. Eaton, D.A. (1974) in Immobilized Biochemicals and Affinity Chromatography, (Dunlap, R.B., ed.) pp. 241-258, Plenum Press.
3. Weetall, H.H. and Filbert, A.M. (1974) Methods Enzymol. 34B, 59-72.
4. Goldstein, L., Levin, Y. and Katchalsi, E. (1964) Biochemistry 3, 1913.
5. Goldstein, L. and Katchalski, E. (1968) Anal. Chem. 243, 375.
6. Weetall, H.H., Havawala, N.B., Garfinkel, H.M., Buehl, W.M. and Baum, G. (1974) Biotechnol. Bioeng. 16, 169.
7. Dixon, J.E. and F.E. Stolzenbach, (1974) ISr. J. Chem. 12, 529.
8. Weetall, H.H. (1974) Anal. Chem. 46, 602A.
9. Weetall, H.H. (1971) Res. Dev. 22, 18.
10. Royer, G.P. and Andrews J.P. (1973) J. Biol. Chem. 248, 1807.
11. Kleine, R., Spengenberg, P. and Flemming, C. (1976) Hoppe-Seyler's Z. Physiol. Chem. 357, 629.
12. Lasch, J. and Koelach, R. (1978) Eur. J. Biochem. 82, 181.
13. Cart, P.W. and Bowers, L.D. (1980) in Immobilized Enzymes in Analytical and Clinical Chemistry pp. 148-196, Wiley.
14. Dixon, J.E., Stolzenbach, F.E., Ching-Lun, T.L. and Kaplan, N.O. (1974) Isr. J. Chem. 12, 529.
15. Neurath, A.R. and Weetall, H.H. (1970) FEBS Lett 8, 253.
16. Royer, G.P. and Uy, R. (1973) J. Biol. Chem. 248, 2627.
17. Weetall, H.H. and Vann, W.P. (1976) Biotechnol. Bioeng. 18, 105.
18. Weetall, H.H. (1985) Appl. Biochem. Biotechnol. 11, 25.
19. Weetall, H.H., Vann, W.P., Pitcher, W.H. Lee, D.D., Lee, Y.Y. and Tsao, G.T. (1976) Methods Enzymol. 54, 776.

Thank you to Ms. Ruth Dhionis for her aid in preparing this manuscript.

Portions of this article were previously published in Trends in Biotechnology, Vol. 3, 11, 276-280, (1985).

IMMOBILIZATION OF BIOAGENTS BY RADIATION

Ayşe Kevser Pişkin
Hacettepe University, Faculty of Medicine, Medical Biology Department, Ankara, Turkey

ABSTRACT: One of the contributions of radiation to biotechnology is immobilization of bioagents. Immobilization by radiation is a physical entrapment method whereby polymerization of carriers is achieved at very low temperatures. The method is applicable to immobilization of a large variety of agents including cells, enzymes, hormones and drugs. By selecting the proper monomers and/or polymers and adjusting the radiation dose it is possible to obtain carriers in a great variety of shapes and structures. Entrapment into hydrophilic polymers leads to higher recovery of enzyme activity due to augmentation of substrate diffusion. Since, polymerization is achieved without using catalyst or crosslinkers at low temperatures enzyme inactivation is minimized. This paper deals with the details of the method and gives examples of its application in immobilizing bioagents.

INTRODUCTION

There is no doubt that application of radiation to health related fields is one of the most important scientific endeavors of the century. The use of radiolabelled molecules have led to great discoveries in biology. A great number of diseases can be diagnosed with the help of radioisotopes. Radiotheraphy is still a major therapy for cancer. Two recent applications of radiaton to biotechnology are: (1) Surface modification of polymeric materials to improve their biocompatibility (2) immobilization of bioactive substances.

Great potential of immobilized bioagents have led to development of new immobilization techniques. Among these, immobilization by radiation is a rapid method for physical entrapment within polymeric matrices.

MAJOR FEATURES OF THE METHOD

1. Immobilization is achieved usually by means of γ - radiation
2. Starting materials are generally glass forming monomers and/or polymers.
3. Immobilization is carried out at very low temperatures (-80 °C)
4. Chemical catalyzers or cross-linking agents are not regulred
5. Immobilized conjugates in a variety of shapes and structures can be obtained
6. Conjugates produced by this method are already sterile

G. G. Guilbault and M. Mascini (eds.),
Analytical Uses of Immobilized Biological Compounds for Detection, Medical and Industrial Uses, 17–23.

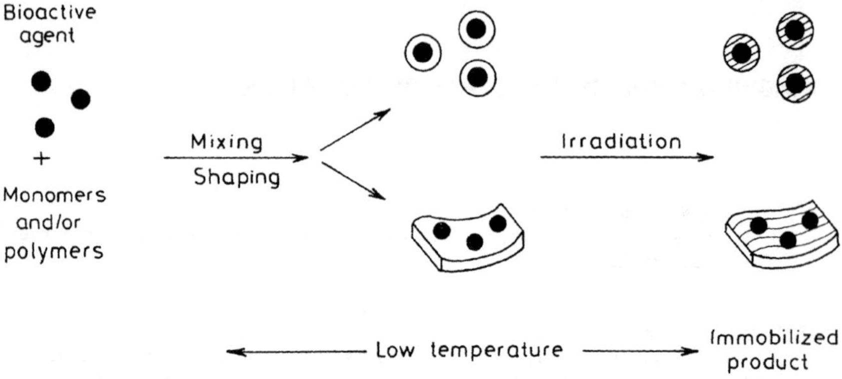

Figure 1. Schematic representations of radiation immobilization technique

7. Method is applicable to a variety of bioagents e. g. cells, hormones, enzymes and drugs
8. It is an easy and rapid method

The possibility of working at low temperatures brings certain advantages. These are: (1) Glass forming monomers are highly viscous in supercooled state. This allows retention of the shape of carrier until the immobilization is complete. (2) The enzyme is protected against thermal inactivation, (3) Radiation damage is minimized in compact, frozen state. A simplified diagram to show the main procedure in radiation polymerization is given in Figure 1.

EXAMPLES OF APPLICATION

Enzymes that are important in food industry, such as, glucose oxydase, aminoacylase and invertase are immobilized by γ - radiation (1-3). It is reported that enzymes retain up to 80 % of their original activity when radiation dose is kept under 4 Mrads.

We have used radiation technique to prepare immobilized urease in bead shaped polymeric gels (4, 5). As shown in Figure 2, urease was first added to the mixture of glass forming monomers, namely acrylic acid and methylmethacrylate and polyethyleneglycol as pore making agent. The mixture was then dropped into supercooled petroleum ether. Frozen spheres were irradiated in a ^{60}Co source to achieve polymerization and copolymerization.

The effects of radiation dose and the ratio of monomers and polyethyleneglycol on enzyme activity were investigated. The results obtained from these studies are given in Table I and II. The optimum dose of radiation was around 2.75 Mrads and the ratio of AA/ MMA and PEG was 40/ 40/ 20 respectively.

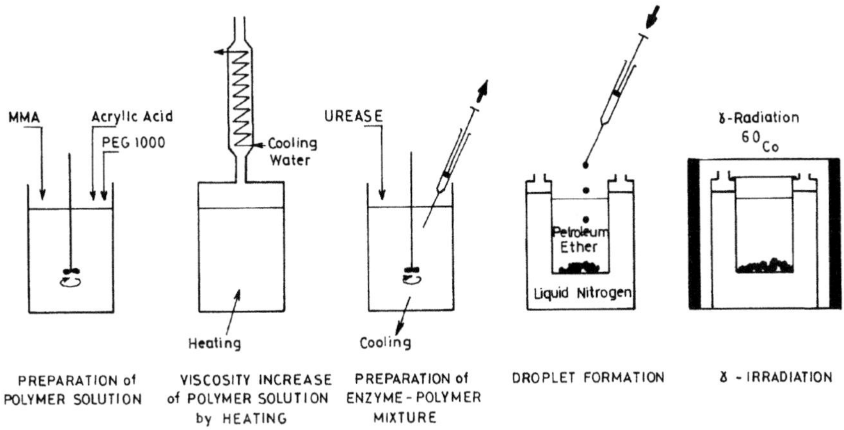

Figure 2. Immobilization of urease by radiation at low temperature.

TABLE I

Effect of Radiation Dose on the Activity Yield of Immobilized Urease

Dose period, h	Activity yield, %
10	70 ∓ 1.0 (there is some leakage)
15	65 ∓ 0.6
30	48 ∓ 0.2
45	35 ∓ 0.2

The urease containing spherical gels had high ability to swell in aqueous solutions. This enabled obtaining high appearent enzyme activity (65 %). Fig 3 (a) and (b) shows the scanning electrone micrographs of gel spheres before and after they are swollen in aqueous media.

As an extension of this approach, an anticancer drug mitomycin-C was immobilized for sustained release by Kiremitçi and Pişkin [6]. Zero order release

TABLE II

Effect of Ratio Components on the Formation of Gel Spheres and Immobilized Enzyme Activity

Sample	Ratio of components (%v/v)			Property of polymer sphere	Immobilized enzyme activity yield (%)	Leakage
	Acrylic Acid	MMA	PEG			
1	60	20	20	white sponge like gel spheres	50 ∓ 2.1	-
2	20	60	20	hard spheres	30 ∓ 1.9	-
3	40	40	20	white sponge like gel spheres	65 ∓ 0.8	-
4	20	40	40	curdy composite	47 ∓ 4.2	+
5	20	20	60	soft gel (nonspherical)	74 ∓ 3.0	+
6	40	20	40	soft gel (nonspherical)	80 ∓ 2.4	+

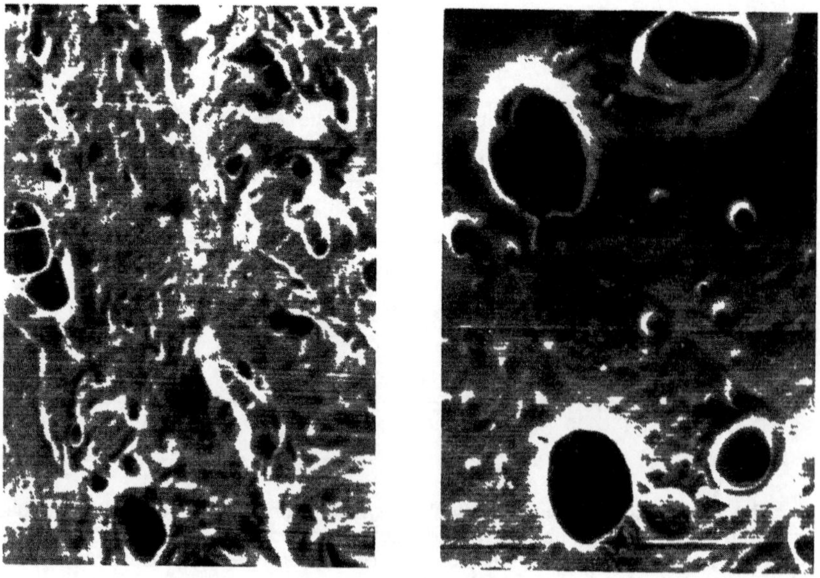

(a) (b)

Figure 3. Scanning electron micrographs of gel spheres: (a) ňon swollen, (ƀ) Swollen

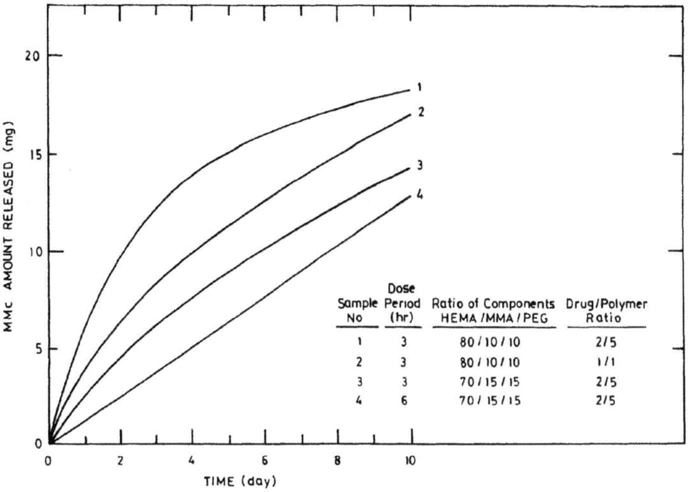

Figure 4. Effect of ratio of components, radiation dose and drug/ polymer ratio on Mitomycin-C release (From Kiremitçi, M., et. al.).

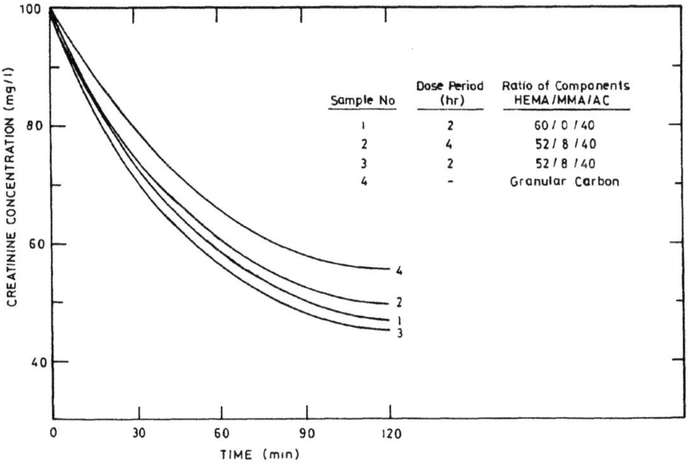

Figure 5. Adsorption spectra of creatinine on granular and immobilized powder activated carbon.

was achived at hydroxyethylmethacrylate (HEMA)/methylmethacrylate (MMA) and polyethyleneglycol (PEG) ratio of 70/ 15/ 15 respectively by 1 Mrads of gamma irradiation (Figure 4).

In another group of experiments we have immobilized powder activated carbon in polymeric beads by radiation. The main purpose of this approach was to make powder activated carbon, which has very high adsorptive capacity, usable in hemoperfusion columns. As shown in Figure 5, these activated carbon preparations exhibited higher adsorptive power compared to granular carbon (7).

Majority of the work on immobilization of bioagents by radiation comes from Kaetsu's group. In an impressive series of studies they have enlarged and elaborated this tecnique for biomedical applications. They have immobilized microbial cells, enzymes, hormones and drugs in polymeric gels. They have also designed systems such as irradiating hydrophobic monomers suspended in an aqueous enzyme solution to yield a porous gel with enzyme coated pores (8, 9). The pore structure has a strong influence on the accessibility of the enzyme thus its activity.

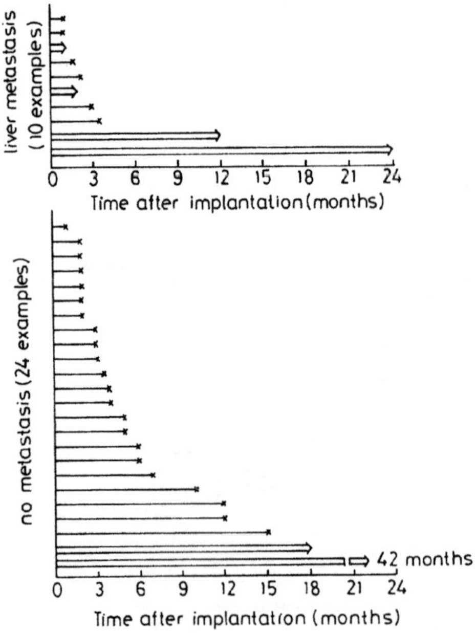

Figure 6. Prolonged survival of pancreas cancer patients implanted with the needle composites including Mitomycin-C. —— died; ==== living (From Kaetsu, I., 10)

The same group applied, anticancer drugs immobilized by radiation polymerization, to the treatment of more than 250 patients over Japan. According to the results of these clinical trials, mainly three kinds of effects were noted. These were: relief of severe pain, decrease in tumor mass and prolonged survival [10]. Needle composites prepared by the same technique containing Mitomycin-C (MMC) was implanted in pancreas cancer patients. Some of these patients have shown better survival rates (Figure 6).

The radiation immobilization technique requires interdisciplinary corporation and it can contribute a great deal to biotechnology and bioengineering.

REFERENCES

1. Maeda, H.,Yamamuchi, A., and Suzuki, H., 1973, Biochem. Biophys.Acta. 315:18
2. Kawashima, K., and Umeda, K., 1974, Biotechnol. Bioeng. XVI: 609
3. Kaetsu, I., Kumakura, M., Yoshida, M., 1979, Biotecnol. Bioeng., 21: 847.
4. Pişkin, K., Arca, E., Pişkin, E., 1984, Applied Biochem. Biotechnol., 10:73.
5. Arca, E., Pişkin, K., Pişkin, E., 1983, Life Support Systems, 1:262.
6. Kiremitçi, M., Öncü, A., Denizli, A., Pişkin, E., 1986, A drug delivery system from the polymer matrices containing PHEMA/PMMA/PEG' in Progress in Artif. Organs, Nose,Y., Kjellstrand, C., Ivanovich, P., eds., ISAOPress, Cleveland
7. Pişkin, E., Kiremitçi, M., Pişkin, K., Arca, E., Evren, V., Mutlu, M.,1985, Proc. 5 th Symp. Hemoperfusion and Artificial Organs, Tianjin.
8. Kumakura, M.,Yoshida, M., Kaetsu, I. 1979, Biotechnol. Bioeng. XXI: 679.
9. Kaetsu, I., 1981, Radiat. Phys. Chem., 18 (1-2): 343.
10. Kaetsu, I., 1984, Proc. 5th Int.Meeting on Radiation Proessing , San Diego.

STRUCTURAL AND TRANSPORT PROPERTIES OF POLYMERIC CARRIERS FOR BIOAGENTS AND IMMOBILIZATION TECHNIQUES AND THEIR APPLICATIONS

Erhan Pişkin
Hacettepe University, Chemical Engineering Department,
Ankara, Turkey

ABSTRACT: A wide variety of biological materials, are incorporated into or onto a carrier, generally a polymeric material, for a wide variety uses. Biotechnology utilizes immobilized enzymes or microorganisms for the production of chemicals and many important pharmaceuticals Bioactive agent-carrier composites are employed in clinical medicine for treating many diseases including enzyme deficiencies, autoimmune diseases, etc. Controlled administration of chemicals or drug, which are usually incorporated with polymeric carriers, is widely publicized in the field of medicine, pharmacy, agriculture, etc. With extraordinary selectivities and specifities, biological compounds which are immobilized onto surfaces of carrier beads, are used in bioaffinity chromatograpy for identification, purification and seperation of many important biological substances. Novel biosensor includes bioactive agent-carrier composites. Synthetic polymers make up by far the broadest and most diverse class of carriers used in these novel applications. This is mainly because synthetic polymers are available with such a variety forms and also because they may be fabricated readily into complex shapes and structures. Here, the structural and transport properties of polymeric carriers for immobilization of biological compounds, immobilization techniques and relevant applications are presented briefly.

1. POLYMERIC CARRIERS

There is a wide variety of materials which are used for immobilization of bioagents for many diverse applications. High polymers are used widely in the construction of carriers for this purpose. Because, polymers are available a wide variety of compositions, properties and forms (see Table I) and low in cost. They may be fabricated readily into complex shapes and structures (see Table II). They exhibit high chemical resistance, and acceptable stability and compatibility within biological environment. They offer acceptable thermal and mechanical resistances.

Table I. Forms of Polymeric Carriers.

- Solid
  - Hard and Glassy
  - Soft and Rubbery
  - Gel
- Liquid-Like
- Suspensions and Emulsions

G. G. Guilbault and M. Mascini (eds.),
Analytical Uses of Immobilized Biological Compounds for Detection, Medical and Industrial Uses, 25–38.

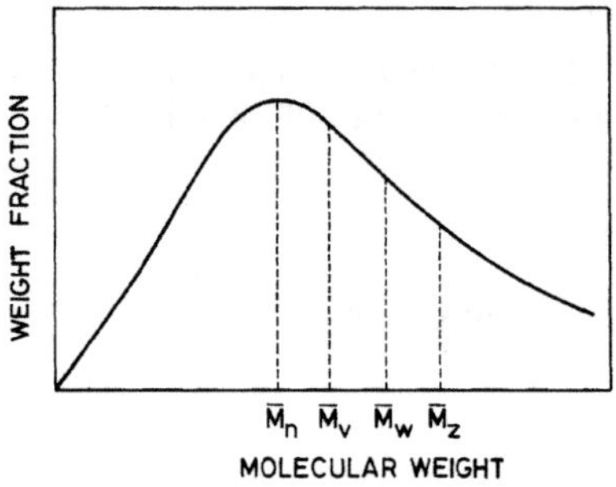

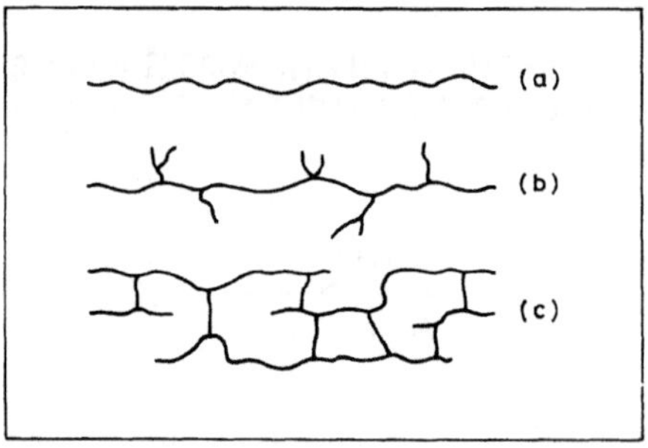

Figure 1. Molecular Weight Distribution.

Figure 2. Polymer Chains: (a) Lineer; (b) Branched; (c) Cross-linked.

Table II. Shapes of Polymeric Carriers.

- Hollow Fibers
- Tubes
- Films, Membranes, Disks
- Micro-to Macro Capsules, Beads
- Fibers, Rods
- Other Complex Shapes

1.1. Characterization of Polymeric Carriers

Polymers are made by joining together thousands of small molecular units known as monomers. Polymers are in general complex mixtures of molecules with different molecular weights (MWs). In other terms, a polymer is a polydisperse mixture of species with a distribution of MWs (Figure 1). Both processing and end-use characteristics of polymeric materials are influenced not only the several MWs but also by the MW distribution.

The chemical nature of the macromolecular chains determines both intermolecular (or intramolecular) interactions and chain flexibility, which are both effective on the state molecular aggregation (i.e. so-called molecular order), and hence on processing and end-use properties.

Intermolecular or intramolecular interactions are due to secondary forces, i.e., dipole-dipole interactions (polarity), dispersion forces, hydrogen bonding and ionic forces. Cohesive energy density gives the strength of secondary forces.

Macromolecules are also characterized by their molecular architecture. Polymer chains may be lineer, branched or cross-linked (Figure 2). Cross-linking involves generally the formation of covalent bonds between polymer chains. If there is enough cross-linking in a matrix, an insoluble three-dimensional network is obtained.

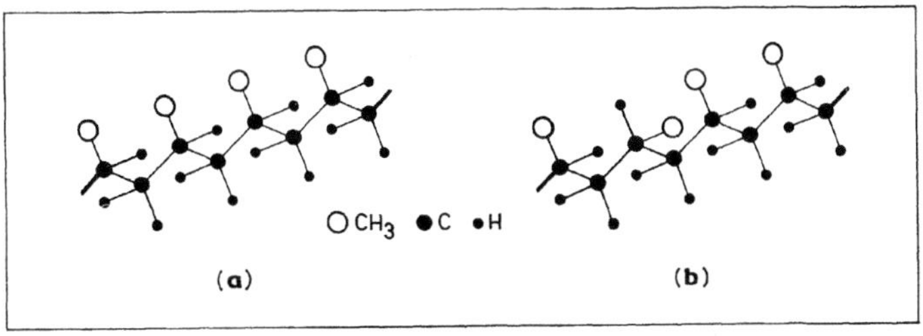

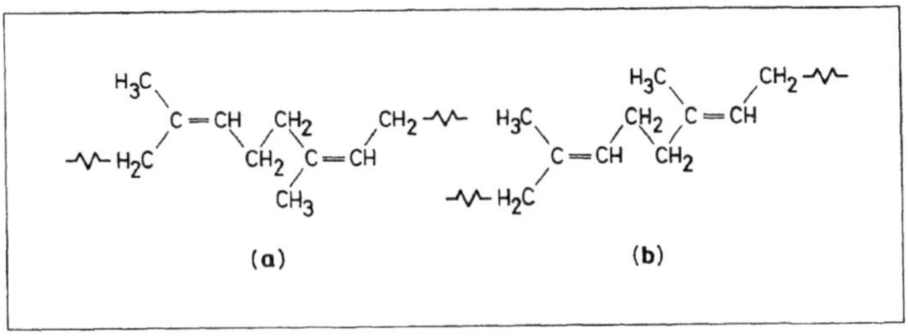

Figure 3. Polypropylene Configurations: (a) Regular (Isotactic); (b) Irregular (Atactic).

Figure 4. Cis-Trans Isomerism of Polyisoprene (a) Cis-(Natural Rubber); (b) Trans-(Gutta-Percha).

Polymers can be made by combining two or more different monomers (i.e., so-called copolymers).

Stereoisomerism is another important feature affecting the state of molecular aggregation. The side groups on a polymer chain may be placed regularly (isotactic or syndiotactic) or randomly (atactic) relative to one another (Figure 3). Cis-trans isomerism is also observed in macromolecules (Figure 4). It should be noted that the molecular or segmental levels prohibit the intermolecular or intramolecular interactions.

Depending upon such factor as steric irregularities, the strength of secondary forces, and temperature, several states of molecular aggregation in polymers are possible: liquid, amorphous (glassy or rubbery), crystalline, semicrystalline and oriented (Figure 5). In the glass state polymers are hard, rigid and often brittle. As it is seen in modules versus temperature curve, on heating a glassy polymer changes to a flexible rubbery solid state at a certain temperature (i.e., so-called glass transition temperature) (Figure 6). In the rubbery state, polymers are soft and flexible , and considerable segmental motions are possible. Note that there is another transition (Tm) from the rubber to the liquid state. In crystalline state, molecular chains were gathered in an ordered fashion. Semicrystalline polymers consist of dense crystalline regions dispersed in amorphous matrix. Stereoregularity and strong secondary forces lead to crystalline structure.

Polymers contain also additives. In some cases impurities are present as a result of polymerization process. In other cases additives such as plasticizers, stabilizers, lubricants, fillers, pigments, etc. are added to enchance the bulk properties of polymers. It should be noted that residues and additives may cause toxic side effects in biological applications.

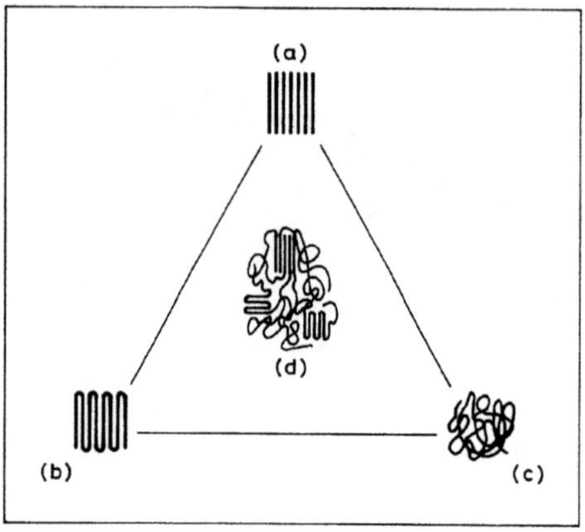

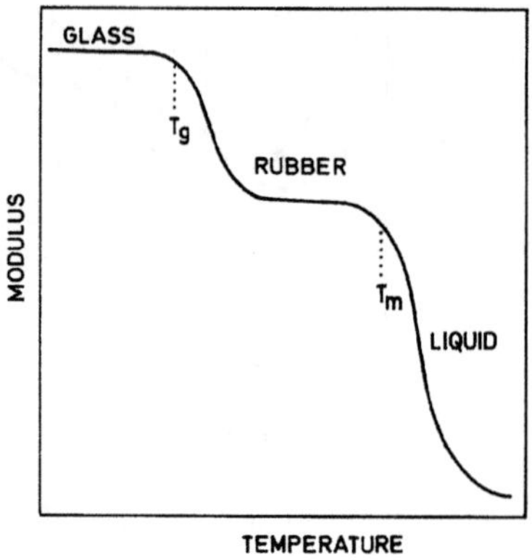

Figure 5. States of Molecular Aggregation (a) Oriented; (b) Crystalline; (c) Amorphous; (d) Semicrystalline.

Figure 6. Modules-Temperature Relation Showing Tg and Tm Transition Temperatures.

1.2. Transport Through Polymers

It has long been recognized that there is close relation between the type of polymer and the principle transport mechanism. From a transport analysis point of view it is convenient to classify the polymeric matrices in two categories.

- Homogeneous (non-porous)
- Heterogeneous (porous)

Homogeneous Matrices : Homogeneous polymeric structures consist of a dense polymer matrix. They have no identifible voidages at colloidal level and species pass through them by diffusion through a solid phase. Transport of species occurs by molecular diffusion in the space (i.e. 10-100 A) available between the macromolecular chains of the carrier. The free volume represents the holes into which diffusing molecules jump as they move a concentration (or chemical potential) gradient.

Diffusion through a homogeneous polymeric layer may be expressed by Fick's law given below [1].

$$J = -D \frac{dc}{dx} \tag{1}$$

Here, J is the flux of diffusant molecule, D and c are the diffusion coefficient and the concentration of diffusant within the polymeric matrix, respectively, t is diffusion time and x is position.

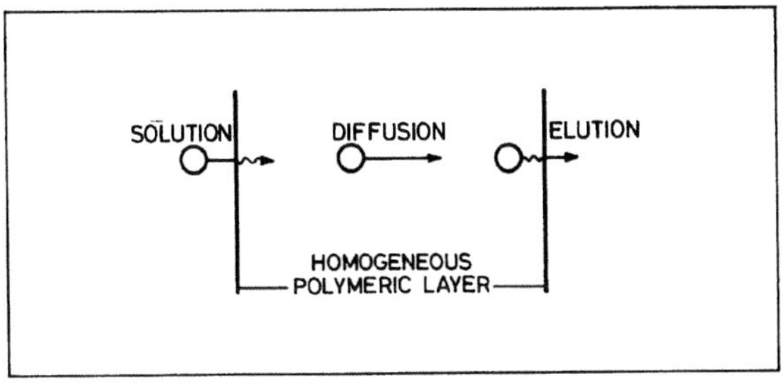

Figure 7. Solution-Diffusion Model for Transport Through Homogeneous Polymeric Layers.

This expression may be used to describe steady-state diffusion through polymers. It should be pointed out that, it is assumed here that diffusivity (D) is independent of concentration and direction. Note that this is not true, especially for liquid or vapor transport through non-porous polymeric layers, where solute-polymer interaction is appreciable [2].

Transmission of species through a homogeneous polymeric layer is generally treated with "solution (or sorption)-diffusion" model (Figure 7) [1,3]. Here, permeation takes place in three steps: the sorption of penetrant molecules within the polymeric matrix, the diffusion of penetrant through the polymeric layer,and elution at the opposite side of the layer. Combination of sorption laws and Fick's law the following expression, which represents the solution-diffusion model, is obtained.

$$J = P \frac{\Delta c}{\Delta x} = D.\,S \frac{\Delta c}{\Delta x} \qquad (2)$$

Here, P and S are the permeability coefficient and the solubility of penetrant, respectively.

There are several parameters which are effective both on diffusivity and solubility, hence on the transport through homogeneous polymeric layers. Some important parameters are discussed below.

- *Temperature* : Diffusion in polymeric matrices is an activated process, and in general both diffusivity and permeability increase with temperature.
- *Nature of Penetrant* : The molecular size of the penetrant is one of the main rate-controlling feature [4]. Small molecules may transfer more easily than bigger ones. The shape of the diffusant molecule is also important, especially in the case of the diffusion of complex molecules. It should be noted that solubility does play a significant role in the later cases and affects the permeability. Sorption and diffusion of complex molecules is rather complicated due to significant interaction between the penetrants and the polymeric material.
- *Nature of Polymeric Matrix* : Diffusional processes through homogeneous polymeric layers occur through the spacing between the macromolecular chains [5,6]. Large differences in permeabilities may be attributed to differences in the state of molecular aggregation, or in other terms, in morphology of the polymeric matrix. Chain flexibility and cohesive energy

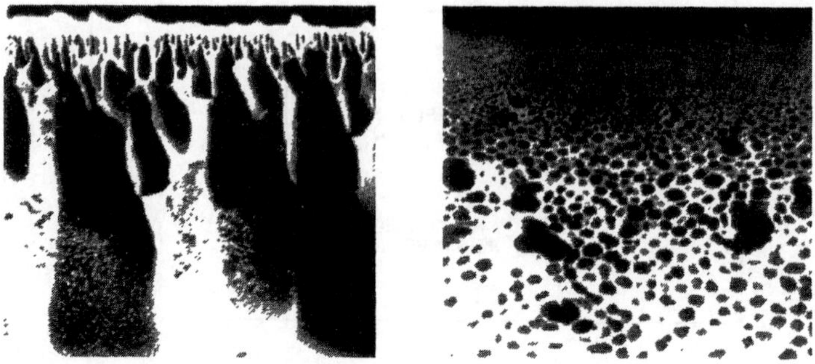

Figure 8. Examples of Porous Strustures of Heterogeneous Polymeric Layers (SEM Photographs) (Ref. 14 with Permission).

density are two important parameters which affect diffusion inversely. Higher diffusivities are obtained by increasing chain flexibility and decreasing cohesive energy. Both changes in polarity and steric effects result in changes in the chain flexibility and/or cohesive energy, hence leading differences in transport properties.

Cross-linking is also effective on the diffusion process. In general as the degree of cross-linking increases, the chain flexibility and the effective area between the chains for diffusion decreases, which in turn leads to reduction of the diffusion coefficient [7,8].

Crystalline regions of polymers are ordered structures of chains of higher density. Therefore, they are usually impermeable to penetrant molecules and act as barriers for diffusional process. Crystalline polymers are usually less permeable than amorphous ones. Percent crystallinity , crystalline size and orientation influence the mass transfer properties [9-11].

Like crystalline regions, fillers and other particle type additives introduce a heterogeneity at the microcrystalline level within homogeneous polymeric matrices. The geometry (shape, size and size distribution) concentration and concentration distribution, orientation, and other properties of the disperse phase significantly affect the transport through polymers.

Usually, addition of a plasticizer to a polymer decreases the cohesive forces and causes increas in both diffusivity and permeability [12,13].

Heterogeneous Matrices: Heterogeneous polymers have interconnected holes or pores into which species can pass through (Figure 8). They may be macroporous or microporous, with pores of average diameter, between 0.1 and 1.0 μ and 100-1000 A, respectively. The pore patway is tortuous and irregular. They exhibit usually a wide pore size distribution (Figure 9).

The models to determine or to explain the transport properties and mechanisms within the heterogeneous polymers, assume sieve-like behaviour, which means species with molecular sizes smaller than the pore size of the matrix may pass through the polymeric layer (Figure 10). Permeability and permselectivity within porous polymer layers depend on the pore size distribution and tortuosity of the matrix as well as the penetrant size, charge and may be the chemistry of the penetrant molecules. In contrast to the homogeneous membranes, the polymer structure plays only a minor role here, especially for macroporous matrices, but may become important only in situations where the penetrant molecules exhibit strong interactions with the polymer.

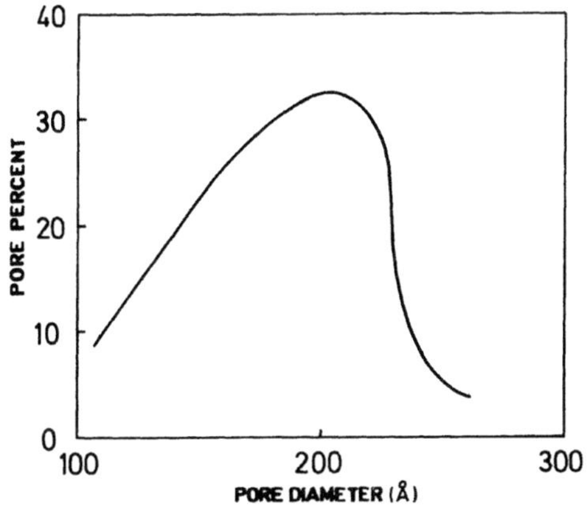

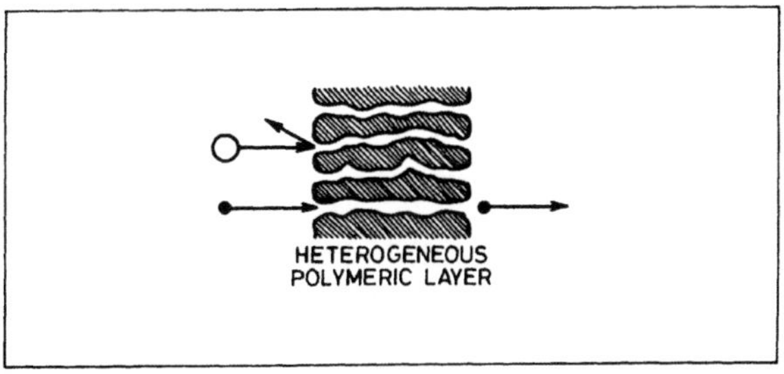

Figure 9. Pore Size Distribution in Heterogeneous Polymers.

Figure 10. Sieve-Like Transport Through Heterogeneous Polymeric Layers.

In general, diffusion through heterogeneous polymeric layers may be described by Fick's law given in Equation 1. As the diffusion coefficient, the effective diffusion coefficient is used here. The effective diffusion coefficient is usually expressed in its most general form given below [15,16].

$$D_{eff} = D \frac{\varepsilon K_p K_r}{\tau} \tag{3}$$

Here, D is the penetrant diffusion coefficient through the pore liquid, ε is the void fraction (i.e. so-called porosity), τ is the tortuosity, K_p is the partition coefficient of the penetrant between the polymer and the pore liquid, and K_r is a restriction coefficient, which is a geometrical factor (the normalized ratio of penetrant molecular radius and the average pore radius) [17].

In addition the molecular diffusion, convective transport may be occured in porous polymer, if there is a driving force, such as pressure difference. There are several models to determine the convective transport through porous polymeric layers. The simplest model for convective transport assumes the heterogeneous polymers as a porous medium. Darcy's Law is generally obeyed for the flow of fluid through a porous medium, and may be expressed by the following equation [18].

$$J_S = P_H . \Delta p \tag{4}$$

Here, J is the fluid flux, P_H is the hydraulic permeability, and Δp is the hydrostatic pressure difference.

2. IMMOBILIZATION TECHNIQUES

"Immobilization" can be defined as any technique that limits the free migration of bioagents. A number of technique have been proposed and evaluated [19-22]. Various terms have been used to describe different types of immobilization. However, they are usually classified into five groups (see Table III).

Table III. Immobilization Techniques.

- Primary (Covalent or Chemical) Bonding
- Secondary (Physical) Bonding (Adsorption)
- Encapsulated Within Polymeric Capsules
- Covalently Cross-Linked to Each Other
- Dissolved or Dispersed in Polymeric Matrix

Bioagents may be covalently bound (primary bound) to a carrier system which may be a variety of solid supports, e.g. synthetic polymers, activated carbon, glass, ceramics, etc. Bioagents are usually attached to the solid support through their functional groups. In some instances, the activity of chain, an "arm" between the ligand and the solid matrix [23]. It was assumed that these arms relieve the steric restrictions imposed by the matrix backbone allow an increased flexibility and availability of the bioagent. Solid support materials in the form of membranes (fibers, films, etc.), or porous beads are used in this technique.

The bioagent may be adsorbed physically to various kinds of sorbents. The forces that cause "physical adsorption" are mostly van der Waals forces (secondary bonding). Ionic forces, dispersion forces and other hydrophilic and hydrophobic attractions of agents to the sorbent may sometimes be considered. Many types of activated carbons, ion exchangers, etc., in the form of porous granules are used as soebents in this approach [19-21].

In some novel biotechnological applications, such as microcarrier cell cultures, carriers are employed in the process without any prior immobilization [24]. Immobilization of the anchorage-dependent cells to the surface of the microcarriers by adhesion and proliferation is achieved just before the procedure. Bead-shaped microcarriers based on, mostly cross-linked dextran with a diameter in the range of 100 - 200 μ are used in these applications [25].

The bioagent may be encapsulated with or without its medium within various forms of semipermeable polymeric membrane systems. Usually, capsules with a diameter in the range of microns to millimeters are used [26]. Depending on the type of application, the substrate molecule or bioagent itself can freely permeate through the surrounding membrane and thus show activity. In some applications, the agents are simply entrapped into the lumen of hollow fibers [27].Numereous methods for preparing this type of carrier systems for biotechnoogical, agricultural, pharmaceutical, medical and other applications have been elaborated.

The bioagents, for instance enzymes, may be covalently cross-linked to each other. Attachment is obtained by using biofunctional cross-linkers like glutaraldehyde, dimethy adipimidate [19-21].

The bioagents may be physically entrapped within natural or synthetic matrices in this method, bioagent is dissolved or throughly dispersed in the polymer structure. The mixture is then treated by injection or compression molding, casting, extrusion, calendering, or by other techniques of polymer technology into the reguired geometrical shapes [19-21,28]. Bioagents may also be dissolved or dispersed in suitable monomer or monomer mixtures and may then be shaped by following a proper polymerization procedure.

3. APPLICATIONS OF BIOAGENT-POLYMER COMPOSITES

A wide variety of bioactive agents (see Table IV) are incorporated onto several solid supports, or entrapped into different matrices, for many diverse applications (see Table V).

Table IV. Examples of Bioactive Agents.

Aminoacids	DNA or RNA Sequences
Antibacterial Agents	Drugs
Antibiotics	Living or Dead Cells
Antibodies	Microorganisms
Anticancer Agents	Peptide Sequences
Antigens	Tissue Slices
Cell Extracts	Others
Chemicals	

Table V. Major Fields of Application of Bioactive Agent-Polymer Composites.

Biotechnology
Bioaffinity Chromatography
Biosensors
Controlled Release
Therapeutic Applications

3.1. Biotechnology

The rapidly growing field of biotechnology encompasses sophisticated fermentation processes using normal and genetically altered microorganisms, biochemical conversions utilizing enzymes and cells, and the production of many important biological substances from mammalian cell cultures [22,29-33].

In advanced biotechnological systems, the biocatalysts are preferentially employed in immobilized forms, which are generally bioactive agent-polymeric carrier combinations. Immobilization avoids the drawbacks associated with the use of bioactive agents in free forms and gives significant advantages (see Table VI).

Table VI. Advantages of Immobilization.

- Allow Reuse of Bioagent
- Bioagent - Free Product
- Bioagent Properties Favorably Altered
- Enhanced Bioagent Stability
- Closer Control over Product
- Reduced Cost
- Continuous Processing

Immobilized bioagents are being used or investigated for many different applications. They are used for the production of chemicals, in food and beverage industries, for waste treatment, etc. Immobilized enzyme reactors, which employ polymers as matrices for enzyme containment, are now gaining increasing popularity for the conduct of continuous biochemical transformation. The major potential use for immobilized agents in the biomedical area is for the production of pharmaceuticals and reagents by novel techniques. Hollow-fiber membrane devices containing mammalian cell cultures, which are so-called artificial capillary bioreactors are now being used for the production of hormones, vaccines, antibodies, nucleic acids, etc. for diagnostic and therapeutic uses [29-33]. Polymeric microcarriers are also used in mammalian cell cultures to facilitate the production of many important biomolecules.

3.2. Bioaffinity Chromatography

Another interesting application of bioagent-polymeric carrier combinations is bioaffinity. There are a number of traditional techniques for identification, purification and seperation of complex molecules, in that processes rely on differences in physical properties (e.g., size, charge, hydrophobicity) of molecules to be treated. Instead, bioaffinity techniques exploit the unique property of extremely specific biological recognition [23,34].

Affinity sorption requires that compound to be isolated is capable of reversible binding to biosorbent. In bioaffinity chromatography, solutions containing compounds to be seperated are circulated through affinity columns filled with biosorbents. One of the compounds is preferentially adsorbed. Then using a proper elutant system, it is desorbed. Elution may be achieved by changing pH, temperature or ionic strength, by using protein denaturants, or by other ways.

Biosorbents are usually spherical beads of bioactive agent (ligand)-polymer (carrier) composites. There are a number of polymeric carriers used for preparation of biosorbents (see Table VII). Their surfaces are activated by using various activating agent with several techniques, for ligand immobilization [34].

A huge variety of ligands (see Table VII) are potentially available, and are used to prepare biosorbents to be tailor-made for very specific affinity sorption of complex biomolecules and other substances.

Table IV. Polymeric Carriers and Ligands for Bioaffinity Chromatography.

Carriers:	Agarose, Cellulose, Dextran, polyalkylacrylates, polyacrylamid, polystyrene, polyvinylpyridine, polyacrolein-agarose, polycynoacrylates, polyvinylalcohol polyglutaraldehydes, etc.
Ligands:	AMP, ATP, Con A, CDP, Cellulose, Choline, Collagen, DNA, Histidine, Glu, Heparin, Glycogen, Lipoprotein, L-Asp, Lysine, NADP, NAD, Phosphoric Acid, Penicillins, Phenylalanine, Sugar Derivative, tRNA, UDP, Vitamin B_{12}

3.3. Biosensors

As it is being intensively reviewed in this volume, biosensors are used for detection and measurement of biologically important substances by means of high specificity.

In these systems the biochemical signal generated from the interaction of a bioactive agent and the compound to be dedected (i. e., so-called analyte) is converted into an electrical signal by a transducer, amplified and recorded by an electronic device. The bioagent is combined to the transducer by using usually a polymeric carrier (i. e., a membrane). Table VIII lists the types of transducers, gives a number of bioagents and examplifies the analytes to be potentially dedected and measured [35].

The major impetus for the development of biosensors has undoubtedly come from the rapid advanced biomedical sensing technology. This development hs been accelerated recently by other areas of application, such as analytical biotechnology, veterinary and agricultural analysis, pollution monitoring and control, dedection of industrial gases and toxic gases, etc.

Table VIII. Transducers, Bioagents and Analytes of Biosensors.

Transducer:	Potentiometric; Amperometric; Optical; Calorimetric; Conductance; Mass Change
Bioagents:	Enzyme and Enzyme Systems; Immunocomplexes; Biological Membrane Components; Microorganisms; Tissue Slices; Living or Dead Cells
Analytes:	Gases; Inorganic Ions; Amino Acids; Amines; Alcohols; Amids; Carboxylic Acids; Carbonhydrates; Antibiotics; Vitamins; Others

3.4. Controlled Release

Control administration of chemicals or drugs is widely publicized in the field of medicine and pharmacy, but it is also applied in other fields including agriculture, veterinary, food industry, etc [28,36-38]. In controlled release systems, the active agent, which is incorporated into a carrier, generally a polymeric material, is released to the surrounding environment by desired mechanisms and rates.

There are a number of techniques of achieving controlled release. Among the others, bio-agent-polymer combinations are mostly given credit. As seen in Table IX controlled release polymeric systems are usually classified based on the mechanism controlling the release of incorporated active agent [37].

Table IX. Classification of Controlled Release Polymeric Systems.

- Diffusion Controlled Systems
- Chemically Controlled Systems
- Swelling Controlled Systems
- Magnetically Controlled Systems

Two basic configurations, namely reservoir and matrices are the most widely used systems. In reservoir systems, a core of agent is surrounded by a polymer film, and diffusion of the agent through this film is the rate-controlling step. These systems include membranes, micro-to macrocapsules, hollow fibers and many other forms.

In matrix systems, bioagent is dissolved or throughly dispersed in a polymer matrix. The rate limiting step of these devices may be pure diffusion of the agent (diffusion controlled systems), chemical reaction at the continuously depleted interface between the polymer and the dissolution medium (chemically controlled systems), or countercurrent diffusion of the agent and the dissolution medium in the polymer (swelling controlled systems). These devices may be any required geometrical shape.

Controlled release formulations of pharmaceutical, agricultural and veterinary agents have been widely accepted at commercial level. Bioagents, such as analgesics, steroids, vitamins, antibiotics, chemotherapeutic agents, enzyme bacteria, viruses, pesticides, fertilizers, herbicides, and many other agents have been incoroporated with polymeric carriers and have been succesfully employed in controlled release formulations.

3.5. Therapeutic Applications

Biofunctional substances like sorbent systems, enzymes, enzyme systems, antigens and antibodies, cells, and cell extracts, tissue slices, etc. are used in therapeutic medicine for the treatment of patients suffering with renal; diseases, drug overdoses, liver failure and other metabolic disorders; in enzyme replacement therapy; removing of immunogenic substances via specifiv sorbent systems;as an artificial pancreas in order to maintain the glucose level in diabetics; in drug delivery devices in order to provide controlled release or to target drugs by specific means, and in many other applications [21,26,37-39].

In order to increase the biological lifes of bioagents, and also to prevent immune reactions and toxic side effects, bioagents are used in immobilized forms in therapeutic applications. Depending on the route of application (see Table X), these bioagent-carrier (i.e., generally polymeric material) systems are prepared in different forms, such as micro-to macro capsules, hollow fibers, granular sorbents, etc.

Table X. Routes of Application of Bioagent-Polymer Composites in Therapeutic Medicine.

- Topical
- Implantation (e.g., Transdermal, Subcutaneous, Intraperitoneal)
- Injection (e.g., Intravascular)
- Extracorporeal
- Gastrointestinal

REFERENCES

1. Crank, J., (1975), "The Mathematics of Diffusion", Clarendon Press, Oxford.
2. Harris, F.W., Seymour, R.B., (1977), "Structure-Solubility Relationships in Polymers", Academic Press, New York.
3. Sweeting, O.J., (1976), "The Science and Technology of Polymer Films", Vols. I and II, Wiley-Interscience, New York.
4. Stannett, V., (1968), In "Diffusion in Polymers", J. Crank and G.S. Park, eds., Chap. 2., Academic Press, New York.
5. van Amerongen, G.J., (1950), J. Polym. Sci., 3:549.
6. Langer, R.S., Peppas, N.A., (1983), J. Macrom. Sci., Revs. Macromol. Chem. Phys., C23:61.
7. Barrer, R.M., Barrie, J.A., Wong, P.S., (1968), Polymer, 9:609.
8. Peppas, N.A., Lustig, S.R., (1987), Ann. N. Y. Acad. Sci., (in press).
9. van Amerongen, G.J., (1964), Rubber Chem. Tech., 37:1065.
10. Barrer, R.M., Chio, H.T., (1965), J. Polym. Sci., 10:11.
11. Michaels, A., Bixler, H., Fein, H., (1964), J. Appl. Phys., 35:3165.
12. Barrer, R.M., Mallinder, R., Wong, P.S., (1967), Polymer, 8:321.
13. Meyers, A., Tammela, V., Stannet, V., Szwarc, M., (1960), Mod. Plastics, 37:139.
14. Strathmann, H., (1981), J. Memb. Sci., 38:117.
15. Cunningham, R.E., Williams, R.J., (1980), "Diffusion in Gases and Porous Media", Plenum Press, New York.
16. Peppas, N.A., (1984), Proc. Symp. Controled Rel. Bioact. Mater., 11:94.

17. Colton, C.K., Scatterfield, C.N., Lai, C.J., (1975), AIChE Journal, 21:289.
18. Bird, R.B., Stewart, W.E., Lightfoot, E.N.,(1960), "Transport Phenomena", John Wiley & Sons., New York.
19. Zaborsky, D.R., (1973), "Immobilized Enzymes", CRC Press, Cleveland.
20. Mosbach, K., (1976), Methods in Enzymology, Vol. 44, Academic Press, New York.
21. Chang, T.M.S., "(1980), "Biomedical Applications of Immobilized Enzymes and Proteins", Vols. I and II, Plenum Press, New York.
22. Scott, C.D., (1987), Enzyme Microb. Technol.,9:66.
23. Wilchek, M., Miron, T., Kohn, J., (1984), In " Methods in Enzymology", Vol. 104, Academic Press, Orlando, Florida.
24. Glacken, M.W., Fleischhaker, R.J., Sinskey, A.J., (1983), Trends in Biotechnology, 1:102.
25. Cytodex TMa, (1978), "Beaded Microcarriers for Cell Culture", Pharmacia Fine Chemical, Technical Booklet Series, Upsala, Sweden.
26. Lim, F., (1980), "Biomedical Applications of Microencapsulation, CRC Press, Boca Raton, Florida.
27. Mattiason, B., (1983), "Immobilized Cells and Organelles, CRC Press, Boca Raton, Florida.
28. Kydonieus, A.F., (1980), "Controlled Release Technologies: Methods, Theory and Applications", Vols. I and II, CRC Press, Boca Raton, Florida.
29. Science, (1980), 209 (4463):1317.
30. Scientific American, (1981), 245(3):66.
31. Science, (1983), 219(4585).
32. Micheals, A.S., (1981), Chem. Tech., 245(3):66.
33. Micheals, A.S., (1984), Chem. Eng.Progress, 80(4):9 and 80(6):19.
34. Dean, P.D.G., Johnson, W.S., Middle, F.A., (1985), "Affinity Chromatography", I.R.L Press, Oxford.
35. Rechnitz, G.A., (1981), Science, 214:287.
36. Baker, R., (1980)" Controlled Release of Bioactive Materials", Academic Press, New York.
37. Langer, R.S., Wise, D.L., (1984), "Medical Applications of Controlled Release", CRC Press, Boca Raton, Florida.
38. Pişkin, E., Hoffman, A.S., (1986)" Polymeric Biomaterials", Martinus Nijhoff,Dordrecht.
39. Pişkin, E , Chang, T.M.S.,(1983), "The Past, Present and Future of Artificial Organs", Meteksan, Ankara.

IMMOBILIZED ENZYMES IN THERAPY

Ayşe Kevser Pişkin
Hacettepe University, Faculty of Medicine, Medical Biology Department, Ankara, Turkey

ABSTRACT: Enzyme therapy takes the advantage of highly specific action of these catalysts in correcting metabolic disorders. However introduction of enzymes to the body usually elicits immune reactions which can range from a mild case of hives to severe, life threatening anaphylactic schock. Another problem with injecting enzymes is that they are liable to be destroyed by the proteolytic enzymes and the reticuloendothelial system. Problems related to the therapeutic applications of solubilized enzymes have resulted in extensive research into the possible therapeutic uses of immobilized enzymes. Blood detoxification with urease is one of the most important applications of immobilized enzymes in therapy. several enzymes are either used in experimentally or applied in therapy including asparaginase for tumor suppression, thrombolytic enzymes i. e. streptokinase and urokinase in making polymeric materials biocompatible in order to be used in contact with blood. One of the most applicable situations is the treatment of inherited enzyme deficiency conditions. This paper reviews these applications.

INTRODUCTION

The use of enzymes in therapy is hampered by a variety of problems. Most enzyme are not stable at body temperature and they can not be kept at sites where they are aimed to act. When introduced to the body, as foreign proteins they are rapidly eliminated by the reticuloendothelial system. Furthermore, when injected repeatedly, they can create immune reactions ranging from a mild case of hives to severe, life threatening anaphylactic shock [1].

These drawbacks have led to intensive research on immobilization of enzymes with therapeutic potential. Immobilization brings about certain advantages such as increased stability, repeated use and elimination of immune reactions. Immobilization may also enable targetting of enzyme to the site of action. Enzymes are immobilized by five major techniques: (1) Covalent binding to insoluble carriers, (2) physical adsorption on solid carriers, (3) cross-linking of enzyme molecules to each other or to carriers, (4) entrapment within synthetic or natural polymeric gels, (5) microencapsulation into semipermeable polymers (Figure 1).

Among these methods covalent attachment is applicable to extracorporeal applications, whereas, gel entrapment and microencapsulation is usually preferred to safely introduce the enzyme into the body. Both techniques prevent the enzyme from immune recognition provided that the carriers are biocompatible

G. G. Guilbault and M. Mascini (eds.),
Analytical Uses of Immobilized Biological Compounds for Detection, Medical and Industrial Uses, 39–46.

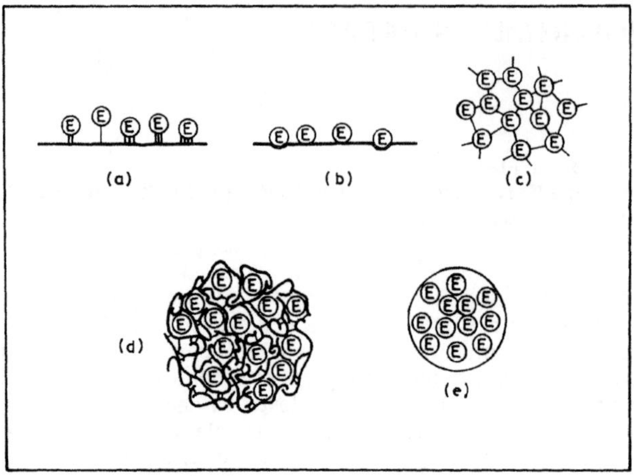

Figure 1. Enzyme Immobilization Techniques:
(a) Covalent binding
(b) Adsorption
(c) Cross-linking
(d) Matrix entrapment
(e) Microencapsulation

WAYS OF INTRODUCING IMMOBILIZED ENZYMES TO THE BODY

1. Topical application: This is used in the treatment of external wounds.
2. Administration into the intestinal tract: By this route enzyme acts on its substrate which equilibrates across the system
3. Implantation: Immobilized enzyme can be implanted by intradermal, intramuscular or intraperitoneal injections.
4. Intraveneous or intraarterial injections: This route is used to maintain a certain level of enzyme activity in the circulation or to target the enzyme at a specific organ.
5. Extracorporeal perfusion system: In this approach immobilized enzyme is placed in an enclosed system through which blood is allowed to circulate. This way enzyme preparation is not introduced into the body and can be removed easily. Figure 2 illustrates this system.

FIELDS OF APPLICATION

Main fields to which immobilized enzymes are applicable, are as follows:

1. Blood purification: Enzymes which hydrolyze endogeneous or egzogeneous toxins are used for blood purification especially in kidney and liver failure.

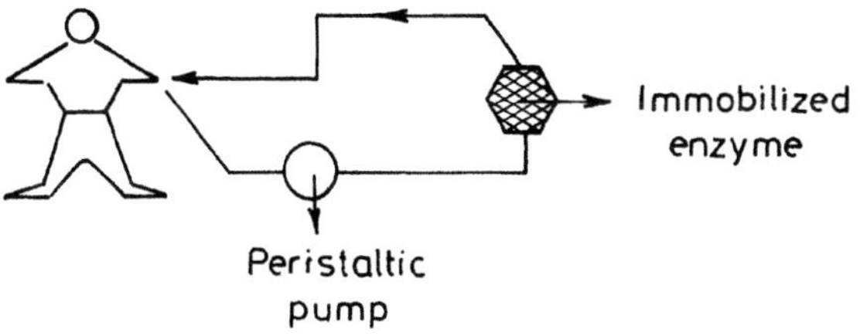

Figure 2. Extracorporeal perfusion system.

2. Treatment of enzyme deficiency: Immobilized enzymes are aimed to correct inborn errors of metabolism mainly due to inherited enzyme deficiency conditions.
3. Cancer therapy: Substrate dependent tumors are suppressed by depleting these substrates using immobilized enzymes.
4. Development of nontrombogenic biomaterilas: Enzymes are incorporated to polymeric devices to be used in contact with blood to help improve their blood compatibility.

EXAMPLES OF THERAPEUTIC APPLICATIONS OF IMMOBILIZED ENZYMES

Urease

Urease is the first enzyme to show the feasibility of using immobilized enzymes in therapy. Urea is the major metabolic end product and the removal of its excess is of viral importance. Chang was the first to show the effective hydrolysis of blood urea by immobilized urease (Figure 3). Microcapsulated urease was implanted into the peritoneal cavity of dogs or it was packed in a hemoperfusion column to hydrolyze urea. The ammonium produced by hydrolysis was removed by ion exchange resins [2]. Gastrointestinal applications were carried out by Chang and by Gardner et. al. where immobilized urease was administered into the gastrointestinal system along with ammonia adsorbent [3, 4]. To introduce urea removal into hemoperfusion columns, urease was adsorbed on activated carbon and then encapsulated with cellulosenitrate membranes by Pişkin and Chang. As shown in Table I, these enzyme preparations provided high conversion rates when packed in a hemoperfusion columns and tested in rats [5].

This form of immobilized urease was also used to hydrolyze urea in the dialysate [6,7]. Dialysate regeneration is a process whereby accumulated toxins or excess metabolic waste molecules are removed continuously to enable using small volumes of dialysate. This helps decrease the size of hemodialysis machines in dialysate regenerators, immobilized urease is used along with other adsorbents and commercially available systems utilize immobilized urease for this purpose [8]

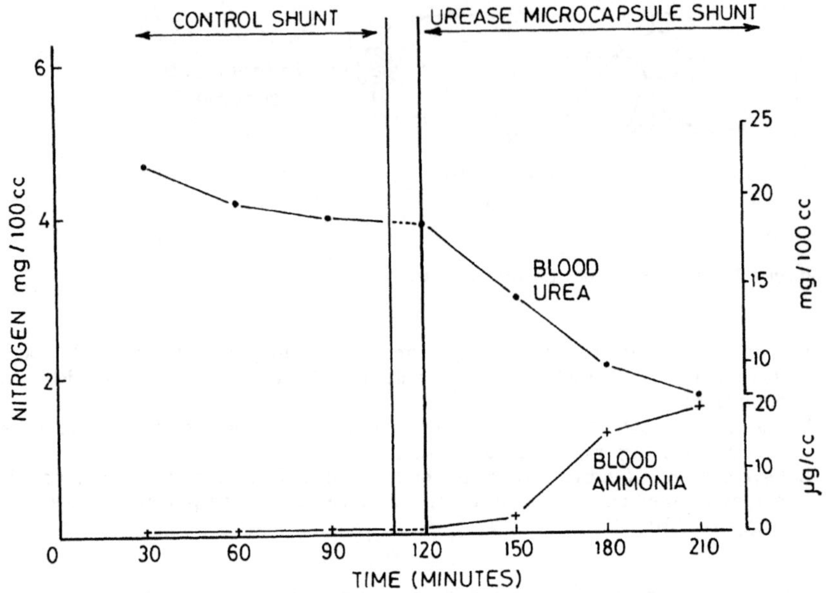

Figure 3. Hydrolysis of blood urea by microencapsulated urease (From Chang 2)

TABLE I

In-vitro and In-vivo Conversion of Blood Urea (Mean ∓ S.D.)

	% CONVERSION ($\frac{A-E}{A}$ X 100)	
Time (min)	In-vitro (blood)	In-vivo
5	84.5∓5.0	94.8∓13.2
15	86.6∓3.7	87.3∓17.1
30	86.4∓3.7	87.8∓13.0
60	89.5∓2.8	81.9∓18.6
90	88.0∓3.5	74.1∓15.2

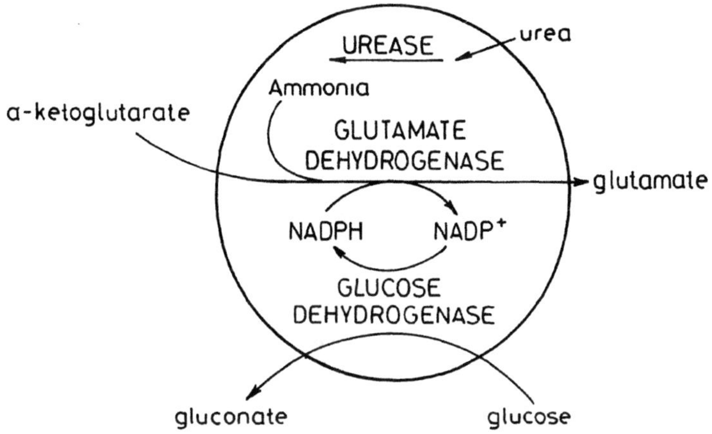

Figure 4. Urea conversion in multienzyme, cofactor recycling system [From Campbell and Chang, (10)].

An alkalinization-gas desorption technique developed by Pişkin and Chang was used for ammonium removal in the form of ammonia [9]. An interesting approach proposed by Campbell and Chang was immobilized multienzyme and cofactor recycling system which converts urea to glutamate [10] (Figure 4)

Asparaginase

Immobilized asparaginase was used to suppress asparagine dependent lymphosarcoma in mice. Microencapsulated asparaginase was applied intraperitoneally by Chang, et. al. and liposome entrapped asparaginase was targetted by injection to the liver by Gregoriadis to breakdown asparaginase [11, 12]. Important advantage of immobilized asparaginese such as increase stability and reduced immune reactions are reported, however for complete tumor suppression sustained release formulations may be necessary for continuous delivery of asparaginase.

Catalase

In acatalasemia, which is a congenital enzyme deficiency condition, the catalase activity in the body is very low. Major problem of acatalasemics is the oral lesions. As blood lacking catalase reaches these lesions hemoglobin is oxydized by bacterial hydrogen peroxide. The lack of oxygen in the injected region results in necrosis. This creates great peridontological problems.

In order to use catalase safely without immune reactions and greater stability to treat these lesions, catalase was immobilized in semipermeable microcapsules by Chang. Microencapsulated catalase suspension applied topically to oral lesions of patients worked efficiently to remove hydrogen peroxide [13].

Heparinase

Extracorporeal perfusion systems are widely applied. Millions of extracorporeal treatment procedures are carried out each year. These are heart-lung machines, hemodialysis, hemoperfusion, plasmapheresis and others where blood is circulated through synthetic surfaces. For operation of these devices anticoagulation is necessary. Systemic heparinization brings about the risk of increased bleeding tendency. Therefore protamin sulphate is used at the outlet of the circuit to neutralize heparin, however, protamin sulphate itself has variable anticoagulant activity and can also lead to hypotension [14]. Heparinase immobilized on a variety of supports was used to break down heparin to nontoxic saccharides eliminating its anticoagulant activity [15,16].

Plasminogen Activators

Polymers used in contact with blood create reactions leading to emboli as shown in Figure 5. One approach to eliminate these reactions is to use thrombolytic agents e.g. tripsin, brinase, plasmin. Since these agents can be highly toxic when used in vivo because of their proteolytic activity another approach is taken which aims to form plasmin, plasminogen activators, such as urokinase and streptokinase are immobilized by Oshiro, and others to increase blood compatibility of polymeric devices [17]. Oshiro's group have reported that emboli formation is eliminate by using plasminogen activator adsorbed cannulas and sutures [18].

FUTURE PROSPECTS

There are many enzymes which have therapeutic potential. Immobilization of these enzymes is rather promising for treatment of many diseases and can lead to production of effective clinical tools.

Microencapsulated urease can be used to remove uric acid in kidney failure and also in the treatment of Gout [19]

An enzyme reactor containing biluribin oxydase immobilized on sepharose can be used to degrade biluribin to treat neonatal jaundice [20].

Phenylalanine ammonia lyase is encapsulated in microcapsules [21] and immobilized on conventional hollow fiber hemodialyzers [22]. This enzyme has great potential in the treatment of phenylketonuria, an inborn error of metabolism which is treated by taking dietary measures. Other enzymes worth immobilizing for therapeutic applications are proteolytic enzymes for breaking down

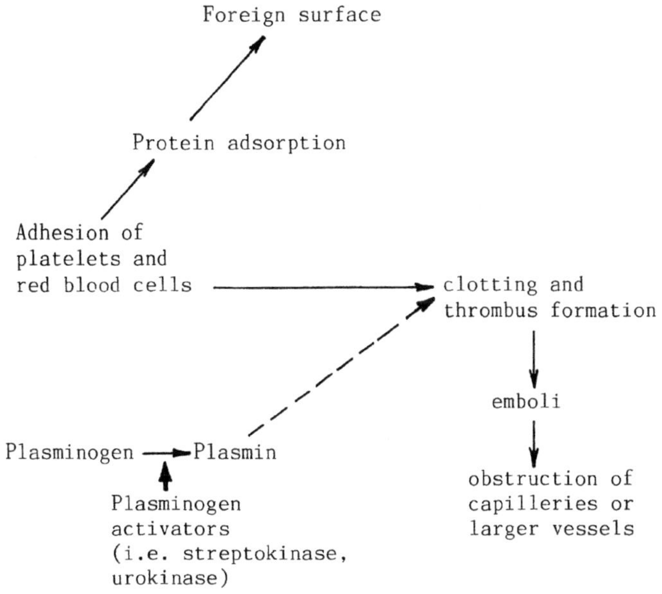

Figure 5. Blood-material interactions and inhibition by plasminogen activators.

eschar tissue, carbonic anhydrase to enchance carbon dioxide transfer in blood oxygenators and alpha 1, 3 glucanase, mutanase or dextranase for combating dental plague, thus preventing cavities.

In conclusion, immobilized enzymes as examplified in this rewiev have great potential in therapy. As more enzymes will be isolated in good purity and availability this potential will increase. Todays technology offers a wide range of targetting agents i.e., highly specific antigens, this should also help a great deal in targeting immobilized enzymes. Accumulating knowledge in enzyme methodology during the past few decades may soon result in development of highly specific and effective clinical tools.

REFERENCES

1. Hasselberger, F.X., 1978, 'Uses of Enzymes and Immobilized enzymes'. F. X. Hasselberger, ed., Nelson Hall Inc., Chigago.
2. Chang, T.M.S., 1966, Trans. Amer. Soc. Artif. Intern. Organs , 12:13.
3. Chang, T.M.S., 1977, 'Biomedical Applications of Immobilized Enzymes and Proteins', Vol I, T. M. S. Chang, ed., Plenum Press, New York.

4. Gardner, D.L., Falb, R.D., Kim,B.C., and Emmerling, D.C., 1971, Trans. Amer. Soc. Artif. Organs, 17: 239.
5. Pişkin, K., and Chang, T.M.S., 1980, Int. J. Artif. Organs, 3:344.
6. Pişkin, K., 1981, Urea Hydrolysis in Hemoperfusion Columns with Immobilized Urease, Ph. D. Thesis, Biology, Ankara.
7. Pişkin, K., Pişkin, E., Noyan, A., 1982, Proc. 4th. Int. Symp. Hemoperfusion Artificial Organs, p. 134, Ankara.
8. Gordon, A., Better, O.S., Greeenbaum, M. A., Marantz, L.B., Gral, T., Maxwell, M. H., 1971, Trans. Amer. Soc. Artif. Intern. Organs, XVII: 253.
9. Pişkin, E., and Chang, T.M.S., 1979, The Int. J. of Artificial Organs, 2(4): 211.
10. Campbell, J., and Chang, T.M.S., 1976, Biochem. Biophysics Acta, 397:101
11. Siu Chang, E.D., and Chang, T.M.S., 1974, Enzyme, 18:218.
12. Gregoriadis, G., 1976, Immobilized Enzymes in 'Methods in Enzymology', K. Mosbach., Vol XII, Academic Press, Inc., New York
13. Chang, T.M.S., 1972, Journal of Dental Research, 51(2):319
14. Swartz, R.D., Port, F.K., 1979, Kidney Int., 16: 513.
15. Ebert, C.D., Kim, S.W., 1982, Thrombosis Res., 26: 43.
16. Langer, R., Linherdt, R.J., Hoffberg, S., Larsen, A.K., Cooney, C.L., Tapper, D., Klein, M., 1983, Science, 217:261.
17. Senatore, F.F., Bernath, F.R., 1986, Biotechnol. Bioeng., 28:58.
18. Oshiro, T., 1983, Enzyme Engineering, Vol 6, p.443, Plenum Press, New York.
19. Segal, A.W., Gregoriadis, G., and Black, C.D.V., 1975, Clin.Sci. Mol. Med., 49:99.
20. Sung,C., Lavin, A., Klibanov,A., Langer, R., 1986, Biotechnol. Bioeng., XXVII:1531.
21. Bourget,L., and Chang, T.M.S., 1985, FEBS Letters, 180 (1):5.
22. Larue,C., Munnich, A., Charpentier,C., Saudubray, J.M., Frezal, J., Reny, M.M., Rivat, C, M., 1986, Dev. Pharmacol.Ther., 9:73.

USES OF IMMOBILIZED BIOLOGICAL COMPOUNDS IN PHARMACEUTICAL ANALYSIS

G. J. Patriarche, J-M. Kauffmann, J-C. Vire, B. J. Vincke'
Free University of Brussels (U.L.B.)
Institute of Pharmacy
Campus Plaine 205/6, Bd Du Triomphe
Brussels, BELGIUM

ABSTRACT. The purpose of this work is to illustrate the increasing interest of biosensors in the field of pharmaceutical analysis. Several pharmacologically important compounds, such as amino-acids, nicotinamide and ascorbic acid have been determined using enzyme or bacterial membranes immobilized on gas sensing electrodes. The analytical parameters involved in the optimization of the electrode response, and also microbiological factors, are discussed for each electrode. Performances of the electrodes and interferences which may occur are presented. Opportunities of using bacteria to produce specific unstable enzymes are pointed out.

1. INTRODUCTION

The history of immobilization of enzymes dates of a long time (1971) but it is since 1954 that a large development in this field has been carried out. The use of these techniques presents many advantages. They offer a catalytic efficiency over a relatively long period of time. They may exhibit selectively modified properties and serve as models for enzymes in their own environment. Many review papers and books were devoted to this field(1-6).

The techniques of immobilization are various and many classifications have been proposed such as: matrix entrapped immobilized enzymes, adsorption, covalent binding or microencapsulation. Also, following the technique of immobilization, the enzymes are much more stable; they can be heated till 50°C with only a little loss of activity and this activity is durable.

There is now an exciting revolution currently underway in industrial and university laboratories around the world and focused on the development of new biosensors essentially for use in the analytical and clinical area. New developments are also interesting in fluorescence optical sensors and piezoelectric-based mass detectors.

The design of enzyme electrode probes is also in an increasing development and is now a part of current commercial analyzers.

Table I non-exhaustively summaries some developments and

G. G. Guilbault and M. Mascini (eds.),
Analytical Uses of Immobilized Biological Compounds for Detection, Medical and Industrial Uses, 47–69.

descriptions for enzyme electrodes applied to pharmaceutical analysis.

Although more than two thousand enzymes are now well identified, most of them are unstable or not commercially available. In addition, the use of enzyme electrodes implies that the catalyst will be physically or chemically immobilized.

More recently, bacterial and tissular electrodes were proposed, which solve these problems and offer several advantages such as a wider choice of enzymes, a longer life time and a better tolerance towards physico-chemical parameters. Some recent applications of baterial electrodes to the determination of pharmacological compounds are shown in Table II.

We will illustrate our paper and our work in the field by the determination of two amino-acids, asparagine and tryptophan, by the determination of nicotinamide using two bacterial electordes and also by an ascorbic acid sensor for which four types of biocatalysts have been compared.

2. Experimental

An ammonia gas sensing electrode (Universal Sensors Inc., type 019733582 or Tacussel pNH_3-1) is used for the construction of the enzyme or bacterial electrodes. For the ascorbic acid sensor, a pO_2 Orion 97-08-00 electrode was used.

Potentials are monitored with a Tacussel Minisis 6000 millivoltmeter in conjunction with a Goertz Servogor 120 recorder. Measurements are performed in a thermostatic cell ($\pm$ 0.2°C).

All chemicals are of analytical or pharmacopoeia pure grade (Difco and Merck). The strains used are : Serratia marcescens ATCC 13880 (asparagine) and Escherichia coli B/It 7-A ATCC 27553 (tryptophan), Escherichia coli ATCC 27195 and wild isolates of Bacillus pumilus (nicotinamide) and wild isolates of Enterobacter

TABLE I: Analytical applications of enzyme electodes

SUBSTRATE	ENZYME	ELECTRODE	REFERENCE
Acetylcholine	Acetylcholinesterase	pH	(7)
		Choline	(8)
D-amino acids	D-amino acid oxidase	NH_4^+	(9,10)
L-amino acids	L-amino acid oxidase	NH_4^+	(11,12)
		NH_3	(13)
		$Pt(H_2O_2)$	(14,15)
		$Pt(O_2)$	(16)
		$C(H_2O_2)$	(17)
	L-amino acid oxidase + peroxidase	I^-	(12)
	L-amino acid decarboxylase	CO_2	(19)
L-glutamic acid	L-glutamate dehydrogenase	NH_4^+	(20)
		Pt (NADH)	(21)
	L-glutamate dehydrogenase + NADH oxidase	O_2	(22)
	Glutamate decarboxylase	CO_2	(23)
L-lactic acid	Lactate dehydrogenase	C (NADH)	(24,25)
		Pt (NADH)	(26)
	Lactate dehydrogenase + NADH oxidase	O_2	(22)
	Cytochrome b_2	Pt or GC ($Fe(CN)_6^{=}$)	(27-29)
	Lactate oxidase	O_2	(30-32)

Table I (continued)

SUBSTRATE	ENZYME	ELECTRODE	REFERENCE
Pyruvic acid	Glutamate dehydrogenase + lactate dehydrogenase	NH_4^+	(20)
	Lactate dehydrogenase	C (NADH)	(25)
Uric acid	Uricase	Pt (O_2)	(33)
		O_2	(34)
		CO_2	(35)
	Uricase + Peroxidase	C($Fe(CN)_6^{\equiv}$)	(36)
Adenosine	Adenosine deaminase	NH_3	(37)
L-alanine	L-alanine dehydrogenase	NH_3	(38)
	L-alanine dehydrogenase + NADH oxidase	O_2	(22)
Biological amines (adrenaline, tyramine, serotonine)	Monoamine oxidase	O_2	(39)
		NH_3	(40)
L-arginine	Arginase + urease	NH_4^+	(41,42)
	Arginine decarboxylase	$CO_3^=$	(44)
		CO_2	(45)
	Arginine decarboxylase + diamine oxidase	O_2	(46)
Cholesterol and cholesterol esters	Cholesterol oxidase	Pt (O_2)	(47)
		Pt (H_2O_2)	(48,49)
	Cholesterol esterase + cholesterol oxidase	Pt (O_2)	(50,51)
		Pt (H_2O_2)	(52-54)
Choline	Choline oxidase	O_2	(55)

Table I (continued)

SUBSTRATE	ENZYME	ELECTRODE	REFERENCE
Creatinine	Creatininase	NH_4^+	(56)
		NH_3	(57,58)
L-cysteine	β-cyanoalanine synthetase	Ag_2S	(59)
L-dopa	Tyrosinase	O_2	(60)
Flavine adenine dinucleotide	Alkaline phosphatase + adenosine deaminase	NH_3	(61)
Galactose	Galactose oxidase	Pt (H_2O_2)	(49,62, 63)
Glucose	Glucose oxidase	Pt(O_2)	(16,64, 65)
		Pt(quinone)	(66)
		Au(quinone)	(67)
		Pt(H_2O_2)	(15,49, 62,68, 69)
		Au(H_2O_2)	(70)
		C(Fe(CN)$_6^{\equiv}$	(71)
		O_2	(18,72, 73)
	Glucose oxidase + peroxidase	I^-	(74)
		F^-	(75)
		Pt or GC (Fe (CN)$_6^{\equiv}$)	(36,76)
	Glucose oxidase + catalase	Pt (H_2O_2)	(77)

Table I (continued)

SUBSTRATE	ENZYME	ELECTRODE	REFERENCE
Glucose-6-phosphate	Glucose-6-phosphate dehydrogenase	C(NADH)	(25)
L-gutamine	Glutaminase	NH_4^+	(23)
		NH_3	(78)
L-histidine	Histidine ammonia-lyase	NH_3	(79)
	Histidine decarboxylase	CO_2	(80)
Lactose	β-galactosidase +	O_2	(81)
	glucose oxidase	Pt(H_2O_2)	(49)
	Galactose oxidase	Pt(H_2O_2	(62)
Leucine	Leucine dehydrogenase + NADH oxdiase	O_2	(22)
Lipids	Phospholipase D + Choline oxidase	Pt(H_2O_2)	(82)
L-lysine	Lysine decarboxylase	CO_2	(83)
	Lysine decarboxylase + diamine oxidase	O_2	(46)
Maltose	Glucoamylase +	Pt(H_2O_2)	(49,84)
	glucose oxidase	O_2	(73,81,85)
L-methionine	Methionine ammonia-lyase	NH_3	(86)
NAD^+	Hexokinase + glucose oxidase or glucose dehydrogenase	O_2	(73,87)
NADH	Alcohol dehydrogenase	Pt	(88)
	Peroxidase	O_2	(89)

TABLE I (Continued)

SUBSTRATE	ENZYME	ELECTRODE	REFERENCE
Penicillines (ampicilline, nafcilline, penicillines G and V, cyclicilline, dicloxacilline)	Penicillinase	pH	(90-95)
L-phenylalanine	L-phenylalanine ammonia-lyase	NH_3	(11)
	L-phenylalanine decarboxylase	CO_2	(19,96)
Triglycerides	Lipase + glycerol dehydrodenase + peroxi ase	O_2	(97)
L-tyrosine	Tyrosine decarboxylase	CO_2	(19,96,98)
	Tyrosinase	O_2	(60,99)
Urea	Urease	pH	(90)
		CO_2	(18)
		NH_4^+	(100,101)
		NH_3	(102-107)

TABLE II: Analytical applications of bacterial electrodes

SUBSTRATE	BIOCATALYST	ELECTRODE	REFERENCE
Lactic acid	Hansenula anomala (cytochrome b_2)	Pt or GC ($Fe(CN)_6^{=}$)	(108)
Arginine	Streptococcus daecium faecalis or lactis (multienzyme sequence)	NH_3	(109-111)
Aspartate	Bacterium cadaveris (aspartate ammonia-lyase)	NH_3	(112)
Cholesterol	Nocardia erthropolis (cholesterol oxidase)	O_2	(113)
Cysteine	Proteus morganii (crysteine desulf-hydrase)	H_2S	(114)
Creatinine	Creatininase + Nitro-somonas sp. + Nitro-sobacter sp.	O_2	(115)
Glucose	Aspergillus niger (glucose oxidase)	$C(Fe(CN)_6^{=})$	(71)
	Hansenula anomala(BOD)	O_2	(116)
Glutamate	Escherichia coli (glut-amate decarboxylase)	CO_2	(117)
Glutamine	Sarcina flava (glutaminase)	NH_3	(118)
Histidine	Pseudomonas sp. (histidine ammonia-lyase)	NH_3	(119)
Nystatine	Saccharomyces cerevisiae (BOD)	O_2	(120)
Pyruvate	Streptococcus faecium (pyruvate dehydrogenase)	CO_2	(121)
	Hansenula anomala (BOD)	O_2	(116)
Tyrosine	Aeromonas phenologenes (tyrosine phenol-lyase)	NH_3	(122)
Saccharose	Hansenula anomala (BOD)	O_2	(116)
Steroids(cortisol, testosterone, methyltestosterone	Nocardia opaca (Δ^1-dehydrogenase	GC(dichloro-phenolindo-phenol)	(123)
Urea	Proteus mirabilis (urease)	NH_3	(124)

agglomerans (ascorbic acid). Tryptophanase is a Sigma Chemical Co. product and ascorbate oxidase is a Boehringer Co. product.

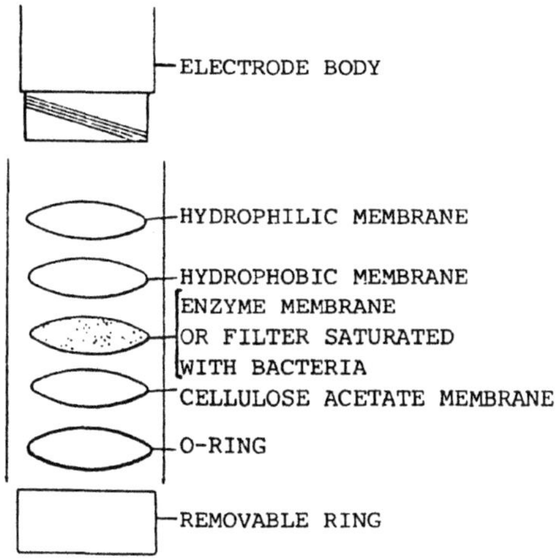

Figure 1. Setting of the membranes in an enzyme or a bacterial electrode

The ammonia gas sensing electrode is modified by setting the enzyme or the bacterial membrane against the hydrophobic membrane (Fig. 1). The catalytic layer is protected by a cellulose acetate membrane.

The bacterial membrane is prepared as follows: the stock strain, conserved at 4°C on Heart Infusion Agar slants, is cultured at 37°C in the liquid growth medium. The third subculture is filtered on a microporous cellulose acetate filter. This one, coated with the bacteria, is dried 30 minutes at 37°C and the electrode membrane is cut out (Fig. 2).

Two growth media have been tested for each bacterial strain: the first, said "poor medium", contains yeast extract and tryptone. The second, or "rich medium", is the Brain Heart Infusion medium (BHI Difco)(125).

3. RESULTS AND DISCUSSION

3.1. Asparagine

Asparagine has been determined according to the following reaction:

$$\text{L-asparagine} + H_2O \xrightarrow{\text{bacterial}} \text{L-aspartic acid} + NH_3$$

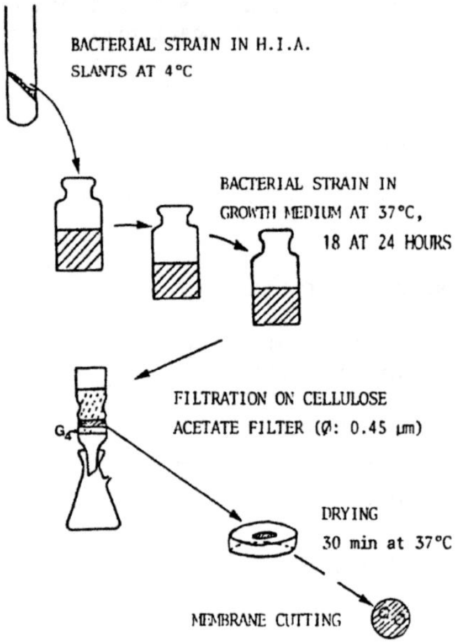

Figure 2. Preparation of the bacterial membrane.

Studying the influence of the growth medium on the electrode response, we have pointed out that the better response was obtained when the substrate, asparagine, was added to the growth medium (1.5 g per 100 ml). This procedure induces or increases the synthesis of asparaginase by the bactria(126). It has been observed that the BHI medium increases the proliferation of the cells, giving rise to a better sensitivity, but a wider linear range is obtained with the first or poor medium which will be preferred for analytical purposes.

Asparaginase is very unstable and enzyme electrodes equipped with a liquid membrane of purified asparaginase lose the half of their activity after three or four days. Chemically bound enzyme electrodes are still more stable(9).

The enzyme activity of bacterial electrodes decreases if they are stored in their growth medium at 4°C or in the buffer at room temperature (Fig. 3 A). The slope and the linear range decrease. In such conditions, the growth of the cells is inhibited. But if the electrode is stored in its growth medium at 30°C under slow shaking, proliferation of the bacteria is maintained at the surface of the electrode and its stability is enhanced up to more than three weeks (Fig. 3 B). During the fourth week, the slope decreases slowly as the linear range and the response time increases.

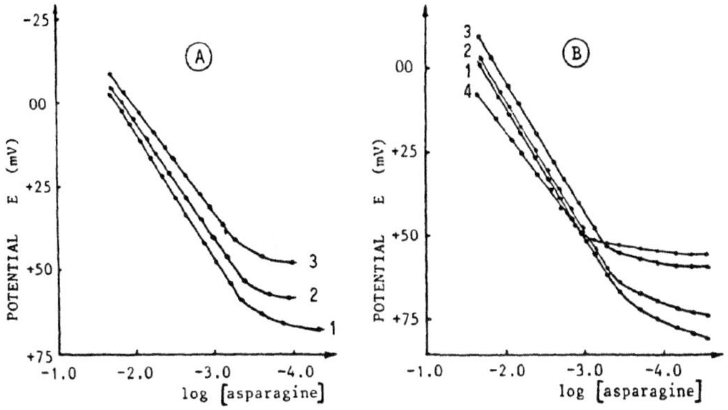

Figure 3. Stability of the asparagine bacterial electrode.
A) conserved in the poor medium at 4°C.; 1) 1st and 2nd days; 2) 3th day; 3) 5th day.
B) conserved in the poor medium at 30°C under shaking. 1) 1st day; 2) 16th day; 3) 22nd day; 4) 27th day.

TABLE III: Influence of the nature of the buffer, its concentration and of the pH on the slope and the linearity range of the electrode response

BUFFER AND CONCENTATION	pH	SLOPE (mV/dec)	LINEARITY RANGE (mole $.1^{-1}$)
Tris - HCl 0.1 M	8.5	35	2.10^{-2} to 8.10^{-4}
Phosphate 0.1 M	8.5	48	3.10^{-2} to 3.10^{-3}
Tris - HCl 0.1 M	7.8	36	2.10^{-2} to 6.10^{-4}
Phosphate 0.1 M	7.8	50	4.10^{-2} to 2.10^{-3}
Tris - HCl 0.05 M	7.8	38	2.10^{-2} to 4.10^{-4}
Tris - HCl 0.1 M	.4	37	2.10^{-2} to 6.10^{-4}

The physico-chemical parameters are also of great importance (Table III). If a phosphate buffer increases the sensitivity, the tris-hydrochloride buffer allows the attainment of a linear range. A variation of pH from 7.4 to 8.5 does not strongly affect the slope and the linear range, allowing studies with bacterial electrodes in a wider pH range than with enzyme electrodes, for which this parameter is more critical(9).

The best response time is obtained at 30°C but, again, the maximum is not so sharp as that observed with an enzyme electrode.

Although microbial electrodes have sometimes a longer response time than enzyme electrodes, due to the thickness of the bacterial membrane, this parameter does not exceed 4 to 9 minutes, the longer time being observed with the lower concentrations.

From this study, the best operating conditions for the asparagine electrode can be summarized as follows: 0.05 M tris-HCl buffer at pH 7.8, 30°C, storage of the electrode in the growth medium at 30°C under shaking. The characteristics of the electrode are then: a slope of 38 mV per decade, a linear range from 2.10^{-2} to 4.10^{-4}M, a response time of 4 to 9 minutes and a lifetime of about 26 days.

Serratia marcescens contains several enzymes able to deaminate other amino-acids, thus some interferences may occur. From a study covering 18 amino-acids and urea, it appears that only L-glutamine and L-histidine give a relative response of 20%, the others having a relative response lower than 1%.

3.2 Tryptophan

Tryptophan can be determined using the tryptophanase synthetized by Escherichia coli. This enzyme catalyses the following reaction:

$$\text{L-tryptophan} + H_2O \xrightarrow[\text{or bacterial tryptophanase}]{\text{tryptophanase}} \text{indole} + \text{puyruvic acid} + NH_3$$

The bacterial electrode is constructed following the same procedure as for asparagine, the cells being cultured in a poor medium with indole as inducer (674 B/It-7a medium)(125).

The characteristics of the bacterial electrode were compared with those obtained for the enzyme electrode covered with a crude extract of tryptophanase. 30 mg of the enzyme and 0.5 mg of pyridoxal phosphate as cofactor are put on a cellophane membrane and fixed against the hydrophobic membrane of the electrode. The electrode is then immersed during one hour in the buffer at 37°C in order to dissovle the enzyme. Physical immobilization in agar gel and chemical immobilization with glutaraldehyde on a cellulose acetate membrane were also tested but the stability of such electrodes is not better than that of solubilized enzyme. This will be due to the impossibility for the immobilized enzyme to react with its cofactor, the pyridoxal phosphate. This cofactor of tryptophanase must always be added to the solution but at a higher concentration for the enzyme electrode (20 mg for 100 ml) than for the bacterial electrode (2.5 mg for 100 ml), the bacteria containing it partially.

In order to maintain the best linearity range, the working temperature must stay below 35°C for both electrodes. While the sensitivity of the bacterial electrode is not affected by a change of temperature between 25°C and 35°C, the slope of the enzyme electrode decreases with this parameter. As a compromise, we have choosen a working temperature of 30°C for both electrodes (Fig. 4 A).

The effect of pH is also critical for the enzyme electrode where the maximum slope appears at pH 8.7 but with a restricted linear range. This effect is not pronounced with the bacterial electrode and a pH of 8.3 was selected (Fig. 4 B).

The tris-HCl buffer employed for asparagine cannot be used in this case, the electrode response being inhibited. A 0.1 M phosphate buffer gives the best results.

The bacteria electrode is also stored in its growth medium at 37°C in order to regenerate the enzyme synthesis. The enzyme electrode is stored at 4°C in the working buffer.

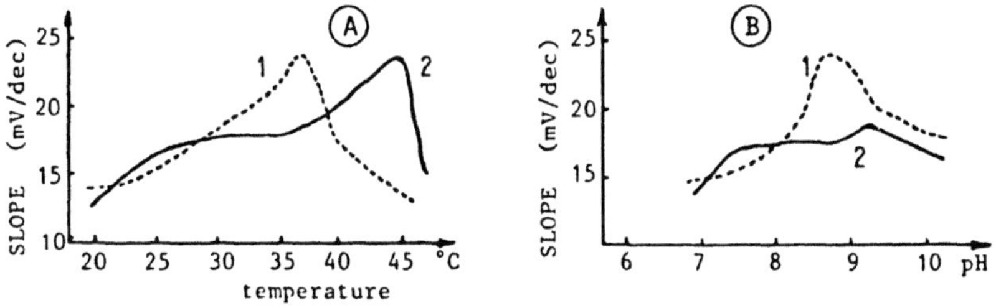

Figure 4. Effect of temperature (A) and of pH (B) on the slope of the tryptophan electrodes. 0.1 M phosphate buffer (A) pH 8.3: (B) T = 30°C
1) Enzyme electrode; 2) Bacterial electrode.

Figure 5 shows that both electrodes exhibit a maximum of sensitivity at the fourth day. This probably may be attributed to the complex part taken by pyridoxal phosphate in the steric modification of the enzye(127). The slope of the bacterial electrode

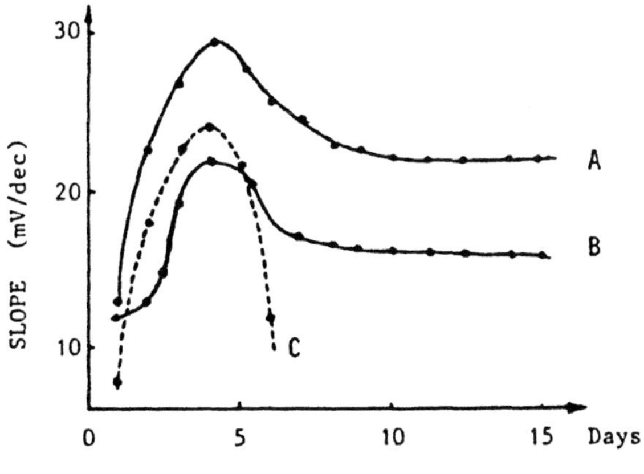

Figure 5. Stability of the tryptophane electrodes. 0.1 M phosphate buffer, pH 8.3; T = 30°C.
A Bacterial electrode regenerated in poor medium
B Bacterial electrode regenerated in BHI medium
C Solubilized enzyme electrode.

is then stabilized, giving a lifetime of three weeks. The sensitivity of the enzyme electrode falls rapidly after the maximum and this sensor cannot be used more than four days. Immobilization of the enzyme does not increase the stabiltity.

Several compounds may interfere, due to the fact that tryptophanase catalyses other reactions, the substrates being tryptophan, cysteine and serine. Molecules which are structually related to tryptophan also interfere: tryptamine or 5-hydroxytrytophan. In addition, Escherichia coli has also other deaminating enzymes, able to react with asparagine, glutamine and aspartic acid. These interferences also occur with the enzyme electrode because it is a crude extract not well purified.

Table IV compares the characteristics of both sensors.

TABLE IV: Comparison of the characteristics of enzyme and bacterial electrodes for tryptophan determination.

Criteria	Bacterial electrode	Solubilized enzyme electrode
Linearity range (mole. l^{-1})	1.10^{-2} to 3.10^{-4}	1.10^{-2} to 5.10^{-4}
Slope (mV/dec)	22	18
Response time (min)	10 to 15	10 to 15
Stability	3 weeks	4 days
Cofactor concentration	2.5 mg/100 ml	20 mg/100 ml
Interferences	Asparagine Glutamine Aspartic acid	Asparagine Glutamine Aspartic acid
Stability of stock material	Several years on HIA slants at $4^{o}C$.	Less than 2 months (lyophilized form)

3.3 Nicotinamide

Nicotinamide deaminase (nicotinamide amiohydrolase EC 3.5.1.19), a hyrolyzing enzyme involded in nicotinamide adenine dinucleotide (NAD) metabolism, was found to be present in microorganisms such as Aspergillus niger, Escherichia coli, Bacillus pumilus as well as in vertebrate tissues (rabbit and rat liver). The role of nicotinamide deaminase in the maintenance of the cellular NAD level has been studied by Sarma et al.(128). According to these investigations, the enzyme deaminates nicotinamide to nicotinic acid and the nicotinic acid formed is utilized for NAD-biosynthesis via the Preiss and Handler pathway.

The interest of the enzymatic deamination, following the reaction:

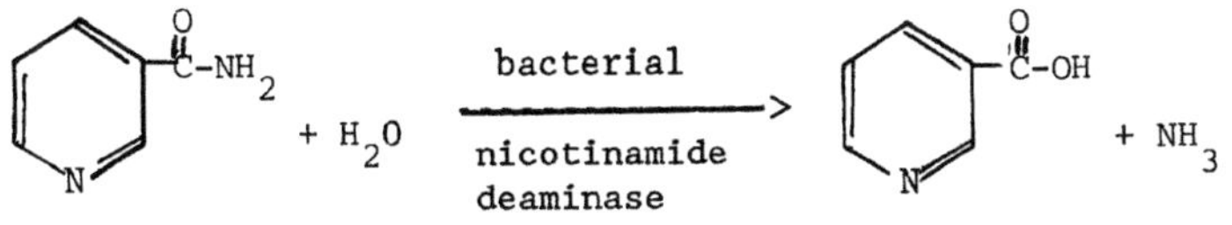

is to present a greater selectively than the chemical hydrolysis.

Two electrodes have been constructed using either Escherichia coli ATCC 27195 (mutated strain) after three subcultures in BHI medium, or Bacillus pumilus with first an induction step in poor medium followed by two subcultures in BHI medium in order to obtain the best growth rates and enzymes activities.

The best slope and linearity were obtained in a 0.01 M tris-HCl buffer, pH 7.8 at 30°C, supplemented with 4.10^{-3}M magnesium chloride as metallic cofactor for the enzyme activity.

It is well known that the stability of the purified enzyme at 30°C is only about 20 minutes. The way we work strongly increases the analytical lifetime by regenerating the electrode in the growth medium. Table V compares the stability of two different types of electrodes and their corresponding analytical characteristics.

TABLE V: Stability and analytical characteristics of E. coli and B. pumilus electrodes used for the determination of nicotinamide.

Type of electrode	Slope (mV/dec)	Linearity range (mole. l^{-1})	Stability (days)
E. coli ATCC 27195	38 ± 1	$2.8x10^{-4}$ to $2.0x10^{-2}$	5
B. pumilus after induction	38 ± 2	$2.8x10^{-4}$ to $1.4x10^{-2}$	3

The decrease of stability after 3 days in the case of Bacillus pumilus is caused by the appearance of a progressive sporulation observed by microscopy in spite of the regeneration. This fact appears also in the study of three other enzymatic deamination pathways of B. pumilus, such as asparaginase, glutaminase and histidine ammonia-lyase, where a reduction of the slope and the linearity at higher substrate concentration values is also reached after the third day. On the contrary, in E. coli, the activity is maintained during more than 10 days.

Quantitative determinations of nicotinamide in aqueous solutions have been realized on the two proposed electrodes (E. coli and B. pulilus). The results, as shown in table VI, demonstrate a perfect correlation between the two electrodes.

TABLE VI: Determination of nicotinamide in tris-HCl 0.01 M pH 7.8 buffer + $MgCl_2$ 4 x 10^{-4} M at 30.0 $\pm$ 0.2°C.

Nicotinamide added (mg/l)	Nicotinamide found* (mg/l)		Coefficient of variation (%)	
	E. coli	B. pumilus	E. coli	B. pumilus
34.1	34.5	34.0	3.2	3.8
58.3	57.9	58.5	3.1	2.7
82.5	82.7	83.1	2.4	2.5
130.5	130.5	129.9	2.4	2.2
225.3	224.5	226.1	2.2	2.1
410.6	411.1	409.3	2.4	2.6
849.8	845.1	853.2	1.6	1.8

* Average of 5 measurements

3.4 Ascorbic acid

L-ascorbic acid membrane electodes based upon the use of four classes of biocatalysis immobilized at an oxygen electrode were evaluated and compared in terms of electrode properties and operating requirements. Isolated ascorbic oxidase enzyme in soluble form and in covalent binding matrices, peel of cucumber and living bacterial cells of Enterobacter agglomerans strain, respectively, are employed as biocatalytic layers.

The ascorbate oxidase is able to catalyse the following reaction:

$$\text{L-ascorbic acid} + 1/2\ O_2 \xrightarrow[\text{oxidase}]{\text{ascorbate}} \text{dehydroascorbic acid} + H_2O$$

Depletion of oxygen is measured with a pO_2 electrode.

The growth of the bacteria is carried out in a poor medium containing 0.2% of the substrate as inducer.

The tissular electrode is prepared by cutting 500 to 600 µm peels of cucumber mesocarp and immobilizing it onto the electrode surface by means of a nylon net.

The solubilized enzyme is maintained by means of a dialysis membrane (cellophane 50 µm).

Two types of covalent binding have been used for the immobilized enzyme electrodes: the first is a direct binding of the enzyme on a cellulose acetate support with glutaraldehyde (type I), the second being a coreticulation binding with an inert protein on the same support(15) (type II).

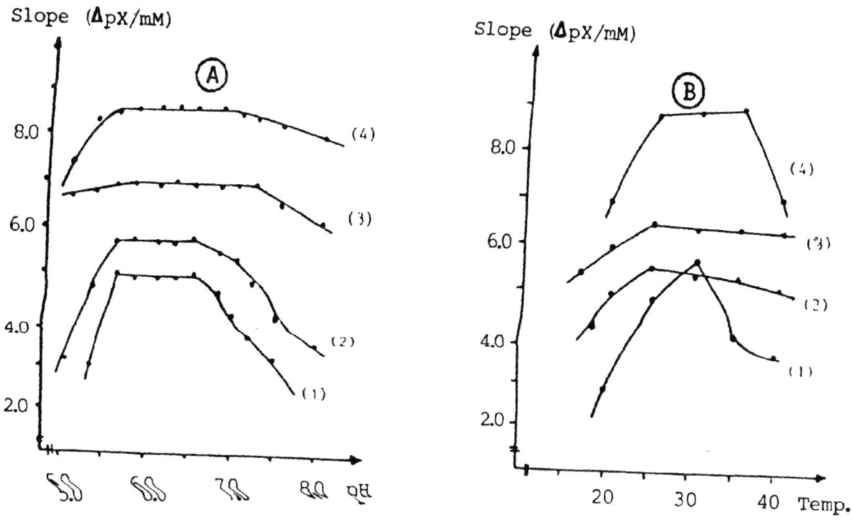

Figure 6. Effect of pH (A) and of temperature (B) on the slope of the ascorbic acid electrodes. 0.1 M phosphate buffer (A) T = 25°C; (B) pH 6.5; (1) Solubilized enzyme electrode (2) Tissular electrode; (3) Immobilized enzyme electrode (4) Bacterial electrode.

Figure 6 describes the influence of pH (A) and temperature (B) on the slope of the four investigated electrodes. These parameters however do not affect their linearity range. From these results, a 0.1 M phosphate buffer, pH 6.5 at 25°C has been selected. The linearity range decreases above this buffer concentration.

Figure 7 shows that the stability of the electrodes is closely related to the immobilization mode of the enzyme or to its synthesis by the living cells, especially when the catalytic membrane is regenerated.

The analytical charateristics of these electrodes, which are summerized in table VII, are relatively similar. As can be expected, the solubilized enzyme electrode exhibits a very short lifetime but also a lower slope and a longer response time. For these reasons, this type of sensor has not been considered for analytical purposes.

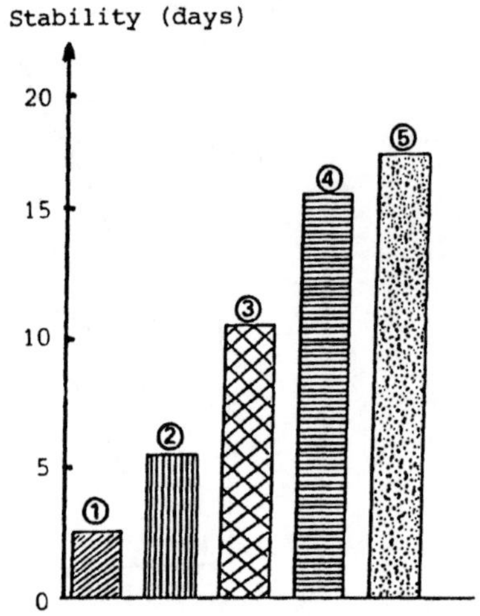

Figure 7: Stability of the ascorbic acid electrodes.
(1) Solubilized enzyme electrode
(2) Tissular electrode
(3) Bacteria electrode without regeneration
(4) Bacterial electrode with regeneration
(5) Immobilized enzyme electrode (type I and II)

TABLE VII: Charateristics of the ascorbic acid electrodes.
0.1 M phosphate buffer pH 6.5, 25.0°C

Type of electrode	Slope (DpX/mM)	Linearity range (mole . 1^{-1})	Response time (min)	Stability (days)
Bacterial electrode	8.6±0.3	4.10^{-6}-7.10^{-4}	2.5to3.5	10 to 11[a] 15 to 16[b]
Tissular electrode	5.8±0.8	4.10^{-6}-1.10^{-3}	1.5to2.0	5 to 6
Solubilized enzyme elec.	4.9±0.4	4.10^{-6}-5.10^{-4}	4.0to5.0	2 to 3
Immobilized enzyme elec. (type I & II)	6.7±0.2	4.10^{-6}-9.10^{-4}	2.0to2.5	16 to 18

(a) conserved at 4°C.
(b) regenerated in its growth medium at 27°C.

4. CONCLUSIONS

This study shows that there are many similarities between enzyme and bacteria electrodes. The linearity range, the slope and the response time are quite similar, but the stability of the bacterial electrode is enhanced in most cases, owing to the regeneration of the living cells during the storage of the electrode. This possibility is of great interest when the enzyme is unstable or cannot be immobilized. The selectivity, which is often better for the enzyme, depends on the purification of the catalyst or on the metabolism of the bacteria. In addition, the enzyme is not always available or may be unstable, even in its lyophil zed form, whereas the strain is available and may be stored at 4°C during several years.

5. ACKNOWLEDGEMENT

Thanks are expressed to the Fonds National de la Recherche Scientifique (Belgium) for help to one of us (G.J.P.) and to the SPPS (Belgium Politic Research, ARC), contract n° 86/91-89.

6. REFERENCES

1. G.G. Guilbault, Enzyme Electrodes in Analytical Chemistry, in Wilson and Wilson's Comprehensive Analytical Chemistry, vol 8. G. Svehla (Ed.), Elsevier, Amsterdam, 1977.
2. G.G. Guilbault, Analytical Uses of Immobilized Enzymes, M. Dekker, Inc., New York, 1984.
3. P. Vagdama, Enzyme Electrodes, In A.K. Covington, Ion-Selective Electrode Methodology, vol 2, C.R.C., Boca Raton, 1979.
4. M.M. Fishman, Anal. Chem. 52, 185 R, (1980).
5. M.A. Arnold, M.E. Meyerhoff, Anal. Chem., 56, 20 R, (1984).
6. G.G. Guilbault, G. de Olivera Neto, Immobilized Enzyme Electrodes, in J. Woodward, Immobilized Cells and Enzymes, a Practical Approach, I.R.L. Press, Oxford, 1985.
7. P. Durand, A. David, D. Thomas, Biochim. Biophys. Acta., 527, 277, (1978).
8. G. Braum, M. Lynn, Anal. Chim. Acta, 65, 385, (1973).
9. G.G. Guilbault, E. Hrabankova, Anal. Chim. Acta, 56, 285, (1971).
10. R. Lienado, G.A. Rechnitz, Anal. Chem., 46, 1109, (1974).
11. C.P. Hsiung, S.S. Kuan, G.G. Guilbault, Anal. Chim. Acta,90, 45, (1977).
12. G.G. Guilbault, G. Nagy, Anal. Letters, 6, 301, (1973).
13. G. Johannson, K. Edstrom, L. Ogren, Anal. Chim. Acta, 85, 55, (1976).
14. G.G. Guilbault, G.J. Lubrano, Anal. Chim. Acta, 69, 183, (1974).
15. G.G. Guilbault, G.J. Lubrano, Anal. Chim. Acta. 97, 229, (1978).
16. M. Nanjo, G.G. Guilbault, Anal. Chim. Acta., 73, 367 (1974).

17. R.M. Ianniello, A.M. Yacynych, Anal. Chim. Acta, 131, 123, (1981).
18. C. Tran-Minh, G. Broun, Anal. Chem., 47, 1359, (1975).
19. C. Calvot, A.M. Berjonneau, G. Gelf, D. Thomas, FEBS Letters, 59, 258, (1975).
20. P. Davis, K. Mosbach, Biochim. Biophys. Acta, 370, 329, (1974).
21. A. Malinauskas, J. Kulys, Anal. Chim. Acta, 98, 31, (1978).
22. H. Huck, A. Schelter-Graf, J. Danzer, P. Kirch, H-L. Schmidt, Analyst, 109, 147, (1984).
23. G.G. Guilbault, F. Shu, Anal. Chim. Acta, 56, 333, (1971).
24. L. Thomas, G.D. Christian, Anal. Chim. Acta, 78, 271, (1975).
25. N. Cenas, J. Rozgaite, J. Kulys, Biotechnol. Bioeng., 26, 551, (1984).
26. W.J. Blaedel, R.C. Engstrom, Anal. Chem., 52, 1691, (1980).
27. H. Durliat, M. Comtat, A. Baudras, Clin, Chem., 22, 1802, (1976).
28. H. Durliat, M. Comtat, Anal. CHem., 52, 2109, (1980).
29. T. Clerbaux, P. Vanhove, L. Brasseur, Path. Biol., 29, 636, (1981).
30. F. Mizutani, K. Sasaki, Y. Shimura, Anal. Chem., 55, 35, (1983).
31. M. Mascini, D. Moscone, G. Palleschi, Anal. Chim. Acta, 157, 45, (1984).
32. E.B. Makovos, C.C. Liu, Biotechnol, Bioeng., 27, 167, (1985).
33. G.G. Guilbault, M. Nanjo, Anal. Chem., 46, 1769, (1974).
34. W. Dritschilo, M.K. Weibel, Biochem. Med., 11, 242, (1974).
35. T. Kawashima, G.A. Rechnitz, Anal. Chim. Acta, 83, 9, (1976).
36. J.J. Kulys, V-S. A. Laurinavicius, M.V. Pesliakiene, V.V. Guerviciene, Anal. Chim. Acta, 148, 13, (1983).
37. I. Deng, C. Enke, Anal. Chem., 52, 1937, (1980).
38. D.P. Nikolelis, Anal. Chim. Acta, 167, 381, (1985).
39. K. Kaminiski, J. Bodziony, W. Matuszewski, R. Winter, W. Waronski, Arch. Geshw-Forsch., 46, 360, (1976).
40. L.R. Meyerson, K.D. McMurtrey, V.E. Davis, Anal. Biochem., 86, 287, (1978).
41. T. Anfalt, A. Granelli, D. Jagner, Anal. Letters, 6, 969, (1973).
42. T.A. Neubecker, G.A. Rechnitz, Anal. Letters, 5, 653, (1972).
43. P. Valle-Vega, C.T. Young, H.E. Swaisgood, J. Food Sci., 45, 1026, (1980).
44. S.L. Tong, G.A. Rechnitz, Anal. Letters, 9, 1, (1976).
45. P. Valle-Vega, C.T. Young, H.E. Swaisgood, J. Food Sci., 45, 1003, (1980).
46. L. Macholan, Coll. Czech. Chem. Commun., 43, 1811, (1978).
47. I. Satoh, I. Karube, S. Suzuki, Biotechnol. Bioeng., 19, 1095, (1977).
48. C. Bertrand, P.R. Coulet, D.G. Gautheron, Anal. Letters, 12, 1477, (1979).
49. C. Bertrand, P.R. Coulet, D.G. Gatheron, Anal. Chim. Acta, 126, 23, (1981).

50. A. Noma, K. Nakayama, Clin. Chem., 22, 336, (1976).
51. J.M. Drietschy, L.E. Weeks, J.J. Delenk, Clin. Chim. Acta, 73, 407, (1976).
52. H.S. Huang, S.S. Kuan, G.G. Guilbault, Clin. Chem., 23, 671, (1977).
53. H.S. Huang, S.S. Kuan, G.G. Guilbault, Clin. Chem., 21, 1605, (1975).
54. P.R. Coulet, L.J. Blum, Anal Letters, 16, 541, (1983).
55. K. Matsumoto, H. Seijo, I. Karube, S. Suzuki, Biotechnol. Bioeng., 22, 1071, 1980.
56. H. Thompson, G.A. Rechnitz, Anal. Chem., 46, 246, (1974).
57. G.G. Guilbault, S.P. Chen, S.S. Kuan, Anal. Letters, 13, 1607, (1980).
58. G.G. Guilbault, P.R. Coulet, Anal. Chim. Acta, 152, 223, (1983).
59. G.G. Guilbault, W.F. Gutknecht, S.S. Kuan, R. Cockran, Anal. Biochem., 46, 200, (1972).
60. J.L. Iborra, E. Vilanova, J.A. Lozano, Biochem. Educ., 9, 51, (1981).
61. A.R.M. Yatim, T.L. Riechel, Anal. Letters, 17, 835, (1984).
62. T.A. Grooms, L.C. Clark Jr., B.J. Weiner, Enzyme Eng., 5, 217, (1980).
63. J.M. Johnson, H.B. Halsall, W.R. Heineman, Anal. Chem., 54, 1394, (1982).
64. M. Nanjo, G.G. Guilbault, Anal. Chim. Acta, 75, 169, (1975).
65. M. Koyama, Y Satoh, M. Aizawa, S. Suzuki, Anal. Chim. Acta, 116, 307, (1980).
66. D. Williams, A. Doig, A. Korosi, Anal. Chem., 42, 118, (1970).
67. L. Gorton, K.M. Bhatti, Anal. Chim. Acta, 105, 43, (1979).
68. G.G. Guilbault, G.J. Lubrano, Anal. Chim. Acta., 64, 439, (1973).
69. D.R. Thevenot, R. Sterngerg, P.R. Coulet, J. Laurant, D.G. Gautheran, Anal. Chem., 51, 96, (1979).
70. E. Lobel, J. Rishpon, Anal. Chem., 53, 51, (1981).
71. B.J. Vincke, J-M. Kauffman, M.J. Devleeschouwer, G.J. Patriarche, Analusis, 12, 141, (1984).
72. J.L. Romette, B. Froment, D. Thomas, CLin. CHim. Acta, 95, 249, (1979).
73. D. Pfeiffer, F. Scheller, M. Janchen. K. Berteermann, Biochimie, 62, 587, (1980).
74. G. Nagy, L.H. Von Storp, G.G. Guilbault, Anal. Chim. Acta, 66, 443, (1973).
75. I.W. Siddiqi, Clin. Chem., 28, 1962, (1982).
76. J.J. Kulys, M.V. Pesliakiene, A.S. Samalius, Bioelectrochem. Bioenerg., 8, 81, (1981).
77. C.C. Liu, L.B. Wingard Jr., S.K. Wolfson Jr., S.J. Yao, A.L. Drash, J.G. Schiller, Bioelectrochem, Bioenerg., 6, 19, (1979).

78. M.A. Arnold, G.A. Rechnitz, Anal. Chem., 52, 1170, (1980).
79. R.R. Walters, P.A. Johnson, R.P. Buck, Anal. Chem., 52, 1684, (1980).
80. P.M. Kovack, M.E. Meyerhoff, Anal. Chem., 54, 217, (1982).
81. F. Cheng, G.D. Christian, Analyst, 102, 124, (1977).
82. S. Suzuki, I. Karube, Enzyme Eng., 55, 205, (1980).
83. G.G. Guilbault, C. White, Anal. Chem., 50, 1481, (1978).
84. P.R. Coulet, C. Bertrand, Anal. Letters, 12, 581, (1979).
85. L.B. Marschall, G.D. Christian, A. Kumar, Analyst, 102, 424, (1977).
86. G.G. Guilbault, S.S. Kuan, K.W. Fung, H.Y. Sung, Anal. Chem., 51, 2319, (1979).
87. D. Pfeiffer, F. Scheller, M. Janchen, K. Bertermann, H. Weise, Anal. Letters, 13, 1179, (1980).
88. H. Jaegfeldt, A. Torstensson, G. Johnansson, Anal. Chim. Acta, 97, 221, (1978).
89. D. Kirstein, F. Schubert, F. Scheller, Anal. Letters, 15, 1479, (1982).
90. H. Nilsson, A. Akerlund, K. Mosbach, Biochem. Biophys. Acta, 320, 529, (1973).
91. G.J. Papariello, A.K. Mukherji, C.M. Schearer, Anal. Chem., 45, 790, (1973).
92. H. Nilsson, K. Mosbach, S.O. Enfors, M. Molin, Biotechnol. Bioeng., 20, 527, (1978).
93. S.O. Enfors, H. Nilsson, Enzyme Microb. Technol., 1, 260, (1979).
94. S. Caras, J. Janata, Anal. Chem., 52, 1935, (1980).
95. L.F. Cullen, J.F. Rusling, A. Schleifer, G.J. Paperiello, Anal. Chem., 46, 1955, (1974).
96. A.M. Berjonneau, D. Thomas, G. Broun, Path. Biol., 22, 497, (1974).
97. T.A. Kelly, G.D. Christian, Analyst, 109, 453, (1984).
98. J. Havas, G.G. Guilbault, Anal. Chem., 54, 1991 (1982).
99. A. Kumar, G.D. Christian, Clin. Chem., 21, 325, (1975).
100. G.G. Guilbault, g. Nagy, S.s. Kuan, Anal. Chim. Acta, 67, 195, (1973).
101. G.G. Guilbault, G. Nagy, Anal. Chem., 45, 417, (1973).
102. M. Mascini, G.G. Guilbault, Anal. Chem., 49, 795, (1977).
103. G.G. Guilbault, W. Stokbro, Anal. Chim. Acta, 76, 237, (1975).
104. N.R. Larsen, E.H. Hansen, G.G. Guilbault, Anal. Chim. Acta, 75, 9, (1975).
105. P. Eliard, A. Laudet, J-C. Viré, G.J. Patriarche, Analusis, 10, 182, (1982).
106. M. Fischer, A. Laudet, J-C. Viré, G.J. Patriarche, S.P. Dufrane, J. Pharm. Belg., 37, 202, (1982).
107. G.G. Guilbault, M. Tarp, Anal. Chim. Acta, 73, 355, (1974).
108. B.J. Vincké, M.J. Devleeschouwer, G.J. Patriarche, Anal. Letters, 18, 593, (1985).

109. G.A. Rechnitz, R.K. Kobos, S.J. Riechel, C.R. Gebauer, Anal. Chim. Acta, 94, 357, (1977).
110. S.R. Grobler, N. Basson, C.W. Van Wyk, Talanta, 29, 49, (1982).
111. C.A. Corcoran, R.K. Kobos, Anal. Letters, 16, 1291, (1983).
112. R.K. Kobos, G.A. Rechnitz, Anal. Letters, 10, 751, (1977).
113. V. Wollenberger, F. Scheller, P. Atrat, Anal. Letters, 13, 825, (1980).
114. M.A. Jensen, G.A. Rechnitz, Anal. Chim. Acta, 101, 125, (1978).
115. I. Kubo, I. Karube, S. Suzuki, Anal. Chim. Acta, 151, 371, (1983).
116. J.J. Kulys, Anal. Letters, 14, 377, (1981).
117. M. Hikuma, H. Obana, T. Yasuda, I. Karube, S. Suzuki, Anal. Chim. Acta, 116, 61, (1980).
118. G.A. Rechnitz, T.L. Riechel, R.K. Kobos, M.E. Meyerhoff, Science, 199, 440, (1978).
119. R.R. Walters, B.E. Moriarty, R.P. Buck, Anal. Chem., 52, 1680, (1980).
120. I. Karube, T. Matsunaga, S. Suzuki, Anal. Chim. Acta, 109, 39, (1979).
121. C.L. di Paolantonio, G.A. Rechnitz, Anal. Chim. Acta, 148, 1 (1983).
122. C.L. di Paolantionio, G.A. Rechnitz, Anal. Chim. Acta., 141, 1, (1982).
123. V. Wollenberger, F. Scheller, P. Atrat, Anal. Letters, 13, 1201, (1980).
124. B.J. Vincke, M.J. Deveeschouwer, G.J. Patriarche, Anal. Letters, 16, 673, (1983).
125. American Type Culture Collection, Catalogue of Strains I, 15th ed., 12301 Paklawn Drive, Rockville, Ma 20852, U.S.A., 1982.
126. J. Brisou, Techniques d'Enzymologie Bacterienne, Masson, Paris, 1971.
127. Y. Morino, E.E. Snell, J. Biol. Chem., 242, 5591, (1967).
128. D.S.D. Sarma, S. Rajalakshmi, P.S. Biochem. Biophys. Res. Commun., 6, 389, (1961).

APPLICATION OF IMMOBILIZED ENZYMES IN DIAGNOSTICS

Angelo Manzoni
Research and Development Department
Instrumentation Laboratory S.P.A.
Viale Monza 338
20128 Milan, Italy.

ABSTRACT. Immobilized enzymes in conjunction with a detector have led to the development of highly specific electrode systems for diagnostic purposes. The great steps that have been made in the field of immobilization research techniques afford us many analytical configurations each year. The use of electrochemical sensors to directly monitor the products of enzymatic reactions eliminate the need for additional reagents (i.e. dyes, other enzymes) and allows the application of simple analytical equipment.

INTRODUCTION

In clinical chemistry analyzers, sodium and potassium I.S.E. technology is replacing the conventional flame photometric method.
Moreover,
- chloride (1)
- ionized and total calcium (2)
- total carbon dioxide (3)
- and lithium (4)

sensors are commercially available and versatile enough to be included in modern analyzers for blood serum electrolyte analyzers and lastly the emerging metabolite sensors for glucose, urea, creatinine and other substrates (uric acid, cholesterol) obtained by immobilizing the enzyme or the biocatalyst

a) directly on the surface of the sensor (5), (enzyme, bacterial or tissue electrodes)

b) in a reactor connected with electrochemical sensors (6)

G. G. Guilbault and M. Mascini (eds.),
Analytical Uses of Immobilized Biological Compounds for Detection, Medical and Industrial Uses, 71–82.

can be considered the last stage of the industrial application of the ion selective electrodes. Experience with the use of electrochemical sensors, their reliability, adaptability and high resolution will lead to the continuous improvement of the industrial development of these emerging electrochemical devices combined with enzymatic systems to provide highly selective and sensitive procedures for the determination of a given substrate.

Moreover, the development of enzyme immobilization techniques on or within a solid matrix (synthetic, water-insoluble, polymeric support), in conjunction with a detector, have led to the development of highly specific electrode systems for diagnostic purposes.

Nevertheless, wide scale commercialization of enzyme electrodes has not yet been achieved, probably due to the complexity of manufacturing the multilayer composite membrane.

Approaches based on any combination of discrete reactors (nylon tube bound enzymes) with different sensors seem to be more easily incorporated into systems for analytical purposes.

ANALYTICAL USE OF IMMOBILIZED ENZYMES

Most of the current commercial systems based on enzymes require individual sample volume, specialized reagents, individual fluidic and detector channel : the enzymes that are utilized are not consumed during the enzymatic reactions. However, since they are used in soluble form, they cannot be reused and are sent to the waste container with each sample.

The multichannel analyzer "Beckman Astra 8' (3) performs urea analysis by measuring the increase in conductivity obtained producing ammonium and bicarbonate ions from urea in urease catalyzed reaction and, a reagent containing glucose oxidase reacts with glucose to consume oxygen, the depletion of which is detected by an oxygen electrode.

An alternative to utilizing enzymes in soluble form is the immobilization of the same onto some solid support.

Sometimes the high cost of a particular enzyme system makes its routine use impractical.

If we consider the enzymatic conversion of creatinine

$$\text{Creatinine} \xrightarrow{\text{Creatinine iminohydrolase}} NH_3 + \text{N- methyl-hydantoin}$$

to a measurable specie (NH_3) with a commercially acceptable cost/test, it is necessary that the enzyme be immobilized on a solid support.

Thus, by coupling the enzyme to an electrochemical (ammonium sensor (6)) or optical sensor (NH_3 + bromophenol blue → blue dye λ max 600 nm) (7), a simple, highly specific and sensitive measuring device can be obtained.

Furthermore, where the enzyme is mechanically or chemically attached to the sensor, the assay methodology can be essentially reagentless and require little, if any, pretreatment of the sample.

Many instruments incorporating immobilized enzymes to measure a variety of substrates are now available on the market (8).

Figure I shows the schematic diagram of an enzyme electrode for glucose.

An oxidase enzyme sensor. Substrate (S) reacts with O_2 in the oxidase layer to form product (P) and H_2O_2. The decreased O_2 concentration in the enzyme layer is measured by recording the signal from the oxygen electrode. An increase in concentration of substrate causes a decrease in O_2 concentration at the electrode

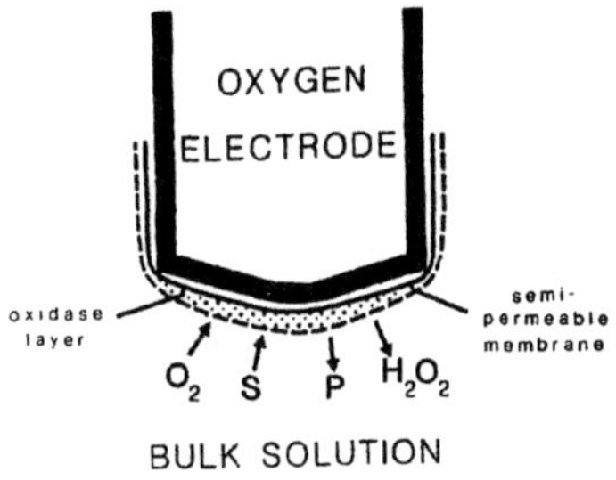

From a practical standpoint, the constitution of an enzyme sensor measurement is uncomplicated.

The essential components are :

a) transducer
b) circuit to measure potential or current generated at the transducer
c) thin layer of an appropriate enzyme, immobilized next to the active surface of the transducer.

Modern enzyme immobilization techniques allow many chemistries to be directly coupled to electrochemical sensors. In recent years, in fact, various methods for urea and glucose determinations have been based on using gas probes (pCO_2 or pNH_3 and pO_2) fitted with enzymes immobilized on their surface.

One of the most common enzymes used for glucose determination is glucose oxidase (GOD) which catalyzes the oxidation of glucose to gluconic acid and peroxide.

$$\beta\text{- D glucose} + O_2 \xrightarrow{\text{GOD}} \text{gluconic acid} + H_2O_2$$

The gas sensors used in conjunction with a GOD membrane monitor the rate of oxygen consumption or peroxide formation through an amperometric technique (9).

Urea nitrogen determinations are commonly based on the

hydrolysis of urea with the enzyme urease.

$$\text{Urea} + H_2O \xrightarrow{\text{Urease}} CO_2 + 2NH_3$$

Urease-based membranes were placed on the surface of a gas sensor to monitor the amount of carbon dioxide or ammonia produced from this reaction (10, 11). The enzyme electrode for glucose was first introduced by Clark and Lyons (12) : Glucose oxidase was trapped between two cellulose acetate membranes held closely to the surface of a Clark oxygen electrode and small molecules such as glucose diffuse into this layer (figure 1).

An inner hydrophobic membrane allows the selective permeation of oxygen to a platinum cathode.

The consumption of oxygen in the enzyme layer is monitored amperometrically providing the degree of glucose concentration.

An alternative approach is to measure the H_2O_2 evolved in the enzyme reaction : In this case a thin cellulose acetate membrane is placed in front of a platinum anode and H_2O_2 will permeate (9). An analytical system based on this approach was first commercialized by Yellow Spring Instruments (13). An enzyme electrode for urea based on an ammonia gas sensor was introduced by Guilbault : The ammonia generated in the enzyme reaction passes through a gas permeable membrane to a thin layer of solution where the range in pH is monitored with a pH electrode or ammonium ions are detected at an NH_4^+ ISE with nonactine ionophore in PVC membrane (14).

Mascini and Palleschi succeeded in demonstrating the simultaneous determination of glucose and urea based on oxygen and ammonia gas sensors (5).

The enzymes were covalently immobilized onto nylon nets which were attached to the gas sensor. The slow diffusion processes associated with these sensors severely limited the response/recovery time of this approach.

Both enzyme electrodes were fixed in a single flow cell and the wash period was two minutes for recovery to the baseline. Serum was diluted tenfold with 0.1M trisbuffer pH 8.3 and samples were pumped into the flow cell for one minute.

The long washout times required, which directly affected the throughout of samples ($<$ 20 samples/ hr), and probably the complexity of manufacturing the multilayer composite membrane acts as a detriment to wide scale commercialization of analytical systems based on these kinds of enzyme electrodes.

In fact, in order to incorporate these so-called 'enzyme sensors' into automated flowthrough systems, the membrane to which the enzyme would be attached should be strong and reliable. Most procedures for immobilizing enzymes for electrode construction have resulted in fragile membranes, and the mechanical integrity of the enzyme containing membranes becomes impaired after a short period of use. To develop automated enzyme based systems with a high sample throughput, fast washouts and detectors would be needed.

Washout problems could be reduced by removing the enzymes from the physical proximity of the gas sensor and placing them into a tubular reactor. By combination with different sensors, in automated analysis, the enzyme tubes commercialized by Technicon, Miles and Carlo Erba, specific and sensitive substrate detection has been obtained (15). As the nylon tube enzyme reactors are not largely affected by native biological liquids, the reactors seem to be suitable for automated devices in routine substrate assay.

Nylon tubing was first employed by Hornby and Morris as a support for immobilized glucose oxydase (16). Several types of immobilization exist and may be divided into two main categories (17).

Physical Methods	Chemical Methods
Absorption of enzyme on matrix insoluble in water	Crosslinking with multi-functional agents.
Entrapment of enzymes inside polymeric lattice	Immobilization by means of covalent linking technique
Entrapment of enzymes inside microcapsules	

The immobilization can be obtained on

Synthetic materials	Natural Materials
Acrylamide based polymers	Agarose
Maleic anhydride polymers	Dextran
Methacrylic and polymers	Cellulose
Polypeptides	Glass
Styrene based polymers	Nickel Oxide
Dacron	Titanium
Nylon	Manganese Oxide
	Collagen
	Alumina

As previously described, Mascini demonstrated the similtaneous determination of glucose and urea with the same buffer and the enzymes were covalently immobilized onto nylon nets attached to the sensors ; moreover, the immobilization of urease and glucose oxidase onto the inner surface of nylon tubes enables the development of a more reliable analytical system which would require one sample volume, common diluent/reagent and a common fluidic.

In fact, one of the advantages of the immobilized enzymes is a pH shift : The optimum pH can be shifted to that region on which one wants to make a measurement by choosing the right support for immobilization and consequently the enzymes are stabler.

By combination of discrete reactors and detectors we can obtain analytical systems that can be easily incorporated into a flowing stream whose essential components are : sampling device (diluted sample), enzyme reactor and electrochemical detector.

In order to have such a reactor with high conversion efficiency, the nylon tubing developed by Hornby and Morris is one of the best approaches to incorporating immobilized enzymes in the fluidic scheme of an automatic system.

The enzyme immobilization on nylon tubes consists mainly in five phases (figure II).

FIGURE II

Enzyme Immobilization Diagram on Nylon Tubes

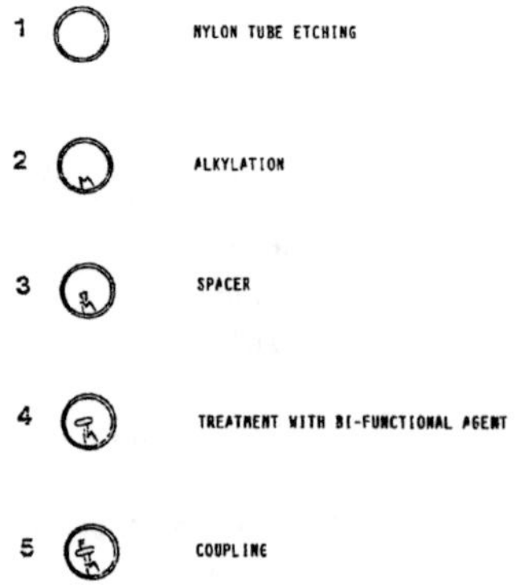

Phase 1 : The nylon tube is treated with 18.6% W/V calcium chloride in 20% W/W methanol water and left at 45°C for 25' to etch the nylon tube and to increase the usable surface.

Phase 2 : Alkylation : It is a process which allows linkings. The one chosen by us is O-Akylation with triethyloxonium tetrafluoroborate, which converts secondary amides into imido-esters.

The nylon structure is not greatly modified.

Phase 3 : Spacer : The nylon O-Alkylation is treated with a solution of a bifunctional molecule. In our case hexamethylendiamin, which, when linked to the nylon acts as a spacer.

Advantages : The introduction of a spacer permits a much higher activity : probably the fact that the enzyme is attached to a fairly long and flexible molecule diminishes the possibility of a steric impediment of the support.

Phase 4 : Consists of treating the tube with a bi-functional molecule such as glutaraldehyde which then permits attachment of the enzyme.

Phase 5 : The nylon tube is then filled with a buffer solution containing the enzyme to be immobilized.

FIGURE III. Experimental-Set-Up

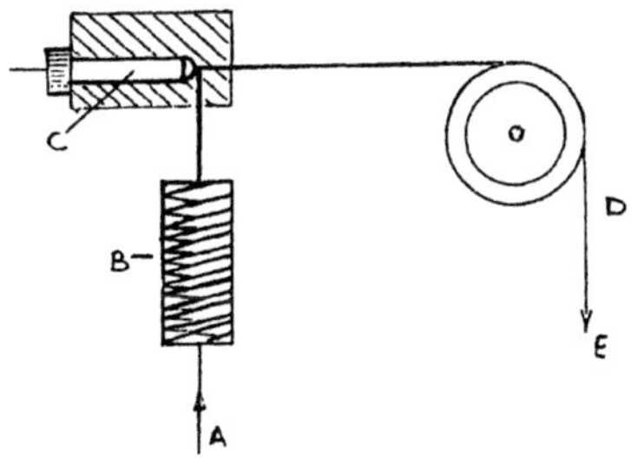

A Diluted sample in

B Immobilized GOD reactor

C pO_2 sensor

D Peristaltic pump

E WASTE

Figure III shows the experimental configuration for the use of immobilized enzymes in conjection with an electrochemical detector for amperometric measurement of glucose based on the detection of oxygen depletion.

The use of a conventional pO_2 sensor for glucose measurement is one of the most reliable approaches, however, the dependence of oxygen levels on partial pressure and termperature can be an added complication.

An alternative to this approach is the detection of hydrogen peroxide, which can be oxydized at a platinum anode : the major problem with the hydrogen peroxide electrode is that the operating voltage (+0.6 V vs. Ag/AgCl) other species present in serum can be oxidized as well.

Consequently, the oxygen electrode benefits greatly in selectivity by its gas permeable membrane. In figure IV may be seen the results obtained :

FIGURE IV.

Relationship of Results for Glucose
Obtained by I. E. and I. L. 919

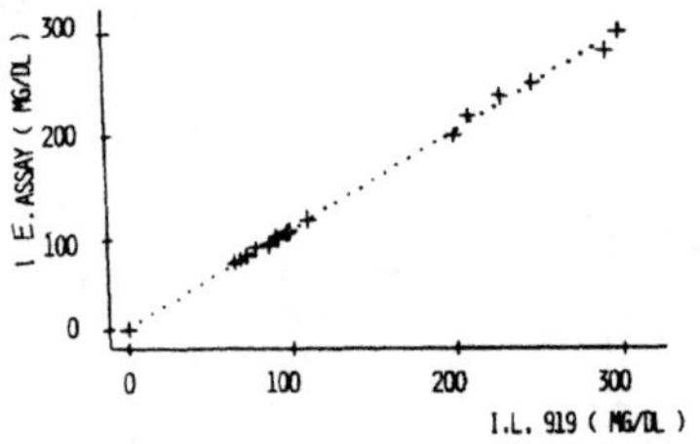

N = 17
SLOPE = 1.018
INTERC. = - 0.08
CORR. COEFF. = 0.999

In the enzymatic analysis of urea via immobilized enzymes the reaction of urea in the presence of the urease enzyme generates ammonia that can be typically monitored with potentiometric gas sensors (Orion, HNU).

These sensors, based on the passage of ammonia through a gas permeable membrane into a thin layer of fluid where the change in pH is monitored with a glass pH electrode, exhibit poor sensitivity and a fairly slow response.

Schindler et al (18) proposed a flow-through system for urea determination, using soluble urease and consecutive determination of liberated ammonium ions by an ammonium ion selective electrode in which nonactin acts as an active component.

This approach, employing soluble urease, can be improved by immobilizing the enzyme onto nylon tube in conjunction with an ammonium electrode (figure V) with an experimental setup very close to the one presented for the glucose assay in diluted serum in order to have the possibility of exploring the simultaneous glucose/urea assay with immobilized enzymes onto nylon tube following the experiment of Mascini and Palleschi using enzyme electrodes (5).

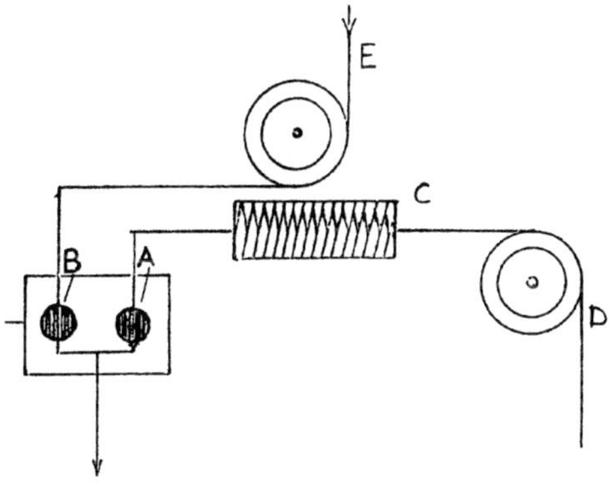

A = Ammonium ion-selective electrode
B = Reference electrode
C = Immobilized urease reactor
D = Diluted sample in
E = Reference solution in
F = WASTE

FIGURE V. Nylon Tube Experimental Set-Up

However, the system based on urea detection through ammonium, very suitable for being incorporated into an automatic analyzer to obtain a high throughput, suffers from the interfering presence of potassium in the sample.

In fact, ammonium ISE exhibits only moderate selectivty overpotassium ($K_{NH_4^+/K^+} = 10$). Due to this interference a background correction is commonly required.

Improved performances could be obtained with blank measurement, calibration adjustment or mathematical correction based on selectivity coefficients.

The results obtained by Schindler (18) encourage this approach ; in fact, he demonstrated that it is possible to work with the presence of potassium interference without sacrificing linearity and analytical performances.

Finally, it should be mentioned that immobilized enzymes, together with electrochemical devices will become popular in the near future due to their compatibility with automatic analyzers with a high throughput.

It has been demonstrated that immobilized creatinine iminohydrolase could convert creatinine into measurable amounts of ammonia in seconds (6).

Other substrates of clinical interest can be measured by connecting immobilized enzymes to an O_2/H_2O_2 electrode such as uric acid (through uricase), lactic acid, (lactate oxidase), cholesterol (cholesterol esterase + cholesterol oxidase) and other redox enzyme assisted reactions.

REFERENCES

1. D. Amman. K. Hartman, P.C. Meier, H. Oehme, F. Osswald and W. Simon
Mikrochimica Acta (Wein), 235-246 (1978)

2. NOVA 7 Instrument Total Clacium/ionized calcium:pH Analyzer by NOVA BIOMEDICAL, 200 Prospect Street, Waltham, MA 02254-9141 U.S.A.

3. P.R. Finley, R.J. Williams, D.A. Lichti and A.C. Thies
Evaluation of a New Multichannel Analyzer, "Astra 8"
Clin. Chem 24/12, 2125-2131 (1978)

4. Lytening 2 Instant ISE
AMDEV INC, 35 Cherry Hill Drive, Danvers MA 01923 U.S.A.

5. M. Mascini, G. Palleschi
Anal. Chim Acta 145 (1983) 213-217

6. M. Mascini, G. Palleschi
Anal. Chim. Acta 136 (1982) 69-76

7. E. Tanganelli, L. Prencipe, D. Bassi, S. Cambiaghi and E. Murador
Clin. Chem. 28 (1982) 1461-1464

8. D.N. Gray, M.H. Keyes and B. Watson
Anal. Chem 50 (8) 1026 (1978)

9. G. Sittampalam and G.S. Wilson
Journ. Chem. Ed. 59 (1) 70-73 (1982)

10. G.G. Guilbault and F. Shu
Anal. Chem. 44 2162, 1972

11. G.G. Guilbault, J. Montalvo, R. Smith
Anal. Chem 41 (1969) 600

12. L. Clark and C. Lyons
Ann. N.Y. Acad. Science 102, 29, 1962

13. D.P. Newman, V.S. Patent Nos. 3, 979, 274 (Sept 7, 1976) and 4, 073, 713 (Feb. 14, 1978). Ass. Yellow Spring Instruments

14. G.G. Guilbault and G. Nagy
Anal. Chem 45, 417, 1973

15. S. Cambiaghi, T. Benecchi, E. Murador
Giorn. It. Chim. Cl. 4(4) 1979

16. Morris, D. L. Campbell, J. and Hornby W.E.
Biochem. J. 147, 593-600 (1975)

17. G.G. Guilbault
Principles of Immobilized Enzymes, pages 77-111
in Analytical uses of Immobilized Enzymes
Marcel Dekker, Inc.

18. J.G. Schindler, R.G. Schindler and O. Aziz
J. Clin. Chem. Clin. Biochem 16, 447-450, 1978

IMMOBILIZED DEHYDROGENASE ENZYMES COUPLED WITH A NADH SENSOR

G. Palleschi
Department of Chemical Science and Technology.
2nd University of Rome.
Via O. Raimondo 8, 00173 Rome. Italy

ABSTRACT. Diverse immobilization techniques have been tested for the enzymes glucose and 3-hydroxybutyrate dehydrogenases.
The coupling of these immobilized enzymes with a previously selected electrochemical carbon probe, leads to the realization of reliable biosensors useful for the determination of metabolites as glucose and 3-hydroxybutyrate and generally for those metabolites whose enzymatic reaction involves the $NAD(P)^+$ molecule as cofactor.
Preliminary studies were devoted to finalizing the best sensor for the NADH reoxidation.
Linear sweep voltammetry as well as constant potential studies were carried out by using NADH in an appropriate buffer solution.
The sensor that gave the best signal/noise ratio was chosen and analysis of the metabolites mentioned above using the enzyme immobilized on the sensor surface were performed.

1. INTRODUCTION

The immobilization of enzymes for the realization of biosensors has been well described (1) and it is becoming increasingly more important in analytical, biological and clinical chemistry (2); moreover, analysis of intermediary metabolites in blood related to some disease states, are also very important(3).

G. G. Guilbault and M. Mascini (eds.),
Analytical Uses of Immobilized Biological Compounds for Detection, Medical and Industrial Uses, 83–93.

There is a group of redox enzymes known as dehydrogenase which are $NAD(P)^+/NAD(P)H$ dependent.
The immobilization of such enzymes and their coupling with an electrochemical NAD(P)H selective transducer, leads to the measurement of substrates as glucose and 3-hydroxybutyrate whose clinical interest, expecially in diabetology, has been well recognized (4).
This paper describes the immobilization of glucose and 3-hydroxybutyrate dehydrogenases directly onto the surface of a carbon probe by using physical as well as chemical techniques. Voltammetric scans and measurements at constant applied potential were previously carried out by using platinum and carbon electrodes.
Results show that carbon is a material more appropriate than platinum for NADH reoxidation, therefore a carbon electrode was selected for assembling a biosensor useful for the determination of glucose and 3-hydroxybutyrate.
The general scheme for the reactions involved is the following:

$$1)\ \text{Substrate} + NAD^+ \xrightarrow[\text{Enzyme}]{\text{Dehydrogenase}} \text{Product} + NADH$$

$$2)\ NADH \xrightarrow[\text{Selected Constant Pot.}]{\text{Selected probe}} NAD^+ + H^+ + 2e$$

According to the reaction 2) the current change, due to the reoxidation of NADH, can be related to the substrate concentration.

2.1 APPARATUS

Platinum (20 mm^2 area) and carbon graphite (8 mm^2 area) electrodes were obtained from Universal Sensors Inc. New Orleans, LA. USA. Carbon graphite rods were obtained from Ultra Carbon Corporation Bay City, Michigan (Purity designation Yu-40).

Glassy carbon rod (4.9 mm^2 area) was from Tokai Co. Japan.
Spectrographic carbon rods (15.9 mm^2 area) were from Electrocarbonium Co. Terni, Italy.
Glassy and spectrographic carbon electrodes were made in our laboratory.
For linear sweep voltammetry and constant potential studies a Tacussel PRGE-DEC and an AMEL 466 Polarographic Analyzer were used.
The reference electrode was Ag/AgCl.
Signals were monitored with a Houston Instrument Omniscribe strip chart recorder and an Amel model 868 recorder.
Dialysis membranes 12,000 m.w.c.o. were from A.H. Thomas CO, Philadelphia PA 19105 USA.
Nylon membranes and preactivate immunoaffinity membranes were from Pall Corporation Milan, Italy .
Polycup resin was from Hercules Inc. 910 Market Street Wilmington, Delaware 19899 USA.

2.2 Reagents

Buffers were a physiological phosphate buffer pH 7.4 consisting of 137mM NaCl; 8.0mM Na_2HPO_4 ; and 1.5mM KH_2PO_4 and a physiological solution (Normosol-R) pH 7.4 from Abbot Aprilia Italy.
The enzymes D+ glucose $NAD(P)^+$ 1-Oxidoreductase EC 1.1.1.47 from Bacillus Species and 3-hydroxybutyrate dehydrogenase (R -3-Hydroxybutyrate: NAD^+ Oxidoreductase EC 1.1.1.30) from Rhodopseudomonas spheroides type II were from Sigma Chemical Co.
Glucose anhydrous RS grade was from C. Erba Italy.
3-hydroxybutyrate was from Sigma Chemical Co.

2.3 Procedure

The platinum electrode was pretreated as previously described (5).
No pretreatment was necessary for carbon electrodes.
NADH was prepared in the aforementioned buffers and stored refrigerated at +5 °C when not in use.
Experiments were performed by dipping the assembled enzyme electrodes in 20 ml of buffer which contained the appropriate concentration of cofactor. The electrode was equilibrated in such solution for 1 minute, then glucose or

3-hydroxybutyrate were injected and changes in current were monitored amperometrically.
3-hydroxybutyrate dehydrogenase was immobilized via physical entrapping by using the dialysis membrane.
For chemical immobilization of glucose and 3-hydroxybutyrate dehydrogenases, procedures via BSA-glutaraldehyde (6), polyazetidine(7) ,carbodiimide (7), and Pall Bia (8) have been performed .

3. RESULTS AND DISCUSSION

3.1 Linear Sweep Voltammetric Studies

Voltammograms from 0 to 900 mV were run both in buffer only and in buffer plus NADH at the rate of 10 mV/sec. by using platinum and carbon electrodes without stirring.
The platinum electrode (Fig.1) gave the highest background current and the lowest reproducibility but it showed a good response to NADH.
Fig. 2 , 3 and 4 show the voltammograms run by using respectively a glassy carbon, a graphite carbon and a spectrographic carbon electrode. Scans were run in buffer and in buffer plus NADH at 0.05 and 0.1 mg/ml of NADH.
The peak current for NADH at graphite carbon electrode was observed at + 450 mV with half-wave potential at + 335 to +340 mV. The peak current at the glassy carbon electrode was at + 500 mV with half-wave potential at +350 mV.
The spectrographic carbon sensor did not show a defined peak but a "plateau" of current from + 300 to + 500 mV was observed.
For all electrodes the peak current is proportional to the NADH concentration.
Constant potential studies were carried out by selecting for each electrode the appropriate potential for the amperometric detection of NADH.
Carbon electrodes were tested by injecting small aliquotes of NADH into the buffer solution containing the electrode, under stirring conditions. Calibration curves of NADH by using carbon graphite and glassy carbon sensors are shown in Fig. 5 and 6. Fig. 7 shows the response time and the stability of NADH solution by performing some calibration curves using a spectrographic carbon sensor. Results show

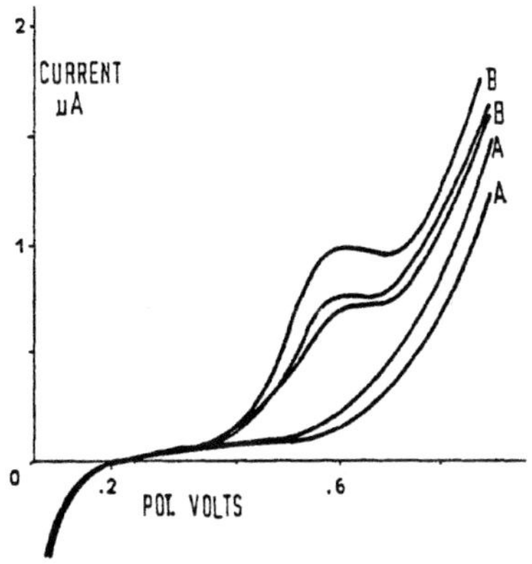

Fig. 1. A = Voltammograms of physiological solution (Normosol-R) pH = 7.4 at Pt electrode (Two scans). B = Voltammograms of physiological solution (Normosol-R) pH = 7.4 plus NADH 0.4 mg/ml.at Pt electrode V scan = 10mV/sec (Three scans).

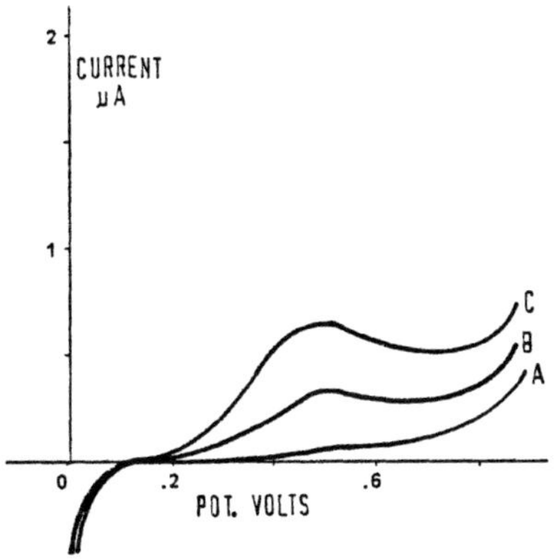

Fig. 2. Voltammograms of physiological solution (Normosol-R) pH = 7.4 and physiological solution plus NADH at glassy carbon electrode; V scan = 10mV/sec. A = Normosol-R; B = NADH 0.05 mg/ml; C = NADH 0.1 mg/ml

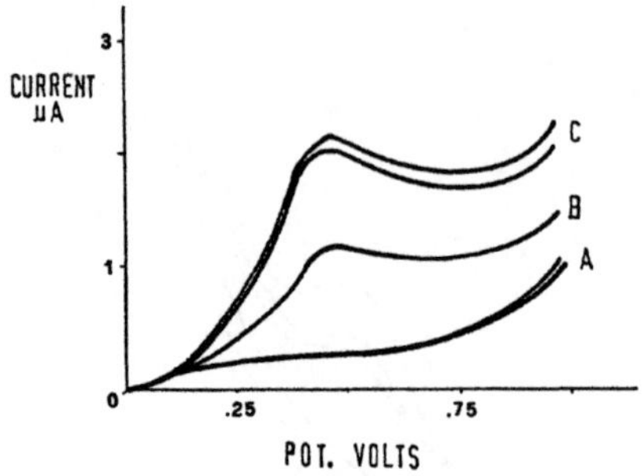

Fig. 3. Voltammograms of phosphate buffer and phosphate plus NADH at carbon graphite electrode; V scan = 10 mV/sec. A=Phosphate buffer pH = 7.4; B = NADH 0.5mg/ml; C = NADH 1.0 mg/ml

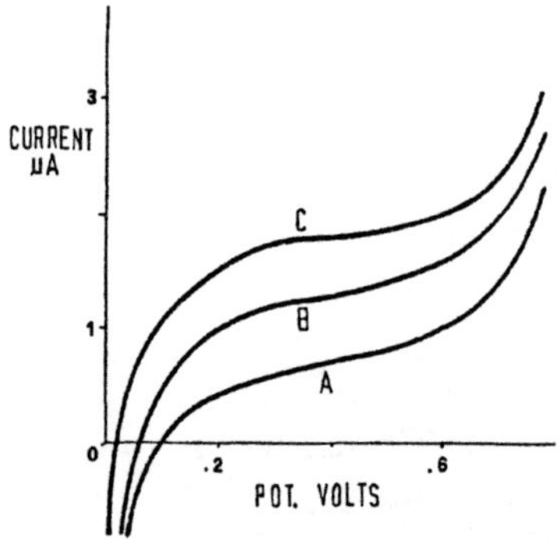

Fig. 4. Voltammograms of physiological solution (normosol-R) pH = 7.4 and physiological solution plus NADH at spectrographic carbon electrode; V scan = 10 mv/sec. A = Normosol-R; B = NADH 0.05 mg/ml; C = NADH 0.1 mg/ml

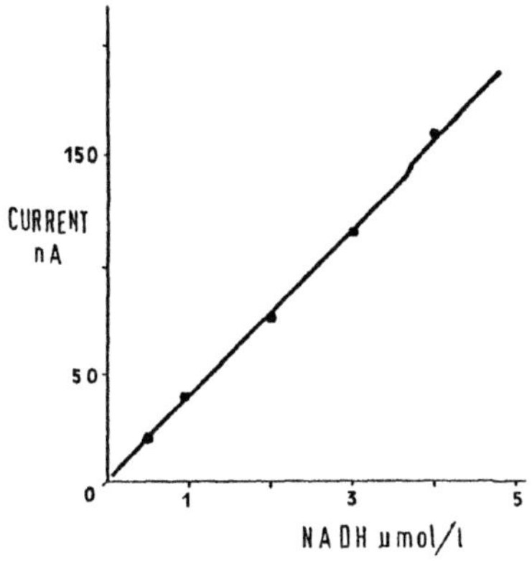

Fig. 5. NADH calibration curve at carbon graphite electrode in physiological phosphate buffer pH = 7.4 at +450 mV applied potential. T = 25 °C

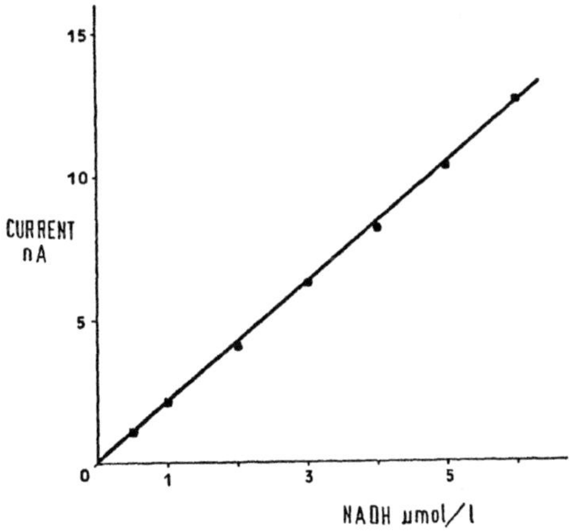

Fig. 6. NADH calibration curve at glassy carbon electrode in physiological solution (Normosol-R) pH = 7.4 at + 500 mV applied potential; T = 25 °C

that these carbon sensors are reliable for the detection of NADH in the range of 1.0 umol/l. - 0.1 mmol/l. with good reproducibility. The response time is less than 1 minute and the NADH solution, stored refrigerated when not in use, is stable for two or three days. Since the spectrographic carbon probe gave the best signal/noise ratio it has been selected for further experiments.

3.2 Glucose and 3-hydroxybutyrate dehydrogenases studies

Two immobilization procedures have been studied for glucose dehydrogenase by using a new nylon support with carboxylic groups activated by carbodiimide and by a preactivated Pall Bia membrane. Results obtained with such assembled probe (Fig. 8), show a linearity for glucose ranging from 0.01 mmol/l. over 0.1 mmol/l. The response time was about 2 minutes and the detection limit was 5 umol/l. The concentration of NAD^+ was 1.0 mmol/l. This concentration has been selected by performing some calibration curves of glucose at different concentrations of cofactor. Results show that when a NAD^+ concentration up to 1.0 mmol/l. was used, no improvements in current change were observed.
For 3-hydroxybutyrate analysis, experiments were carried out by immobilizing the 3-hydroxybutyrate dehydrogenase on the electrode in 5 different ways (Fig. 9). The best calibration curve for this substrate was obtained by using 1 mg of enzyme physically entrapped in the 12,000 m.w.c.o. dialysis membrane (Fig. 9 curve A). In this case the enzyme is in intimate contact with the electrode surface and its specific activity is rather high compared with that of the enzymes chemically immobilized; but, like for all the enzymes immobilized in this way, its activity decreased rapidally.
For chemical immobilizations the procedure by carbodiimide gave the best result (curve B) followed by Pall Bia membrane and polyazetidine (curves C and D).
No good results were obtained by using BSA-glutaraldehyde procedure (curve E). It seems that this enzyme is inhibited by the glutaraldehyde ; this has been also observed for the enzyme pyruvate oxidase (7). The concentration of NAD^+ in the solution has been chosen by using the same procedure as for glucose and was 1.0 mmol/l.
These calibration curves show that the 3-hydroxybutyrate can be determined in concentration 0.05 mmol/l.- 0.5 mmol/l.

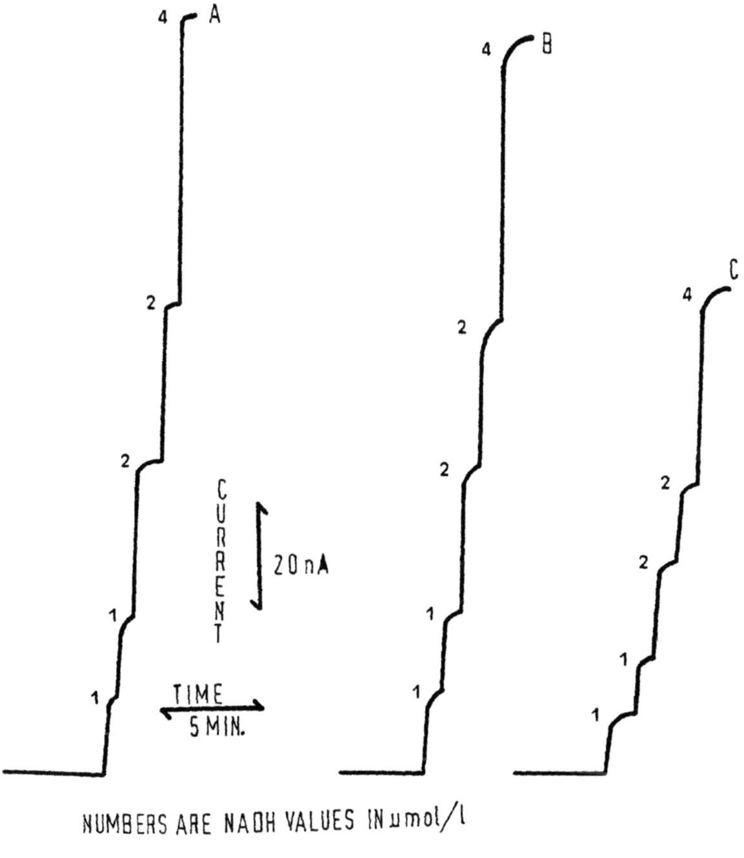

Fig. 7. NADH calibration curves at spectrographic carbon electrode in physiological solution (Normosol-R) pH = 7.4 ; Applied potential + 300mV T = 25 °C. A = First day ; B = Second day ; C = Seven days.

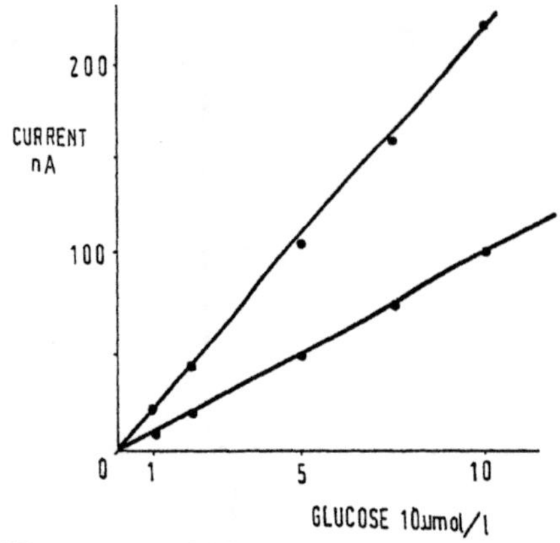

Fig. 8. Glucose calibration curves at spectrographic carbon electrode in physiological buffer (Normosol-R) pH = 7.4 ; T = 25 °C ; NAD^+ = 1.0 mmol/l. A = Immobilization via Pall Bia. B = Immobilization via Carbodiimide.

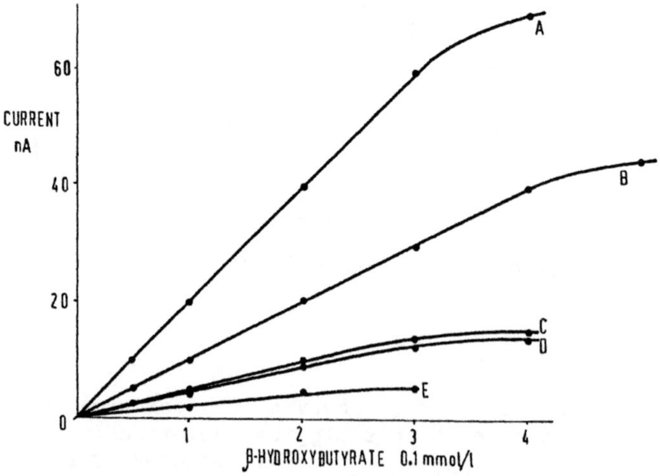

Fig. 9. 3-hydroxybutyrate calibration curves at spectrographic carbon electrode in physiological buffer (Normosol-R) pH = 7.4 ; T = 25 °C ; NAD^+ = 1.0 mmol/l. A = Immobilization via physical entrapment in 12,000 m.w.c.o. dialysis membrane. B = Immobilization via Carbodiimide. C = Immobilization via Pall Bia. D = Immobilization via Polycup. E = Immobilization via BSA-Glutaraldehyde.

The response time is about 2 minutes and the precision of the analysis is about 5% . This concentration range is useful for the determination of 3-hydroxybutyrate in sera or blood.

Some of the main interferences which can be present in blood were then analyzed. Ascorbic acid and uric acid gave serious interference when injected at a concentration of 10 umol/l. Aspirine and cysteine did not give a significant interference at a concentration of 0.1 mmol/l.

In conclusion glucose and 3-hydroxybutyrate can be measured in absence of ascorbic acid or uric acid. The glucose probe is an exellent oxygen or hydrogen peroxide independent glucose sensor, moreover both glucose and 3-hydroxybutyrate sensors gave promising results for analysis in biological media.

4. ACKNOLEDGMENTS

The author wish to thank dr. D.Moscone, dr. R.Pilloton, dr. H.S.Rathore and miss. M.G.Lavagnini for their contribution during the experimental work and prof. M.Mascini for the helpful discussion.

5. References

(1) Guilbault G.G. Analytical Uses of Immobilized Enzymes, Marcel Dekker, New York. (1984)

(2) Mascini M. and Guilbault G.G. Biosensors 2 (1986) 147-172

(3) Lloyd B., Burrin J., Smythe P. and Alberti K.G.M.M. Clin. Chem. 34/10, (1978) 1724-29.

(4) Mascini M., Mazzei F., Moscone D., Calabrese G. and Massi Benedetti M. Clin. Chem. 33, (1987) 591

(5) Palleschi G. Anal. Lett. 19(13&14), (1986) 1501-10.

(6) Lubrano G.J. and Guilbault G.G. Anal. Chim. Acta 97 (1978) 229.

(7) Mascini M. and Mazzei F. Anal. Chim. Acta 192 (1987) 9-16.

(8) Pall Corporation supplier's instructions Pall Biodyne Immunoaffinity Membrane For Biotechnology Applications Pag. 3, Method 1.

TARGETING OF IMMOBILIZED DRUGS

A. Atilla Hincal and H. Suheyla Kas
Hacettepe University Faculty of Pharmacy
Pharmaceutical Technology Department
06 100 Ankara
TURKEY

ABSTRACT. Based on the evidence presented in this investigation, the application of streptomycin sulphate loaded human serum albumin and gelatin microspheres seems advantageous in the treatment of tuberculosis due to the high level of accumulation in lungs when compared to liver, kidneys, spleen and heart. In addition biodegradability, low toxicity and lower antigencity of the matrix materials made formulations an efficient delivery system.

From in vitro release experiments, we can conclude that the degrees of matrix stabilization is varied by duration of exposure time to crosslinkers and by the concentration of the stabilizing agents. It is probable that in vivo microsphere breakdown due to enzyme attack will affect the rate of release of the immobilized drug in a different manner.

1. INTRODUCTION

Several types of carriers intended to modify the systemic distribution of chemotherapeutic agents were recently proposed as possible drug vehicles. Specific delivery of the immobilized chemotherapeutic agents to desired target sites, with minimum systemic side effects, was the aim in drug targeting. The site-specific delivery of the chemotherapeutic agents allows maximum concentration of an agent at a desired body site, thus permitting the use of much smaller doses than normally required with generalized systematic administration. Microspheres showing targeting ability, sustained action and prolonged release provide a new system of immobilized drug delivery to desired target sites.

A chemotherapeutic agent, streptomycin sulphate, is the antimicrobia base produced by certain strains of Streptimyces griseus and is an antibiotic used as a tuberculostatic agent (Waksman 1945, 1953). It was administered intramuscularly, but caused pain and irritation at the site of injection (Valentine and Shooter, 1954) and a systemic side effect on the eighth nerve was predominant. These side effects increased when the drug was given at high doses.

G. G. Guilbault and M. Mascini (eds.),
Analytical Uses of Immobilized Biological Compounds for Detection, Medical and Industrial Uses, 95–113.

The release of the immobilized drug associated with microsphere carriers has been found to be dependent on a number of factors including the size, the extent and nature of crosslinking, the type and amount of the matrix materials, the position of drug in the microsphere, its incorporation level in the microsphere, density of the sphere, physiochemical and physical interactions between the drug and matrix material, the presence of adjuvants in the microspheres and the release environment (Tomlinson, 1984).

Since free streptomycin sulphate distributed in the body tissues and blood and concentrated in the liver, this study was planned to formulate an immobilized streptomycin sulphate and acculate these microspheres at the desired target organ, the lungs, and to reduce the stated systemic effects and also to study the influence of the extent of crosslinking, the type and amount of the stabilizing material and the nature of the matrix material on in vitro release characteristics of streptomycin sulphate from microspheres.

2. Experimental

2.1 Materials

The active substance of the microspheres was streptomycin sulphate (Pfizer) with a mean particle size of 46 μm. The wall and matrix materials used were human serum albumin (HSA) Behringwerke AG) and gelatin (Type B, Merck). The hardening agents were 2,3-butanedione (BASF, and formaldehyde (Merck). The solvents used were diethylether (Merck) and acetone (Merck), ^{99m}Tc was obtained from a generator (Amersham).

Swiss albino mice of about 350-35g in weight were used.

Hand-operated homogenizer, a motor driven glass stirrer (Heidolf), and an automatic well-type gamma counter (Model BF 5300-Berthold) were used in experiments.

2.2 Preparation of Microspheres

Albumin microspheres: Micropheres containing streptomycin sulphate (60.2 percent w/w) were prepared by modification of the method of Scheffel et al (1972) and Widder et al (1978). The aqueous solution containing human serum albumin (HSA), iron hydroxide and streptomycin sulphate was mixed with cottonseed oil and homogenized by passing through a hand-operated homogenizer. The resulting homogenate was added dropwise into continuously stirred cottenseed oil at 25°C. The albumin microspheres were washed with diethylether to remove the oil phase, and microspheres were stablized with 0.2 and 0.4 M concentrations of 2,3-butanedione in diethylether for 15 and 120 min. The stabilized microspheres were washed with diethylether, dried and stored in the dark at 4°C.

Gelatin Microspheres: Microspheres containing streptomycin sulphate (61.4 percent w/w) were prepared by the method of Yoshioka et al (1981). The gelatin solution containing ^{99m}Tc-human serum albumin and the active substance was added dropwise into continuously stirred corn oil containing Tween 80 (0.1 percent v/v) at 40°C. The obtained emulsion was cooled in an ice bath, diluted and washed with acetone to remove the oil phase. The gelatin microspheres were treated with 10 and 20 perent solutions of formaldehyde in acetone to harden the matrix at 25°C. The duration of exposure time to formaldehyde was 10 and 120 min. The stabilized microspheres were washed with acetone, dried and stored in the dark at 4°C.

2 3 <u>Labelling of free streptomycin and its microspheres with ^{99m}Tc</u>

Free streptomycin sulphate. Streptomycin sulphate was labelled with ^{99m}Tc by the tin reduction method (Ercan 1976). The following procedure was adapted for routine labelling. Fifty milligrams of streptomycin sulphate was dissolved in 2ml distilled water, stirred for 5 min after adding 0.1 ml $SnCl_2 2H_2O$ (10 mg. 10 ml^{-1}), and adjusted to pH 7.0 with 0.5 M NaOH. It was filtered through a 0.22 μm Millipore filter into a sterile vial containing $^{99m}TcO_4$ of the desired radioactivity in 1-2 ml saline.

Albumin microspheres: Albumin microspheres were labelled according to the method of Rhodes et al (1969). Human Serum Albumin microspheres were labelled by ^{99m}Tc after preparation. Gelatin microspheres. Gelatin microspheres were labelled using ^{99m}Tc-Human Serum Albumin. HSA was labelled by the method of Richards and Atkins (1967).

Quality control of ^{99m}Tc-Human Serum ALbumin and ^{99m} microspheres and ^{99m}Tc-Streptomycin sulphate. The labelling efficiency of ^{99m}Tc-Human Serum Albumin, ^{99m}Tc-microspheres and ^{99m}Tc-streptomycin sulphate was determined by impreganated thin-layer chromatography (ITLC) using ready plates. Acetone was used as a solvent. The labelled compound stayed at the origin and the free pertecnetate ($^{99m}TcO_4^-$) moved with an Rf of 1.0 (Ercan 1976). The amount of $^{99m}TcO_4^-$ was always less than 3 percent.

2.3.1. <u>Tissue distribution</u>

Free ^{99m}Tc-streptromycin sulphate. To determine the distribution within the body 0.2 ml (∿ 20 μCI) ^{99m}Tc-strptomycin sulphate solution (2 percent w/v) was injected into the tail veins of two mice. The mice were killed after 15 minutes. The organs, such as the lungs, liver spleen, kidneys and the heart, were removed and the radioactivity counted in a well-type automatic counter against a standard solution prepared from a 1/100 dilution of the injection solution. The percentage of radioactivity accumulated in each organ was calculated. Albumin and gelatin microspheres. To determine the distribution within the body, 0.4 mg labelled microspheres suspended in 0.2 ml (∿ 200 μCi) of saline containing Tween 80 (0.1 percent w/v), was injected into the tail veins of 10 mice weighing about 30-35 g. The mice were killed after 10 min, 60 min, 4hr and 24hr. The percentage of radioactivity accumulated in lungs, liver, spleen, kidneys and heart were determined in the previous manner described.

2.3.2 In vitro release procedure

In vitro release studies were run in 25 ml isotonic phosphate buffer (pH 7.4) in a 50 ml flask immmersed in a constant temperature (37± 0.1°C) water bath and shaken at 50 cpm in a horizontal laboratory shaker. At various time intervals, 0.5 ml of the sample solutions were taken and replaced by an equal volume of distribution medium at 37°C.

The samples were assayed by the method reported by Wahbi et al (1977). Spectroscopic determination of the resulting solutions was carrried out at 320 nm.

Microspheres used in the in vitro release experiments were coded as in Table 1.

TABLE 1. Codes of Microspheres Used in Vitro Release Studies.

STABILIZING AGENT		DURATION OF STABILIZATION (min)	CODE
2,3-Butanedione	0.2 M	15	HM_1
		120	HM_2
	0.4 M	15	HM_3
		120	HM_4
Formaldehyde	10 %	10	GM_1
		120	GM_2
	20%	10	GM_3
		120	GM_4

3. RESULTS AND DISCUSSION

Incorporation of drugs into carriers has found increased application and use in therapy in recent years. For successful application, the carrier itself has to be non-toxic and biodegradable. In this investigation human serum albumin and gelatin were chosen as matrix materials because of their low toxicity and biodegradability (Aronsen et al 1979, Baker et al 1984, Bernard et al 1980, Lee et al 1981, Scheffel et al 1972, Sugibayashi et al 1979, Widder.1978). It's existence in blood plasma as a natural circulatory drug carrier, easy control of particle size distribution and its use in medical diagnostic purposes were among the reasons of selection of human serum albumin as a matrix materials to immobilize the drug (Lee et al 1981, Rhodes and Wagner 1969, Rh des et al 1969, Russell 1983, Scheffel et al 1972, Widder et al 1981). Gelatin was chosen as an alternative matrix material due to its protein structure as alubumin (Hashida et al. 1977, 1979, 1980, Yoshioka et al 1981).

Since streptomycin sulphate was not absorbed through the intestines, it was applied intramuscularly. However, intramuscular injection caused pain and irritation at the site of injection and side effects, especially on the nerve (Garrod 1973, Kaymakcalan et al 1979). When given intravenously it accumulated in the liver. The reason for selecting streptomycin sulphate as an immobilized drug was to reduce these side effects and to accumulate the drug in the lungs by intravenous injection.

The microspheres prepared by the emulsion polymerization method were stabilized by the cross linking agents 2,3-butanedione and formaldehyde at two different concentrations and two different duration times. Heat stabilization was not applied because of the possible decomposition of streptomycin sulphate at high temperatures.

Human serum albumin and gelatin microspheres prepared in this investigation were all spherical in shape. Typical photomicrographs of the albumin and gelatin microspheres which immobilize streptomycin sulphate, are shown in Figures 1 and 2,

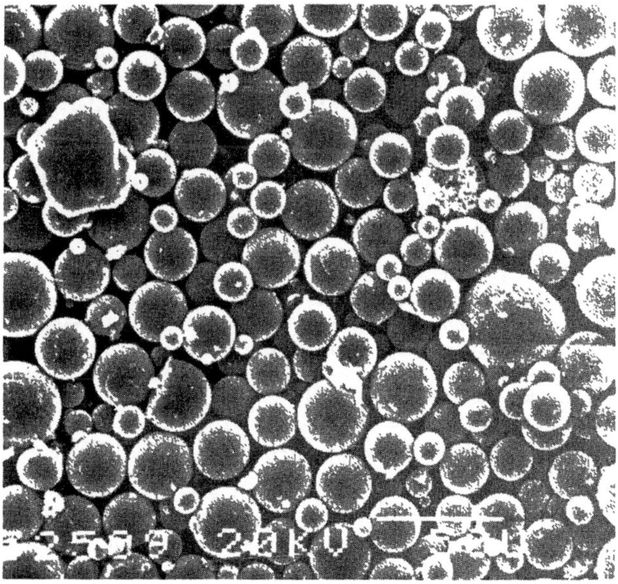

Figure 1. Electron photomicrograph of albumin microspheres

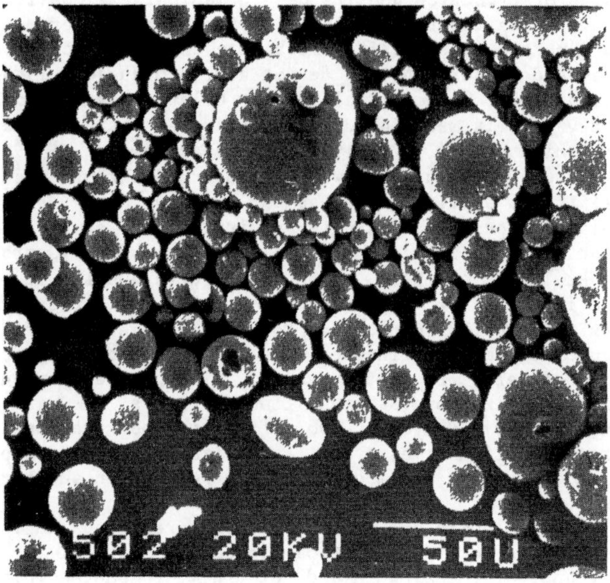

Figure 2. Electron photomicrograph of gelatin microspheres

respectively. The mean particle sizes of the human serum albumin and gelatin microspheres used in animal distribution studies were 32 and 30 μm, respectively. Gelatin microspheres shown in Figure 2 were later passed through appropriate sieves to obtain the desired particle size range. Since the homogenate was passed through a homogenizer this procedure was not necessary for the human serum albumin microspheres.

Following injection, free streptomycin sulphate accumulated initially in the liver (Nelsen et al 1947) while immobilized streptomycin sulphate in human serum albumin and gelatin microspheres accumulated predominantly in the lungs (Figures 3-5). The uptake of microspheres by the lungs is probably due to mechanical filtration caused by capillary blockage before distribution to the whole body can take place (Guyton 1976, Kante et al 1980, Yoshioka et al 1981). This passive targeting to the lungs was due to the particle size distribution of the microspheres, which was around 30 μm (Illum and Davis 1984, Tomlinson 1983, Yoshioka et al 1981).

The radioactivity levels of human serum albumin and gelatin microspheres in the lungs, 10 min after injection were 58.2 percent and 41.3 percent, respectively. The radioactivity levels in the first, fourth and 24th hours showed an appreciable decrease as seen from the histograms (Figures 4,5). The results were in agreement with the other microsphere experiments (Sugibayaski et al. 1977).

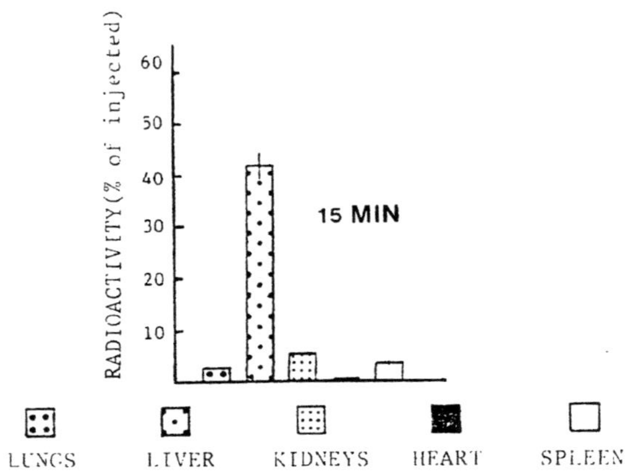

Figure 3. *In vivo* distribution of free streptomycin sulphate labelled with ^{99m}Tc.

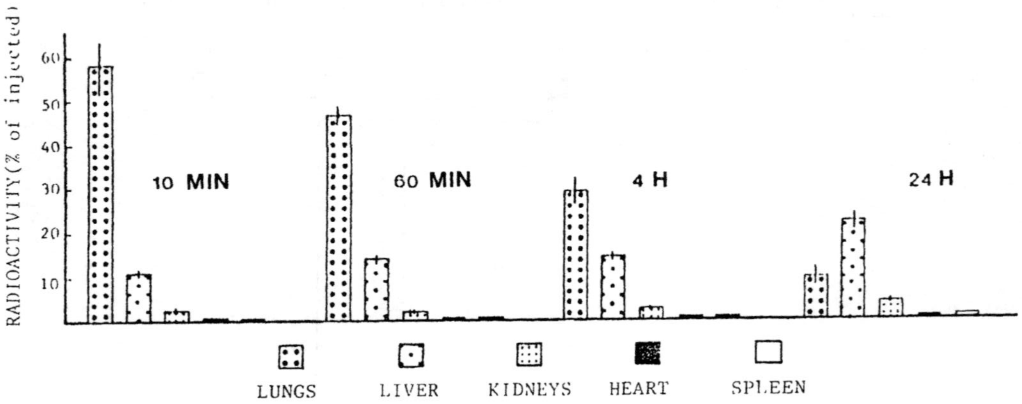

Figure 4. *In vivo* distribution of albumin microspheres.

TABLE 2 Release Kinetics of Streptomycin Sulphate Microspheres According to First Order Kinetics

CODES / Rates Constants	HM_1	HM_2	HM_3	HM_4	GM_1	GM_2	GM_3	GM_4
α	0.182	0.105	0.190	0.0681	0.113	0.161	0.196	0.102
β	0.0261	0.0201	0.0270	0.0354	0.0639	0.0340	0.0396	0.027
r^2	0.997	0.976	0.993	0.991	0.979	0.985	0.975	0.997

α = rate constant of the first phase

β = rate constant of the second phase

r^2 = determiation coefficient

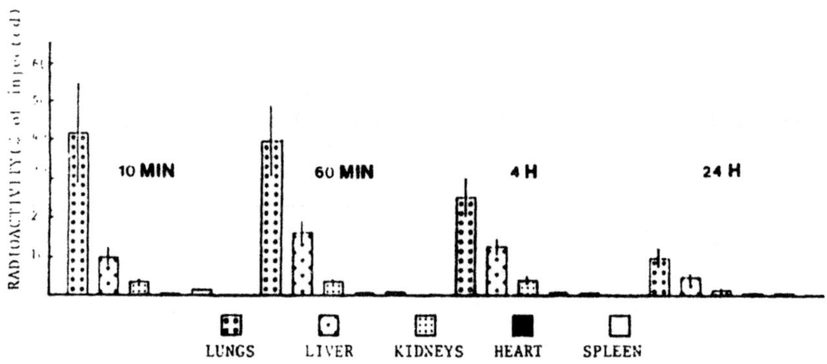

Figure 5. *In vivo* distribution of gelatin microspheres.

The in vitro release kinetics of immobilized streptomycin sulphate from human serum albumin and gelatin microspheres demonstrated a biphasic release pattern. (Table 2) This is in accordance with the results in the literature (Tomlinson et al 1984, Senyei et al 1981, Kojimo et al 1978, Davis 1984, Amsel et al 1984). The release rate profiles and rate constants are shown in Figures 6-9 respectively. It is clear from the figures that in vitro passive release of the immobilized drug from microspheres is biphasic with an intitial large and fast first-order release (burst effect) followed by a slower release. After this burst, release continues to be first order, but is much slower. These release profiles may be modeled using a simple biexponential function as shown by Tomlinson et al. (1984), i.e.

$$Q_t = A_{exp} - {}^{\alpha}{}_t + B_{exp} - {}^{\beta}{}_t$$

Where:

Q_t = the amount of active substance remaining in the microspheres at time t.

A and B = constants

α = rate constant of the first phase
β = rate constant of the second phase

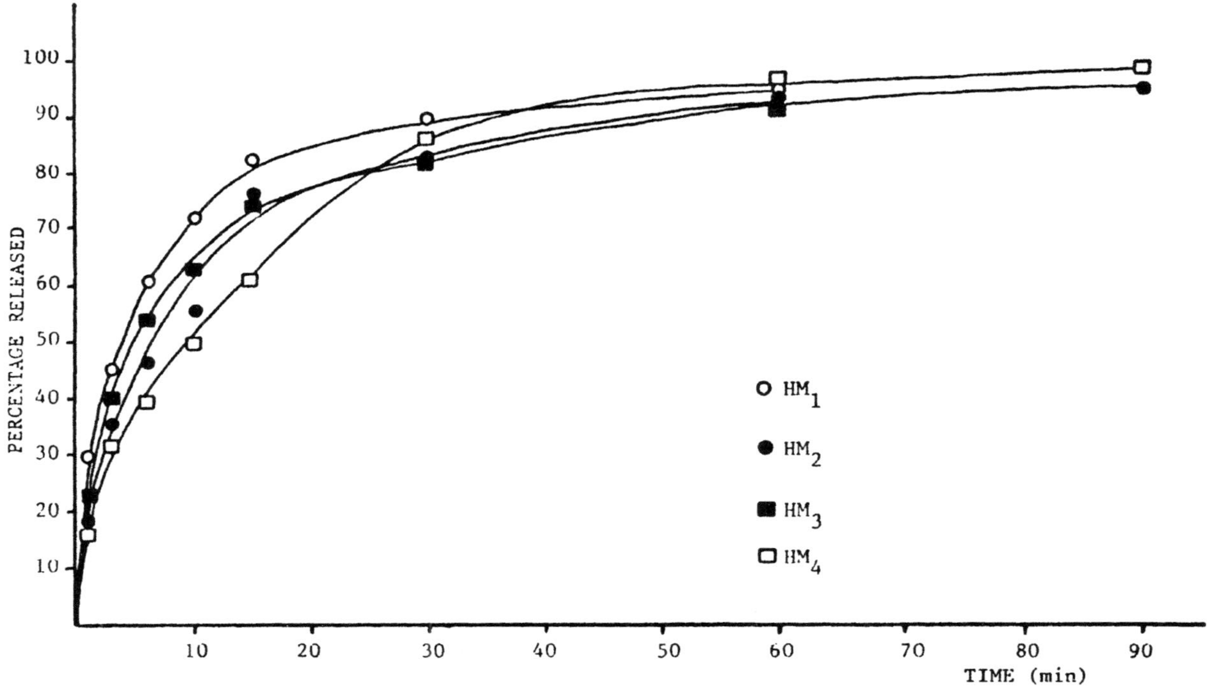

FIGURE 6 : Release profiles of streptomycin sulpnate from 30 μm human serum albumin microspheres in isotonic phosphate buffer(pH 7.4) at 37°C.

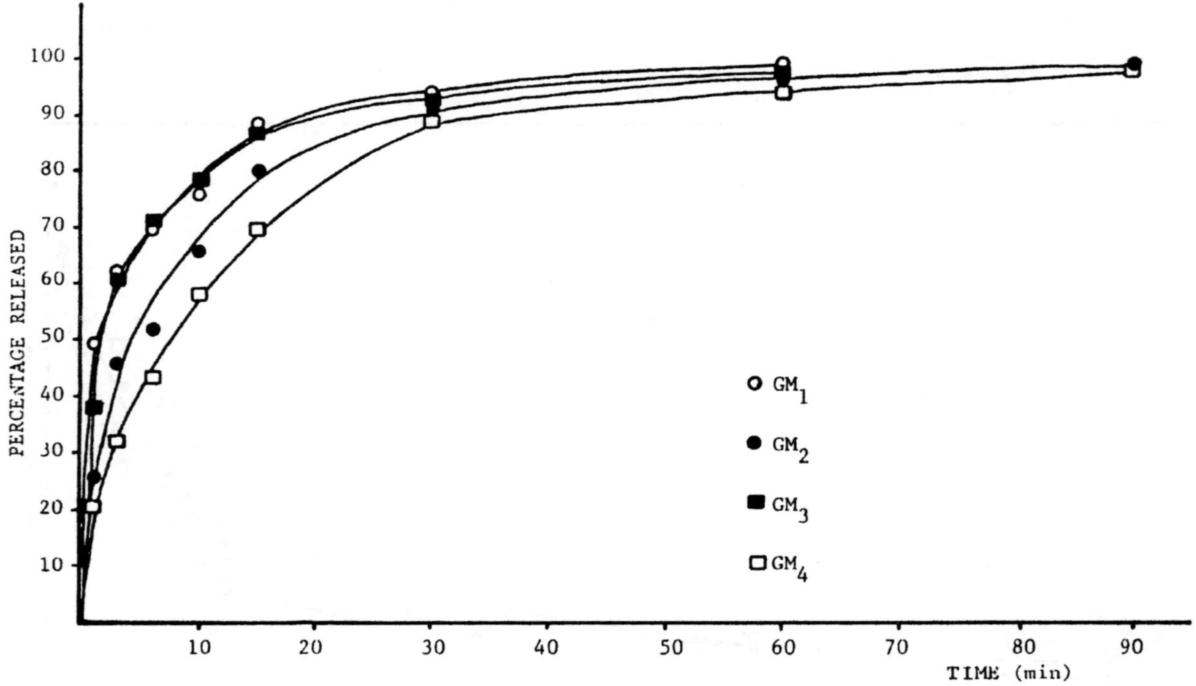

FIGURE 7 : Release profiles of streptomycin sulpnate from 30 μm gelatin microspneres in isotonic phosphate buffer(pH 7.4) at 37°C.

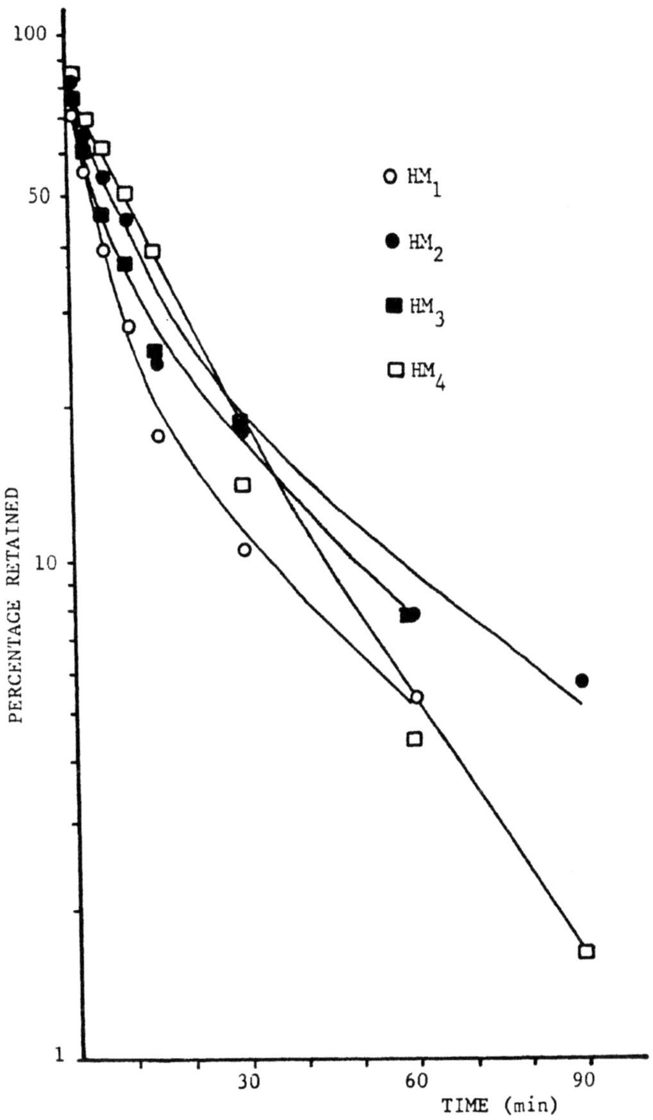

FIGURE 8 : First order release kinetics of human serum albumin microspheres.

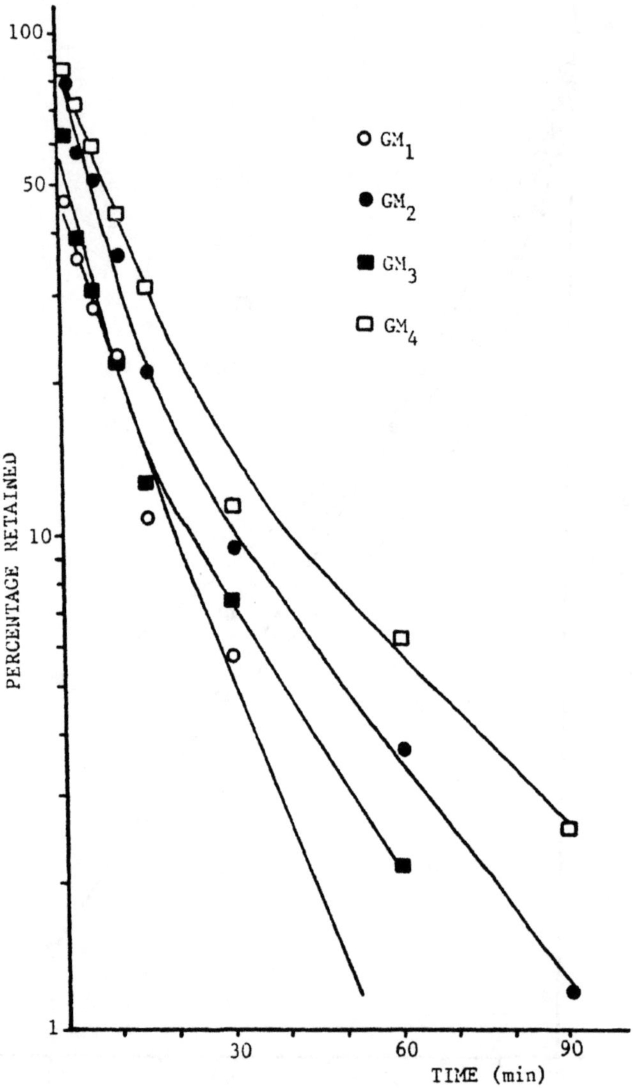

FIGURE 9: First order release kinetics of gelatin microspheres.

The longer duration of crosslinking and the higher concentration of the stabilizing agents reduces the fast release in the first phase, but increases the amount released and time of release in the second phase. This effect arises because more drug remains in the microspheres after the burst phase, so there is more streptomycin sulphate available for release in the second phase (Figures 6-9).

When the rate constants are compared, there seems to be less difference in the release characteristics of microspheres with longer duration of crosslinking. This may be due to the high immobilized drug in the microsphee, the release profile is less affected by the duration of crossliking.

Drug release from gelatin microspheres correspond more closely with the rate of hydration of the microspheres whereas release from human serum albumin microspheres is associated with matrix degradation. The spherical shape of the human serum albumin microspheres observed by the naked eye after dissolution studies are deformed when compared to gelatin microspheres. This deformation in shape is redued as the duration of exposure time to cross-linkers and the concentration of the stabilizing agents are increased.

4. REFERENCES

Amsel, L.P., Hinsvark, O.N., Rotenberg, K. and Sheumaker, J.L. (1984). Recent advances in sustained-release technology using ion-excahnge polymers. Pharmaceutical Technology, 8, 28-48.

Oronsen, K.F., Hellekant, C., Holmberg, J. Rothman, U., and Teder, H., (1979). Controlled blocking of hepatic artery flow with enzymatically degradable microspheres combined with oncolytic drugs. European Surgical Research, 11, 99.

Baker, R.W., Tuttle, M.E., and Helwing, R. (1984). Novel erodible polymers for the delivery of macromolecules. Pharmaceutical Technology, 8, 26.

Bostrom, S. and Brown, D.W., (1971). Spherodized albumin (human) microspheres and lung imaging, Journal of Nuclear Medicine, 12, 474.

Davis, M.A. and Taube, R.A. (1978). Pulonary perfusion imaging. Acute toxicity and safety facators as a function of particle size. Journal of Nuclear Medicine, 19, 1206-1213.

Davis, S.S. (1984). Recent advances in biopharmaceutics. Drug delivery systems, 2nd European Congress on Biopharmacy and Pharmacokinetics, Proceeding Vol. I. Salamanca.

Dispensatory of the United States of America (1960). (Philadelphia: Lippincott), pp. 1321-1334.

Ercan, M.T. (1976). Chromatographic quality control of ^{99m}Tc-labelled radiopharmaceticals. Hacettepe Bulletin of Medicine/Surgery, 9, 85.

Garrod, L.P. (1973). Antibiotic and Chemotherapy (Edinburgh: Churchill Livingstone), pp. 102.

Guyton, A.C., (1976). Textbook of Medical Physiology (Philadelphia: Saunders), pp. 386.

Gurkan, H., Yalabik-Kas, H.S., Hincal, A.A. and Ercan, M.T. (1986). Streptomycin sulphate microspheres formulation and in vivo distribution. Journal of Microencapsulation, 3, 101-108.

Hashida, M. Muranishi, S.H., Tanigawa, N., Satomura, K. and Hisaka, Y. (1979). Increased lymphatic delivery of bleomycin by microsphere in oil emulsion and its effect on lymph node metastasis. International Journal of Pharmaceutics, 2, 245.

Hashida, M., Takahashi, Y., Muranishi, S. and Sezaki, H. (1977). An application of water-in-oil and gelatin-microsphere-in-oil emulsions to specific delivery of anticancer agent into stomach lymphatics. Journal of Pharmacokinetics and Biopharmaceutics, 5, 241.

Hasida, M., Yashioka, T., Muranishi, S. and Sezaki, H. (1980). Dosage form characteristics of microsphere-in-oil emulsions. I. Stability and drug releae. Chemical and Pharmaceutical Bulletin, 28, 1009.

Illum, L. and Davis, S.S. (1984). The kinetics of uptake and organ deposition of colloidal drug carrier particles delivered to rabbits. Proceedings of the Second European Congress of Biopharmaceutics and Pharmacokinetics, Salamanca, Vol. II, pp. 97-106.

Kanke, M., Simmons, G.H., Weiss, D.L. Bivins, B.A. and DeLuca, P.P. (1980). Clearance of 141 Ce-labeled microspheres from blood and distribution in specific organs following intravenous and intraarterial administration in beagle dogs. Journal of Pharmaceutical Sciences, 69, 755.

Kaymakcalan, S., Kayaalp, S.O. and Kiran, B.K. (1976). Tibbi Farmakoloji, Prensipler ve Kavramlar (Ankara: Ankara Universitesi Basimeve) pp. 684-687.

Kojima, T. Hashida, M. Muranishi, S. and Sezaki, H. (1978). Antitumor activity of timed release derivative of mitocycin C Agarose bead conjugate. Chemical Pharmaceutical Bulletin, 26, 1818-1824.

Lee, T.K., Sokoloski, T.D. and Royer, G.P. (1981). Serum albumin beads: An injectable, biodegradable system for the sustained release of drugs. Science, 213, 233.

Merck Index (1976). New Jersey: Rahway Inc., pp. 564.

Nelson, W.E., Forgacs, J. and Kucera, J.L. (1947). Alteration of the distribution and excretion of streptomycin. Proceedings of the Society for Experimental Biology and Medicine, 64, 20.

Ng, B., Shaw, S.M., Kessler, W.V., Landoldt, R.R., Peck, G.E. and Dockerty, G.H. (1980). Distribution and degradation of iodine-125 albumin microspheres and technetium-99m sulfur colloid. Candian Journal of Pharmaceutical Sciences, 15, 30.

Raban, P. Gregora, V., Sindelar, J. and ALvarez-Cervera, J. (1973). Two alternative techniques of labelling iron-free albumin microscopheres with ^{99m}Tc and ^{113}m In Journal of Nuclear Medicine, 14, 344-345.

Rhodes, B.A. and Wagner, H.N., Jr. (1969). Improved Jung scans using ^{99m}Tc-albumin microspheres. Nuclear Medicine, 10, 432.

Rhodes, B.A., Zolle, I., Buchanan, J.W. and Wagner, H.N. Jr. (1969). Radioactive albumin microspheres for studies of the pulmonary circulation nuclear medicine, 92, 1453.

Richards, P. and Atkins, H.L. (1967). Technetium-99m Labelled Compounds, USAAEC Rep. BNL-11831, 1.

Russel, G.F.J., (1983). Starch microspheres as drug delivery systems, Pharmacy International, 4, 20.

Scheffel, U. Rhodes, B.A., Natarajan, T.K. and Wagner, H.N., Jr. (1972). Albumin microspheres for study of the reticuloendothelial system. Journal of Nuclear Medicine, 13, 498.

Senyei, A.E., Reich, S.D., Gonezy, C. and Widder, K.J. (1981). In vivo kinetics of magnetically targetted low-dose doxorubicin. Journal of Pharmaceutical Science. 70. 389-391.

Sugibayashi, K., Akimoto, M., Morimoto, Y., Nadai, T. and Kato, Y. (1979). Drug carrier property of albumin microspheres in chemotherapy III. Effect of microsphereentrapped 5- fluorouracil on Ehrilich Ascites Carcinoma in mice. Journal of Pharmaceutical Dynamics. 2, 350.

Tomlinson, E., Burger, J.J., Schoolderwoerd, E.M.A. (1984). Human serum albumin microspheres for intra-arterial drug targeting of cytostatic compound in microspheres and drug therapy, edited by S.S. Davis, l, Illum. J.G. Mc and E. Tomlinson (Amsterdam: Elsevier), pp. 75, 89.

Tomlinson, E., (1983). Microsphere delivery systems for drug targeting and controlled release. International Journal of Phamaceutical Technology, 4, 49.

Valentine, F.C.O and Shooter, R.A. (1954). Findlay's recent advances in chemotherapy, Vol. III, Antibiotics, J.A. Churchill Ltd., pp. 17-41.

Verdon, Jr., T.A., Anger, H.O., Yano, Y. and Dyke, D.V. (1971). 99m Tc-spherodized albumin (human) microspheres for lung imaging. Journal of Nuclear Medicine, 12, 469-470.

Wahbi, A.M., Abdine, H. and Sadek, A.R.H. (1977). The spectrophotometric determination of streptomycin. Pharmazie, 32, 690-692.

Waksman, S.A. (1953). Streptomycin: Background, isolation, properties and utilization. Antibiotics and Chemotherapy, 3, 33.

Waksman, S.A. and Schatz, A. (1945). Streptomycin-origin, nature and properties. Journal of the American Pharmaceutical Association, 34, 273.

Widder, K., Fluouret, G. and Senyei, A. (1979). Magnetic microspheres: Synthesis of novel parenteral drug carrier. Journal of Pharmaceutical Sciences, 68, 79.

Widder, K.J,, Senyei, A.E., Ovadia, H. aand Paterson, P.Y. (1981). Specific cell binding using staphyloccocal protein A magnetic microspheres. Journal of Pharmaceutical Sciences, 70, 387.

Widder, K.J., Senyei, A.E. and Scarpelli, D.G. (1978). Magnetic microspheres: A model system for site specific drug delivery in vivo. Proceedings of the Society for

Experimental Biology and Medicine, 58, 141.

Yoshioka, T., Hashida, M. Muranishi, S. and Sezaki, H. (1981). Specific delivery of mitomycin C to the liver, spleen and jung: nano and microspherical carriers of gelatin. International Journal of Pharmaceutics, 81, 131.

ENZYME AND MICROBIAL SENSOR

Isao Karube and Koji Sode
Research Laboratory of Resources Utilization
Tokyo Institute of Technology
4259 Nagatsuta-cho, Midori-ku, Yokohama
227 JAPAN

ABSTRACT. Various enzyme and microbial sensors based on amperometric determination were developed. These sensors were composed of enzyme(s) or/and microorganisms immobilized in membranes on an oxygen electrode. Hypoxanthine, fish freshness and free fatty acids were determined by enzyme sensors. Especially, in fish freshness sensors and in free fatty acid sensors, more than three sequential enzyme reactions were utilized. By using CO_2 utilizing microorganisms a CO_2 microbial sensor was constructed. Thermophilic bacteria isolated from Japanese hot springs were used for the microbial sensor for BOD measurement. This microbial sensor showed a good thermostability. By combining creatinine deiminase and nitrifying bacteria, a hybrid type biosensor was constructed for the amperometric determination of creatinine.

1. INTRODUCTION

The determination of organic materials such as nutrients and products in fermentation broths is important for control of fermentation processes. Most of these materials in broth can be determined by spectrophotometric methods. However, when natural materials are used as raw materials in fermentations the broths are not optically clear.

Electrochemical monitoring of these compounds has definite advantages. Many of these methods have utilized enzyme-catalyzed reactions because of their specificity. Many reports on enzyme sensors have been published. However, enzymes are generally expensive and unstable. Therefore, authors have developed microbial sensors using immobilized whole cells. These enzyme and microbial sensors can be classified into two categories. First is amperometric, the second potentiometric determination. In this study, novel enzyme and microbial sensors based on amperometric determination are described.

G. G. Guilbault and M. Mascini (eds.),
Analytical Uses of Immobilized Biological Compounds for Detection, Medical and Industrial Uses, 115–130.

2. Enzyme sensors

2.1. Hypoxantine sensor

Estimation of fish meat freshness is very important for the manufacturer of high quality products. After the death of a fish, the decomposition of ATP in the fish meat sets in and ADP, AMP, and other related compounds are subsequently generated according to the following sequence [1]

$$\mathrm{ATP} \xrightarrow{a} \mathrm{ADP} \xrightarrow{b} \mathrm{AMP} \xrightarrow{c} \mathrm{IMP} \xrightarrow{d} \mathrm{HxR} \xrightarrow{e} \mathrm{Hx} \xrightarrow{f} \mathrm{X} \xrightarrow{g} \mathrm{U}$$

where ATP is adenosine-5'-triphosphate, ADP is adenosine-5'-di-phosphate, AMP is adenosine-5'-monophosphate, IMP is inosine-5'-monophosphate, HxR is inosine, Hx is hypoxanthine, X is xanthine, and U is uric acid. In most cases, the rate-determining step is (e) or (f) depending upon the species of fish, and consequently HxR or Hx accumulates with increase of storage time. Therefore, HxR or Hx may be used as an indicator of the freshness using an oxygen electrode;

$$\mathrm{Hx} + O_2 \xrightarrow{\text{Xanthine oxidase}} \mathrm{X} + H_2O_2$$

$$\mathrm{X} + O_2 \xrightarrow{\text{Xanthine oxidase}} \text{Uric acid} + H_2O_2$$

A flow-type enzyme sensor system has been developed and applied to a rapid and simple determination of hypoxanthine [3] (Fig. 1). Xanthine oxidase was immobilized on a cellulose triacetate membrane and attached to an oxygen electrode. Oxygen

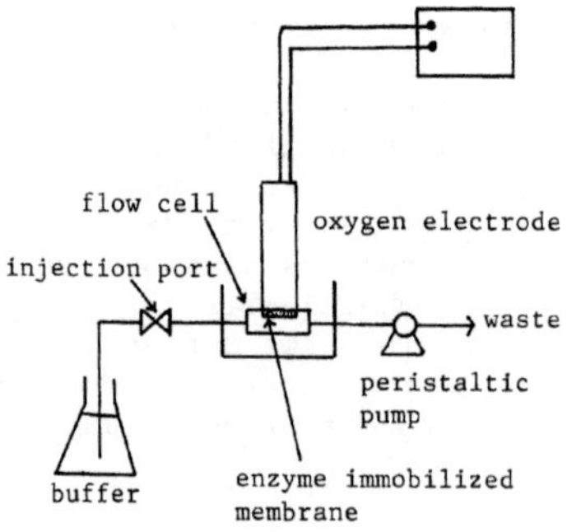

Figure 1. A flow type enzyme sensor system for hypoxanthine determination.

consumption due to the oxidation of hypoxanthine was monitored by this electrode. The enzyme sensor was applied to the determination of Hx in several fish meats.

A linear relationship was obtained for the 62.5 μM to 1.5 mM Hx concentration range (Fig. 2). The minimum concentration that could be determined was 8.5 mg dm^{-3} of Hx. The reproducibility of the current decrease was examined using the same sample and found to the within ± 8% for a sample containing 136 mg dm^{-3} of Hx. The standard deviation was 0.2 mg dm^{-3} for 100 experiments.

Sample solution containing Hx were prepared from sea bass, saurel, mackerel, yellowfish, and flounder. Hx concentration was determined by both the enzyme sensor method [23] and the conventional enzymatic assay method (Fig. 3).

A linear correlation was observed between the value determined by the both methods and the correlation coefficient was 0.98. These results suggest that the sensor proposed is an economical and reliable method for analysis of Hx in the fish meats. However, the Hx content in fish meats depends on the species. Therefore, the Hx content alone cannot be used as universal indicator of fish freshness.

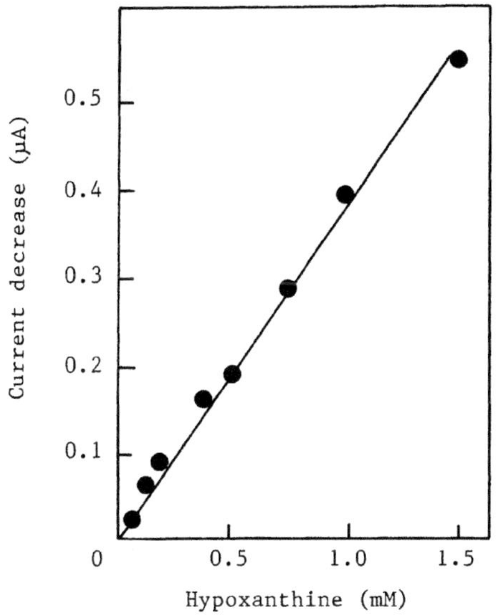

Figure 2. Calibration curve for the hypoxanthine sensor.

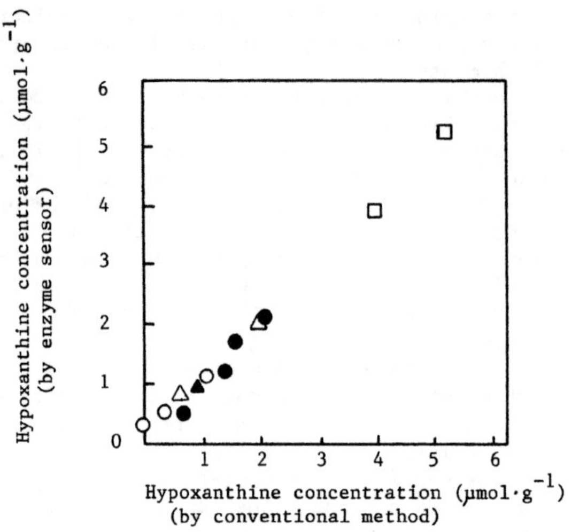

Figure 3. A correlation between hypoxanthine concentration determined by enzyme sensor and by the conventional method.
(▲);seabass, (●); saurel, (○); mackerel, (△);yellowfish, (□); flouder.

2.2. Fish freshness sensor

A multifunctional enzyme sensor system was developed by the authors tfor the estimation of fish freshness [4]. The freshness indicator K_I was represented by following equation;

$$K_I = \frac{HxR + Hx}{IMP + HxR + Hx} \times 100$$

Various enzyme sensors for the detemination of Hx, HxR, IMP, and AMP have been prepared. The enzymes xanthine oxidase, xanthine oxidase-nucleoside phosphorylase, xanthine oxidase-nucleoside phosphorylase-nucleotidase, and xanthine oxidase-nucleoside phosphorylase-nucleotidase-AMP deaminase, respectively, were immobilized on a cellulose triacetate membrane containing 1,8-diamino-4- aminomethyloctane. These immobilized enzyme preparations were attached to an oxygen electrode, and the oxygen consumption due to the oxidation of hypoxanthine was monitored. Each component was determined as the decrease in current. The multifunctional enzyme sensor system consisted of a combination of IMP sensor, ion-exchange resin column and computer (Fig. 4).

Fish freshness was determined simply and rapidly by this sensor system (Table 1). Morever, fine differences among fish freshness, which could not be shown by any indicators so far proposed, were expressed in the freshness pattern (Fig. 5).

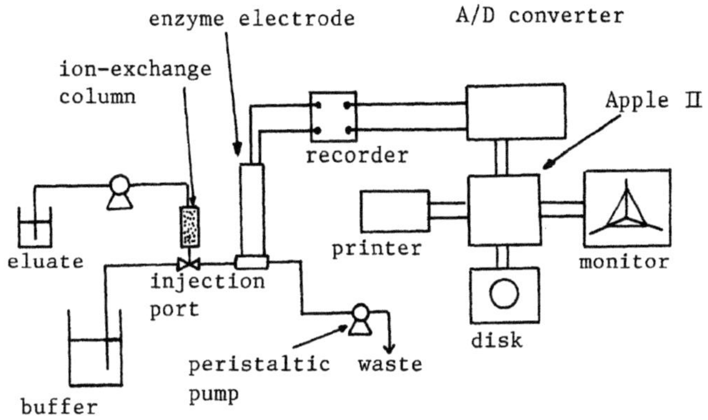

Figure 4. Schematic diagram of fish-freshness sensor system.

Table 1. K_I Values of various fish meat.

Fish	Storage time (days)	K_I Value: Enzyme sensor	K_I Value: Conventional method
Sea bass	1	9.2	10.9
	5	34.8	33.0
	10	55.3	56.0
Bream	1	4.7	6.6
	5	12.4	10.9
	10	17.0	19.9
Flounder	1	10.0	12.9
	5	58.2	55.0
	10	63.0	60.5
Yellowfin tuna	1	8.6	8.0
	5	36.8	35.0

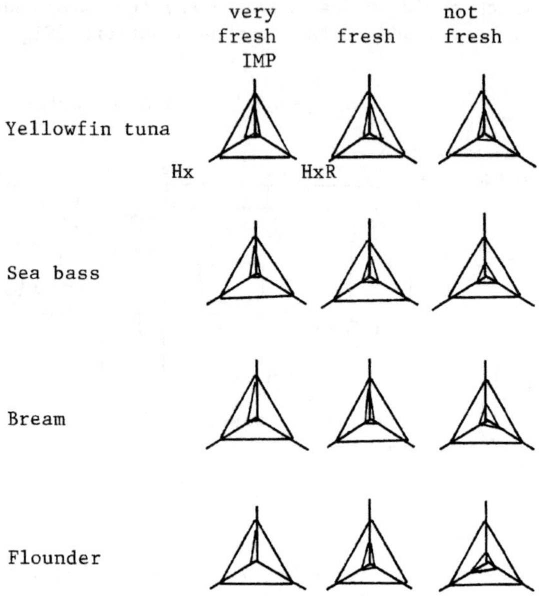

Figure 5. Fish freshness pattern.

very fresh; $K_I < 10$, fresh; $K_I < 40$

not fresh; $K_I > 40$

2.3. <u>Fatty acids sensor</u>

The measurement of free fatty acids in clinical analysis and in food processing are now of great interest. The conventional methods for free fatty acid analysis are mostly based on spectrometry.

Acyl-CoA synthetase is a specific enzyme for free fatty acids having carbon chain length of 8 to 20. This enzyme catalyzes the following reaction, subsequently producing AMP, in the presence of ATP, Co-enzyme A and magnesium.

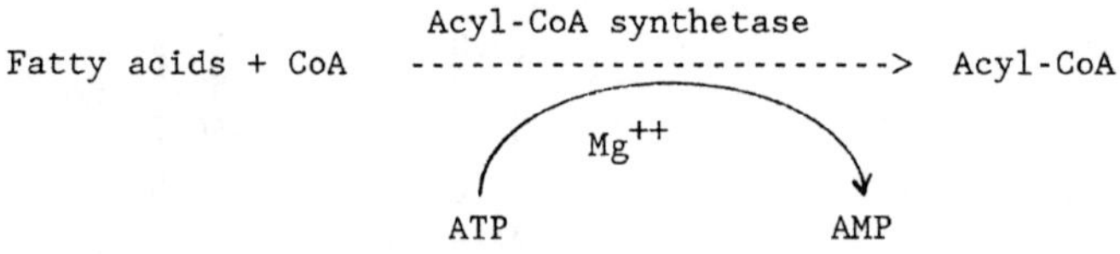

An AMP sensor was developed in our laboratory based on 4 sequential enzyme reactions [5]. Combining acyl-CoA synthetase with the AMP sensor, a free fatty acid sensor was constructed [6].

Acyl-CoA synthetase, xanthine oxidase, nucleoside-phosphorylase, 5'-nucleotidase and AMP-deaminase were immobilized in a porous cellulose nitrate membrane by absorption.

In this experiment, the fatty acid was well emulsified in phosphate buffer solution, by sonication (200w, 60KHz) for 1 min. The emulsion thus formed was stable for more than one day.

The enzyme electrode was immersed in a 30 ml reaction vessel containing phosphate buffer with 3.2 mM of ATP and 6 mM of magnesium chloride. After the output current became steady, the enzyme electrode was taken out from the reaction vessel. Fatty acid was injected to this solution, and emulsified by the method described above. The enzyme electrode was again immersed in this sample solution and the current decrease was recorded. Within 1 minute after the fatty acid sensor was immersed into the sample solution, the output current decreased, and in 3 minutes, the minimum current was obtained. This assay was completed within 5 minutes.

Fig. 6 and 7 show the calibration curve of the sensor towards oleic acid and palmitic acid. In both fatty acids good linearity was observed between the current and the concentration of fatty acid below 15 mM. The minimum concentration for the determination was 0.5 mM, and the rate of incline was 0.11 μ A/mM.

These results suggested that this enzyme sensor appears to be a promising and attractive method for the routine measurement of free fatty acids.

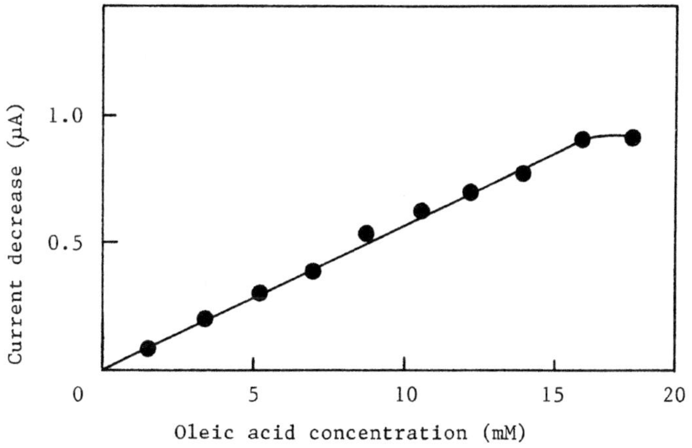

Figure 6. Calibration curve for the free fatty acids sensor. (oleic acid)

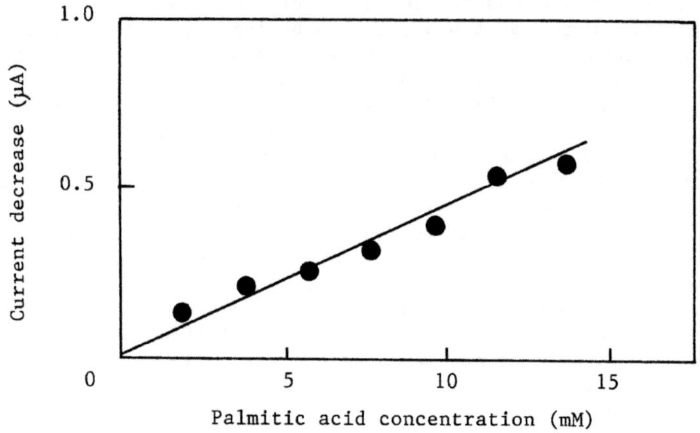

Figure 7. Calibration curve for the free fatty acids sensor. (palmitic acid)

3. Microbial Sensor

3.1. Carbon Dioxide Sensor

Determination of CO_2 is considered to be very important, in industrial processes and environmental and clinical analysis. The pCO_2 electrode was first developed for the measurement of CO_2 in blood. This potentiometric pCO_2 device has found widespread use in biological and industrial application. But major problems are associated with the potentiometric method of detection. Various ions and some organic and inorganic volatile acids affect the potential of the inner pH electrode and the gas permeable membrane of the pCO_2 electrode. Therefore, we have developed a novel microbial CO_2 sensor based on amperometry [7].

The chemoautotrophic bacterium, which can grow with only carbonate as the C-source, used in this study was obtained from The Fermentation Research Institute, Japan. This bacterium was isolated from the soil obtained in Akita prefecture, Japan, and is named S-17. S-17 is supposed to belong to the genus Pseudomonas, and is a 1x4-5 micrometer Gram-negative rod, growing both autotrophically and deterotrophically on common nutrient media. It can also utilize nitrogen, citric acid, acetic acid, butyl alcohol, nitrate and ammonium salts. Optimum temperature and pH for growth are 20-34°C and pH 4.0-8.6, respectively. In this study, S-17 was autotrophically incubated under aerobic conditions at 30°C for approximately one month in an inorganic medium.

Bacterial cells immobilized on a cellulose nitrate membrane (pore size 0.45 μm) were placed in the vicinity of the cathode of a Galvanic type oxygen sensing electrode. The cathode and the immobilized cells were then totally covered with a polytetrafluoroethylene (PTFE) membrane (pore 0.5 μm, Type FH) secured with a rubber O-ring. The PTFE membrane acts as a gas permeable membrane.

Figure 8 shows the current-time response of the CO_2 sensor to the addition of CO_2(K_2CO_3). When a sample solution containing CO_2 is injected into the external buffer solution, CO_2 permeates through the gas permeable membrane, and is assimilated by the bacteria. The respiratory rate of the bacteria consequently becomes elevated, giving rise to the consumption of oxygen. This experiment was executed, injecting 300 μl of 1 M K_2CO_3 solution into 50 ml of KH_2PO_4/NaOH buffer solution (30°C, pH 5.5). As shown, the current decreased rapidly and reached a maximum 2-5 minutes after injection of K_2CO_3, followed by the gradual increase of the current due to the escape of CO_2 to the air. The calibration curve of K_2CO_3 concentration versus current decrease is shown in Figure 9.

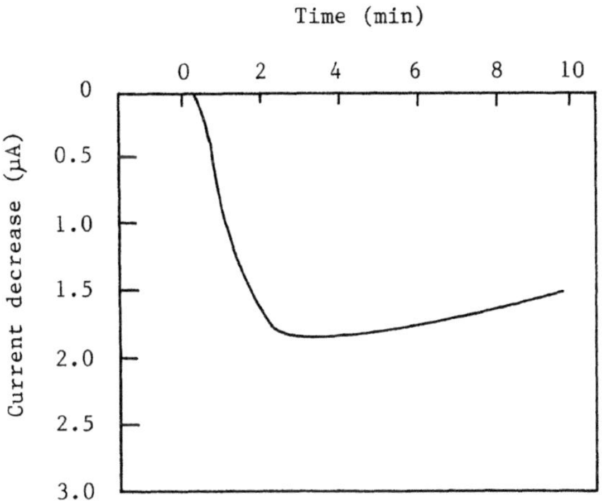

Figure 8. Time course of CO_2 sensor responce.
The curve corresponds to 6mM concentration of K_2CO_3.

For concentrations of K_2CO_3 below 6-8 mM, the current decrease was approximately proportional to the K_2CO_3 concentration, but at higher concentrations the current decrease plateaued because of saturation of the system with CO_2. Initial rate of carbon dioxide utilization does not depend on the K_2CO_3 concentration higher than 8 mM. The rate of CO_2 utilization appears to be limited by the uptake or metabolism of CO_2 by the microbial cells.

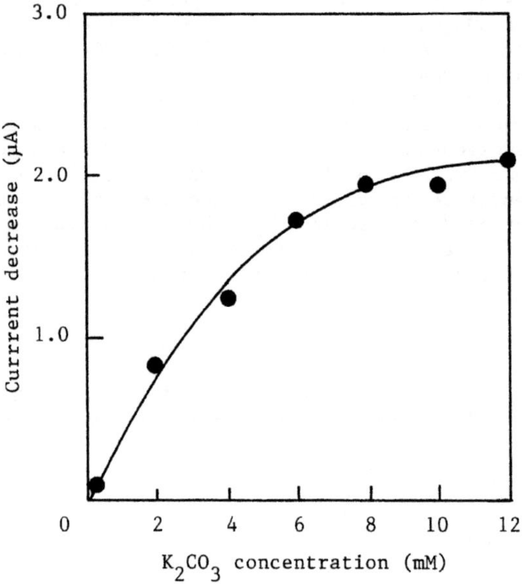

Figure 9. CO_2 sensor responce to K_2CO_3 concentration.

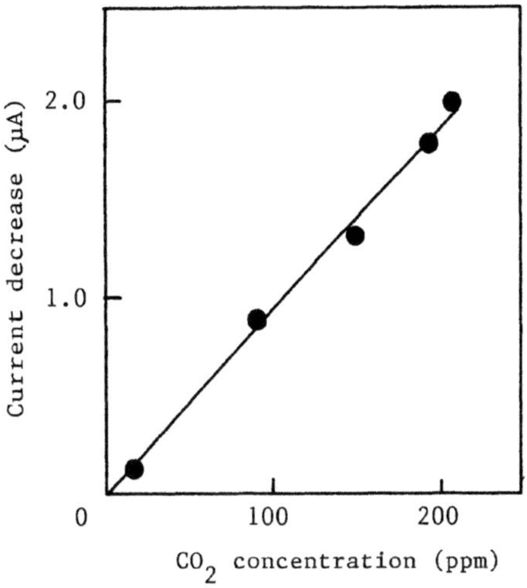

Figure 10. Calibration curve for the CO_2 sensor.

Figure 10 shows the calibration curve for the biosensor. The CO_2 concentration was determined by performing the same experiment as described previously but with a conventional potentiometric pCO_2 electrode. The results of this experiment permitted a linear calibration curve to be constructed relating to the CO_2 concentration and the current decrease of the novel amperometric CO_2 sensor. The linear range of the sensor was found to be below 200 ppm CO_2. The upper limit of the linear range was determined by the saturation of CO_2 in the buffer solution. Considering that the current fluctuation resulting from experimental errors is within 0.5 μA, our novel sensor could easily detect a variation of 5 ppm CO_2 (200 μM K_2CO_3) in the lower concentration range.

Selectivity of the microbial sensor was also examined. Various aqueous solutions of volatile compounds such as acetic acid, citric acid, formic acid, ethyl alcohol, butyl alcohol were added to the sensor system. Though the sensor responded slightly (37% of the response to CO_2) to acetic acid, all the other compounds produced no observable response within the margin of experimental error. Selectivity could also be improved by incorporating a more suitabe material as the gas permeable membrane.

Stability of the sensor was more than three weeks. This was considered to be very satisfactory for its application in practical use, especially in a flow cell configuration.

3.2 Microbial sensor using thermophilic bacteria

We have already developed a BOD sensor using immobilized *Trichosporon cutaneum* and an oxygen electrode, based on measuring their respiration dependent on BOD [8]. This method was simple and not a time consuming procedure. Since this microoorganism used in the BOD sensor was a methophilic microorganism, this BOD sensor was employed at 25-35°C. However, the temperature of waste water is sometimes higher than the optimum temperature for methophilic microorganisms. In such cases, a BOD sensor using a methophilic microorganism could not be applied. Besides, many bio-processes using thermophilic microorganisms have been developed.

Consequently, on-line measurement of these processes by a biosensor for precise control is of great interest.

Regarding these demands mentioned above, we have recently developed a novel microbial sensor using theromphilic bateria [9].

The microorganism used in the biosensor was isolated from hot springs in Japan (Minami Atagawa hot spring, Izu, Shizuoka perfecture). This bacteria was cultured at 65°C with a medium using glucose as a sole carbon source. The bacteria thus cultured was centrifuged and immobilized onto a nitro-cellulose membrane filter (0.45μm pore size) by absorption. This membrane was attached onto an oxygen electrode and covered with a dialysis membrane to construct a microbial sensor.

When this microbial sensor was immersed in the sample solution containing 5.0μM glucose, the output current of the oxygen electrode gradually decreased. Within 6 min, the current decrease showed a steady state.

The effect of temperature on this biosensor response was then examined (Fig. 11). The maximum current decrease was observed at 65°C, while almost no response was observed at 70°C. However, even after 2 hours' incubation at 70°C, this sensor showed a good reproducible respose to glucose at 50°C.

Therefore, this microbial sensor retained a high temperature stability. In further experiments, this biosensor was employed at 50°C. Figure 12 shows the relationship between current decrease and glucose concentration. A good linear correlation was observed between sensor respose and glucose concentration in the range of 2-20 M.

Table 2 summerizes the sensor response to other carbon sources. This table shows that this microbial sensor responds to various carbon sources. The sensor was then applied to BOD measurement. (Fig. 13). A good linear correlation was observed between current decrease and BOD values obtained by conventional methods. Therefore, this microbial sensor using thermophilic bacteria can be used as a BOD sensor.

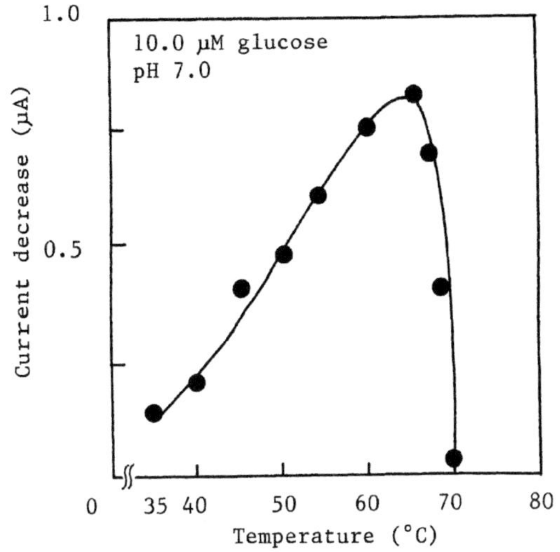

Figure 11. Effect of temperature on microbial-sensor responce.

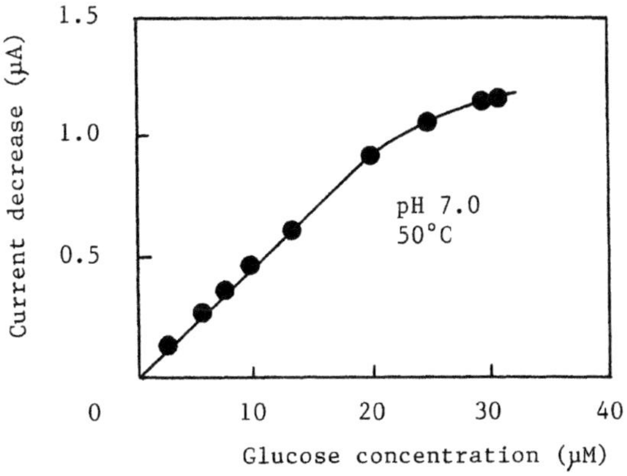

Figure 12. Calibration curve for the microbial-sensor using thermophilic bacteria. (Glucose)

Table 2. The responce of microbial sensor to various organic compounds.

Glucose	100 %
Fructose	29
Galactose	98
Maltose	96
Saccharose	95
Lactose	61
Glycine	25
Glutamate	120
Glycine	96
Ethanol	0
Sodium acetate	0

concentration of each compounds was 10 µM. 50°C, pH 7.0

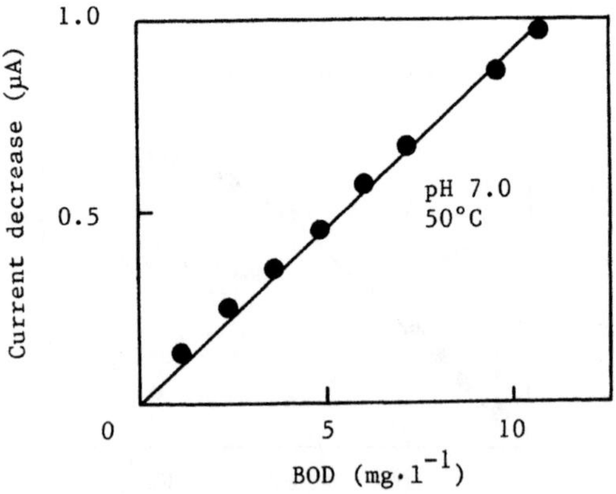

Figure 13. Calibration curve for the microbial sensor using thermophilic bacteria. (BOD)

3.3 Creatinine sensor (hybrid biosensor)

The determination of creatinine in biological fluids is a diagnostically important test. External dialysis (artificial kidney) has become so widely practiced that precise, rapid and simple methods for creatinine determinations are very desirable.

The authors have developed a new sensor for the amperometric determination of creatinine [10]. This is based on amalgamation of an enzyme reaction and bacterial metabolism.

Creatinine deiminase hydrolyzes creatinine to N-methylhydantoin and ammonium ion. The ammonia is successively oxidized to nitrite and nitrate by nitrifying bacteria. The bacteria have been completely characterized, but are known to be mixed culture of Nitrosomonas sp. and Nitrobacter sp. The sequence of reactions is:

$$NH_4^+ \xrightarrow[\text{Nitrosomonas sp.}]{O_2} NO_2^- \xrightarrow[\text{Nitrobacater sp.}]{O_2} NO_3^-$$

The nitrifying bacteria consume oxygen, so that the oxygen decrease can be detected by an oxygen electrode. Therefore, a microbial sensor for ammonia consisted of immobilized nitrifying bacteria and the oxygen electrode [11].

A creatinine sensor consisted of immobilized creatinine deiminase, immobilized nitrifying bacteria, and an oxygen electrode.

A triamine membrane was used to immobilized creatinase. It is a triacetylcellulose membrane containing 1,8-diamono-4-aminomethyl-octane. The acetylcellulose membrane retaining the nitrifying bacteria was attached to the teflon membrane between the two membranes. The immobilized enzyme membrane was attached over the acetylcellulose membrane. These membranes were covered with a dialysis membrane.

When a sample solution containing creatinine was applied to the sensor system, creatinine permeated through the dialysis membrane and was decomposed to ammonia and N-methylhydantoin. The ammonia was assimilated by the immobilized bacteria. At the same time the bacteria consumed dissolved oxygen from around the membrane, so that the current from the oxygen electrode decreased markedly and reached a minimum value within 3 min.

There was a good linear correlation between the current decrease and the concentration of creatinine below 1000 $mg.dm^{-3}$. The minimum detectable concentration of creatinine with this sensor was 50 $mg.dm^{-3}$.

The selectivity of the hybrid biosensor for creatinine was examined with solutions containing other organic compounds. The sensor did not respond to urea, uric acid, citrate, pyruvate, glucose, arginine, glutamine or EDTA. Therefore, the selectivity of the hybrid biosensor seemed to be satisfactory.

The re-usability of the creatinine sensor was examined with a sample solution containing 500 mg.dm^{-3} creatinine. The current output of the sensor eventually decreased but it could be used for more than 3 weeks or 300 assays.

4. REFERENCES

1. T. Saito, A. Arai and M. Matsuyoshi, 'A new method for estimating the freshness of fish'. Bull. Jap. Soc. Sci. Fish. 24, 749 (1959).
2. S. Ehira and H. Uchiyama, 'Rapid estimation of freshness of fish by nucleoside phosphorylase and xanthine oxidase', Bull. Jap. Soc. Sci. Fish. 35, 1080 (1969).
3. E. Watanabe, K. Ando, I. Karube, H. Matsuoka and S. Suzuki, 'Determination of hypoxanthine in fish meat with an enzyme sensor', J. Food Sci..48, 496 (1983).
4. E. Watanabe, K. Toyama, I. Karube, H. Matsuoka and S. Suzuki, 'Multifunctional biosensor for the determination of fish meat freshness', Ann. N.Y. Acad. Sci. 42, 529 (1984).
5. E. Watanabe, K. Toyama, I. Karube, H. Matsuoka and S. Suzuki, 'Determination of inosine-5'-monophosphate in fish meat with an enzyme sensor', J. Food Sci, 49, 114 (1984).
6. Y. Kameda, K. Sode, E. Tamiya and I. Karube, 'Determination of free fatty acids with an enzyme electrode', Anal. Lett. (1987) (submitted).
7. H. Suzuki, E. Tamiya and I. Karube, 'An amperometric carbon dioxide sensor employing immobilized carbon dioxide utilizing bacteria', Anal. Chim. Acta. (1987) (submitted).
8. M. Hikuma, H. Suzuki, T. Yasuda, I. Karube and S. Suzuki, 'Amperometric estimation of BOD by using living immobilized yeasts', Eur. J. Appl. Microbiol. Biotechnol., 8, 289 (1979).
9. I. Karube, K. Yokoyama and E. Tamiya, 'Microbial-sensor using thermophilic bacteria for BOD measurement', Appl. Env. Microb. (1987) (submitted).
10. I. Kubo, I. Karube and S. Suzuki, 'Amperometric determination of creatinine with a biosensor based on immobilized crearininase and nitrifying bacteria'. Anal. Chim. Acta. 151, 371 (1983).
11. I. Karube, T. Okada and S. Suzuki, 'Amperometric determination of ammonia gas with immobilized nitrifying bacteria'. Anal. Chem., 53, 1852 (1981).

REDOX MEDIATORS AND THEIR APPLICATION IN AMPEROMETRIC SENSORS

Anthony P F Turner
Bioelectronics Division
Biotechnology Centre
Cranfield Institute of Technology
Cranfield
Bedfordshire
MK43 OAL
UK.

ABSTRACT. Despite exciting advances in potentiometric, optical and other transducers for use in biosensors, amperometric detectors still predominate in practical configurations. The majority of amperometric devices produced to date rely on expanding the range of conventional electrochemical detectors by monitoring the natural substrate or product of a biochemical reaction. This approach imposes serious limitations on the performance of the sensors and restricts the design opportunities. This brief review considers the advantages offered by replacing the natural redox partner in biological systems with an electrochemically regenerated mediator. Commercial enzyme electrodes exploiting this principle are already emerging onto the market and are likely to be followed by electrochemical immunoassays and mediated systems incorporating intact microorganisms.

1. INTRODUCTION

Enzymes and intact microorganisms have been exploited widely in various forms of biosensors (Turner et al, 1987a). In addition, enzyme labels have become increasingly important in affinity systems such as immunoassay and DNA probes (Kress-Rogers and Turner, 1986). Enzyme activity is normally detected either by monitoring a natural product of the reaction such as ammonium ions, for example:

$$\text{Urea} \xrightarrow{\text{urease}} NH_4^+ + HCO_3^- \qquad (1)$$

or hydrogen peroxide, for example:

$$\text{Glucose} + O_2 + H_2O \xrightarrow[\text{oxidase}]{\text{glucose}} \text{Gluconic acid} + H_2O_2 \qquad (2)$$

or by following the consumption of a cosubstrate such as oxygen in,

G. G. Guilbault and M. Mascini (eds.),
Analytical Uses of Immobilized Biological Compounds for Detection, Medical and Industrial Uses, 131–140.

for example, reaction (2). The most common transducers used in biosensors are electrochemical while colorimetric approaches are still favoured for immunoassay.

2. INDIRECT AMPEROMETRIC BIOSENSORS

Amperometric detectors typically are inexpensive, have a wide dynamic range and are less critically dependent on the provision of an accurate reference electrode than their potentiometric counterparts. They do, however, demand controlled transport of analyte to the electrode (Wilson, 1987). There are many examples of biological redox reactions catalysed by purified oxidoreductase enzymes, cell fragments and intact cells being coupled with amperometric electrodes to form novel analytical systems. In the majority of cases this has been achieved either by using the Clark oxygen electrode to monitor oxygen consumption or by detecting an electroactive product such as hydrogen peroxide at platinum electrodes (Clark, 1987). Given the redox nature of both the catalytic system and the detector it was likely that a more direct method of realising electron transfer between the biological system and the electrode would prove both more sensitive and less prone to interference than previous indirect designs.

3. MEDIATED AMPEROMETRIC SENSORS

In practice a more intimate connection between biological redox systems and electrodes has been achieved by using low molecular weight redox intermediates known as mediators (M) either in solution or to modify electrodes (Cardosi and Turner, 1987a). This approach eliminates the natural electron acceptor from the reaction and can make assays essentially oxygen-independent:

$$\text{Glucose} + H_2O + M_{ox} \xrightarrow[\text{oxidase}]{\text{glucose}} \text{Gluconic acid} + 2H^+ + M_{red}$$

In addition, the potential at which detection is carried out is determined by the redox potential of the mediator, which can be selected to avoid interference from other electroactive species in the sample.

The choice of mediator for such systems is critical and varies depending on the application; an insoluble mediator such as ferrocene (D'Costa _et al_, 1986; Dicks _et al_, 1986) is most appropriate for incorporation into a conventional enzyme electrode whereas an immunoassay may demand a soluble mediator (Cardosi _et al_, 1986). Synthesis of derivatives of a particular mediator is a powerful tool in the manipulation of its relevant properties such as solubility, charge and redox potential (Cass et al, 1984; Green and Hill, 1986). Moreover, derivatives that act, for example, as enzyme substrates may used to monitor enzyme labelled affinity reactions (Kress-Rogers and

Turner, 1986; Higgins and Lowe, 1987).

In addition to the questions of solubility and derivitisation a mediator must satisfy the following criterea:

a) Rapid reaction with the biological system

b) Rapid and reversible electrochemistry with a practical electrode in the assay medium

c) Chemical stability both during storage and use

d) Absence of reactivity of both the oxidised and reduced form to chemicals in the sample or oxygen in the air

e) A redox potential close to zero and away from the redox potential of other electroactive species in the sample

f) A wide pH range or at least an optimum pH which corresponds to that of the biological system

g) Minimal inactivation of the biological system

h) Preferably be inexpensive, amenable to formulation for a mass produced product and non-toxic.

In our hands the best mediators for homogeneous electrochemical enzyme assays are N,N,N',N'-tetramethyl-p-phenylenediamine [TMPD] (Davis et al, 1983), ferricyanide (Cardosi et al, 1986) and ferrocene monocarboxylic acid (Cass et al,1984; Turner et al, 1984; Dicks et al, 1986; D'Costa et al, 1986). For immobilised systems where the mediator is deposited on or in the electrode in a form which is insoluble in the assay medium the most effective mediators have proved to be 1,1'-dimethyl ferrocene [DMF] (Aston et al, 1984; Cass et al, 1984; Turner et al, 1984; Dicks et al, 1986; D'Costa et al, 1986), tetracyanoquinodimethane [TCNQ] (Hendry and Turner, 1987) and tetrathiafulvalene [TTF] (Turner et al, 1987b). In certain specialised cases involving electron transfer from intact microorganisms a cocktail of soluble mediators has proved particularly useful. For example, a synergistic effect was observed between benzoquinone and ferricyanide for use in a broad range microbial activity monitor (Turner et al, 1986).

3.1. Blood and Subcutaneous Glucose Sensors

The most well defined market for a biosensor is undoubtably blood glucose testing. The requirement for simple devices capable of use by the diabetic at home has attracted considerable research effort (Cardosi and Turner, 1987b), but this need is also well serviced by paper test strips supplied predominantly by two major companies. The successful biosensor must either surpass the present technology in convenience, accuracy and cost or provide a service which is beyond the capability of present technology. Mediated amperometric biosensors satisfying both scenarios have been launched or are under development in our laboratories. The original concept of ferrocene-based glucose sensors (Turner, 1983; Cass et al, 1984) has matured into a commercial device in the hands of Genetics International (McCann, 1987). A pen-sized glucose sensor manufactured by this company in the UK was launched in the USA in the Spring of 1987. The device resembles the digital watch/pens which are now ubiquitous in souvenir and trinket shops and upon which the original design was based. A small disposable electrode consisting of mediator-modified carbon and glucose oxidase printed on a plastic support is inserted into the tip of the "pen". A drop of whole blood placed on the electrode can be assayed in 30 seconds without a requirement to clean the test strip. The performance of the sensor compares favourably with conventional laboratory analyses showing high accuracy (correlation coefficient = 0.985, n = 100) and good precision (coefficient of variation for 20 samples with a mean of 4.8 mM glucose = 3.9%). The main appeal of this device and hence its commercial competitiveness lies in its compact design and simple protocol, both of which are greatly facilitated by the use of mediated electrochemistry.

The single-use electrode described above could be constructed using alternative mediators such as TCNQ (Hendry and Turner, 1987) or TTF (Turner et al, 1987b), but for in vivo use ferrocene and its derivatives are favoured due to their low toxicity. Work is in hand in our laboratories, however, to synthesise various derivatives of TTF and TCNQ aswell as investigating the properties of novel polyferrocenes in order to improve the retention of mediators in microminiature devices. A reduction in mediator loss will minimise any risk of toxicity and improve the operating life of the enzyme electrodes. Promising results have already been obtained using poly(ferrocenylenemethylene)- and methoxy-phenyl substituted polyferrocene-modified carbon electrodes (Turner et al, 1987c). Claremont et al (1986) have described the construction of an implantable enzyme electrode based on 1 mm wide DMF-modified carbon strips. Following immobilisation of the glucose oxidase the active surface was protected by a polyurethane membrane. The electrodes responded linearly up to 20 mM glucose with a mean response time of 68 s. Sensor readings correlated well with conventional assays and a proportion of the electrodes (17%) exhibited low or zero drift over 17 h at 37°C. A ferrocene-mediated electrode housed inside a 25 guage needle has been used in Man (Matthews et al, 1987). The needles were well tolerated and gave an unimpaired response for 4.5 h.

3.2. Enzyme Electrodes for Non-medical Applications

Glucose sensors have wide applicability to a range of industrial analyses where it may be either of interest in its own right or used as a indicator of other coupled reactions. In addition to the measurement of absolute concentrations of glucose it is sometimes of interest to measure gradients of concentration. In the microbial deterioration of meat, for example, the aerobic metabolism of the endogenous glucose by bacteria colonising the surface results in an indicative glucose profile being developed. We developed a multiple electrode capable of measuring this profile for the Leatherhead Food Research Association (Kress-Rogers and D'Costa, 1986). The electrodes may be fabricated from laminated carbon or by screen printing techniques. Glucose oxidase was immobilised on the DMF treated carbon electrodes which could then be mounted in a knife shaped holder. The electrodes are capable of operation at the chill temperatures (2-10°C) and low pH (ca 5.5) typically found in stored carcasses.

Many industrial processes would benefit from the improved monitoring offered by biosensors (Brooks and Turner, 1987). Progress in this area, however, has been inhibited by the desire to see on-line devices rather that at line analysers. We have recently shown that it is possible to monitor industrial fermentations using mediated enzyme electrodes housed in a probe configuration (Brooks et al, 1987). The stainless steel housing is sterilised in situ prior to the insertion of a multiple enzyme electrode. The design permits replacement of the sensor, if this becomes necessary, without danger of contamination of the fermentation. The sensor can be calibrated intermittently in situ under computer control and its upper range may be extended by flow dilution within the housing.

Work over the past four years has clearly demonstrated that most oxidoreductase enzymes can be coupled to DMF-, TTF- and TCNQ-modified carbon electrodes. If several enzymes are used to construct an assay including an oxidoreductase at a key site, the overall reaction may be monitored by mediated electrochemistry thus extending the possibilities beyond the substrate range of oxidoreductases alone eg:

$$\text{Cholesterol esters} \xrightarrow[\text{esterase}]{\text{cholesterol}} \text{Cholesterol} + \text{fatty acids}$$

$$\text{Cholesterol} + M_{ox} + H_2O \xrightarrow[\text{oxidase}]{\text{cholesterol}} \text{cholest-4-en-3-one} + M_{red} + 2H^+$$

3.3. Microbial Activity Monitors

Biosensors based on intact microorganisms have received considerable attention over the past decade (Karube, 1987). These sensors have generally been based on the Clark oxygen electrode and can offer increased stability compared to the use of purified enzymes. Microorganism based sensors inevitably tend to be less specific than purified catalysts, but this can be an advantage where a broad substrate range is required, eg pollution monitoring.

We have shown that it is possible to efficiently couple redox reactions of intact microorganisms to an electrode using soluble mediators such as phenazine ethosulphate (Turner et al, 1983), ferricyanide or ferricyanide in combination with benzoquinone (Turner et al, 1986). This principle has been developed into a prototype commercial analyser for the determination of microbial activity (Turner et al, 1986). The device known as the "Biocheck" can detect 10^6 organisms/ml in under 2 minutes, is totally portable and can be operated by unskilled personnel. It responds to a wide range of microorganisms and can accept untreated samples of milk, cutting oil, process water etc.

This mediated technology has also been applied to monitoring metabolic perturbations in indicator organisms. A herbicide monitor has been described exploiting immobilised Synechococcus, a photosynthetic bacteria, illuminated by a light emitting diode (Rawson et al, 1987). Ferricyanide is used as the mediator. Unattended operation for about a week is possible with a sensitivity to herbicides in the order of parts per billion (eg detects 17 ppb Linuron). The detector is now being applied to water intake protection in the field.

3.4. Amperometric Immunoassay and DNA Probes

Electrochemical approaches to immunoassay have been extensively reviewed by Kress-Rogers and Turner (1986), Green (1987) and Heineman and Halsall (1988). Mediators may be used to couple enzyme labels to electrochemical detectors either directly or as a result of an activation step.

Cardosi et al (1986) described a particularly sensitive two stage enzyme amplifier for use in affinity assays with an electrochemical detector. The primary enzyme label was alkaline phosphatase which was used in a conventional enzyme-linked immunosorbent assay (ELISA). However, the label was monitored in an unconventional manner by using it to cleave the phosphate off the cofactor nicotinamide adenine dinucleotide phosphate (NADP). The NAD so formed was then converted to the reduced form (NADH) in the presence of excess ethanol and alcohol dehydrogenase. NAD was then regenerated from the NADH in the presence of diaphorase which was electrochemically coupled using ferricyanide as a mediator. The net result was an amperometric measurement of the amount of label greatly amplified by the NAD recycling step.

Affinity reactions labelled with alkaline phosphatase have also been monitored electrochemically by using novel substrates synthesised from ferrocene (Higgins et al, 1987). Alkaline phosphatase will hydrolyse [N-ferrocenoyl]-4-aminophenyl phosphate to [N-ferrocenoyl]-4-aminophenol, which can be monitored by the change in current at a platinum electrode poised at 180 mV versus SCE. After 15 minutes incubation 8.6 x 10-16 moles of alkaline phosphatase could be detected. A significant advantage of this approach over previous systems is the low operating potential which minimises interference from other electrochemically active compounds that may be present.

4. CONCLUSIONS

The commercial attraction of electrochemical detectors lies in their low cost and simplicity. Although advances in optical devices continue apace they still cannot rival electrochemical approaches for many applications. In the case of electrochemical immunoassay, for example, mediated electrochemical techniques equal the sensitivity of conventional optical plate readers at a fraction of the cost and with two orders of magnitude greater dynamic range (Cardosi et al, 1986). The principal advantage of mediators is in increasing the intimacy of the connection between the biological system and the electrode thus maximising sensitivity and minimising interference. Their role extends beyond the simple enzyme electrode to applications in affinity assays and monitoring membrane bound systems. While only applicable to assays either involving redox reactions directly or coupled to redox systems, mediated electrochemistry is a universal transducer within this broad class of analyses limited only by the imagination of the user.

5. REFERENCES

Aston, W. J., Ashby, R. E., Higgins, I. J., Scott, L. D. L. and Turner, A. P. F. (1984). Enzyme-based methanol sensor. In: Charge and Field Effects in Biosystems (Ed. M. J. Allen and P. N. R. Usherwood). Abacus Press, Tunbridge Wells. pp. 491-498.

Brooks, S. L., Ashby, R. E., Turner, A. P. F., Calder, M. R. and Clarke, D. J. (1987). Development of an on-line glucose sensor for fermentation monitoring. Biosensors, 3, (in press).

Brooks, S. L. and Turner, A. P. F. (1987). Biosensors for measurement and control. Measurement and Control, **20**, 37-42.

Clark, L. C. (1987). The enzyme electrode. In: Biosensors: Fundamentals and Applications (Eds A. P. F. Turner, I. Karube and G. S. Wilson). Oxford University Press, Oxford, UK. pp. 3-12.

Cardosi, M. F. and Turner, A. P. F. (1987a). The realisation of electron transfer from biological molecules to electrodes. In: Biosensors: Fundamentals and Applications (Eds. A. P. F. Turner, I. Karube and G. S. Wilson). Oxford University Press, Oxford. p.257-275.

Cardosi, M. F. and Turner, A.P.F. (1987b). Glucose sensors for the management of diabetes. The Diabetes Annual **3** (Eds. K.G.M.M. Alberti and L.P. Krall). Elsevier, Amsterdam (in press).

Cardosi, M. F., Stanley, C. J. and Turner, A. P. F. (1986). An electrochemical immunoassay using enzyme amplification. In: Chemical Sensors. Proceedings of the 2nd International Meeting on Chemical Sensors, Bordeaux, 1986. (Eds. J.L. Aucoutourier, J.S. Cauhape, M. Destriau, P. Hagenmuller, C. Lucat, F. Menil, J. Portier and J. Saladene). Imprimiere Biscaye, Bordeaux. p.634.

Cass, A. E. G., Davis, G., Francis, G. D., Hill, H. A. O., Aston, W. J., Higgins, I. J., Plotkin, E. V., Scott, L. D. L. and Turner, A. P. F. (1984). Ferrocene-mediated enzyme electrode for amperometric determination of glucose. Analytical Chemistry **56**, 667-671.

Claremont, D. J., Penton, C. and Pickup, J.C. (1986). Potentially-implantable ferrocene-mediated glucose sensor. J.Biomed.Eng. **8**, 272-274.

D'Costa, E. J., Turner, A. P. F. and Higgins, I. J. (1986). Quinoprotein glucose dehydrogenase and its application in an amperometric glucose sensor. Biosensors, **2**, 71-87.

Davis, G., Hill, H. A. O., Aston, W. J., Higgins, I. J. and Turner, A. P. F. (1983). Bioelectrochemical fuel cell and sensor based on a quinoprotein alcohol dehydrogenase. Enzyme and Microbial Technology **5**, 383-388

Dicks, J. M., Aston, W. J., Davis, G. and Turner, A. P. F. (1986). Mediated amperometric biosensors for D-galactose, glycolate and L-amino acids based on a ferrocene-modified carbon paste electrode. Anal.Chim.Acta, **182**, 103-112.

Green, M. J. (1987). New approaches to electrochemical immunoassay. In: Biosensors: Fundamentals and Applications (Eds A. P. F. Turner, I. Karube and G. S. Wilson). Oxford University Press, Oxford, UK. pp. 60-70.

Green, M. J. and Hill, H. A. O. (1986). Amperometric enzyme electrodes. J.Chem.Soc., Faraday Trans.1, **82**, 1237-1243.

Heineman, W. R. and Halsall, H. B. (1988). Electrochemical immunoassays. (this volume).

Hendry, S. P. and Turner, A. P. F. (1987). Glucose sensors incorporating immobilised glucose oxidase with ferrocene, tetrathiafulvalene or tetracyanoquinodimethane as mediator. Hormone and Metabolic Research (in press).

Higgins, I. J., Bannister, J. V. and Turner, A. P. F. (1987). Development and applications of biosensors. In: Biosensors (Ed. R. Schmidt). GBF Monograph Series, Braunschweig, FRG. (in press)

Higgins, I. J. and Lowe, C. R. (1987). Introduction to the principles and applications of biosensors. Proc.Royal.Soc. (B) (in press).

Karube, I. (1987). Micro-organism based sensors. In: Biosensors: Fundamentals and Applications (Eds A. P. F. Turner, I. Karube and G. S. Wilson). Oxford University Press, Oxford, UK. pp. 13-29.

Kress-Rogers, E. and D'Costa, E.J. (1986). Biosensors for the Food Industry. Anal.Proc. **23**, 149-151.

Kress-Rogers, E. and Turner, A. P. F. (1986). Immunosensors based on acoustic, optical and bioelectrochemical devices and techniques. Technical Notes, **49**. Leatherhead Food Research Association Publications, Surrey, UK. pp. 31. and In: Advances in Immunoassays for Veterinary and Food Analysis (Eds. M.N. Clifford, R. Jackman, B.A. Morris and J.A. Morris). Elsevier, Barking, UK. (in press).

Matthews, D. R., Brown, E., Beck, T., Frank, M., Gosden, E., Lock, L., Plotkin, E.V., Wickham, M. and Turner, R.C. A transcutaneous 25 guage amperometric glucose-sensing needle, and its use in Man. (1987). Diab.Med. (in press).

McCann, J. M. (1987). The forthcoming revolution in clinical biochemistry. In: The World Biotech Report 1987, **1**(2). Online Publications, Pinner, UK.

Rawson, D. M., Willmer, A. J. and Cardosi, M. F. (1987). The development of whole-cell biosensors for on-line screening of herbicide pollution of srface waters. Toxicity Assessment (in press).

Turner, A. P. F., Ramsay, G. and Higgins, I. J. (1983). Applications of electron transfer between biological systems and electrodes. Biochemical Society Transactions **11**, 445-448.

Turner, A. P. F., Aston, W. J., Bell, J., Colby, J., Davis, G., Higgins, I. J. and Hill, H. A. O. (1984). CO: Acceptor oxidoreductase from Pseudomonas thermocarboxydovorans strain C2 and its use in a carbon monoxide sensor. Analytica Chimica Acta **163**, 161-174.

Turner, A. P. F., Cardosi, M. F., Ramsay, G., Schneider, B. H. and Swain, A. (1986). Biosensors for use in the food industry: a new rapid bioactivity monitor. In: Biotechnology in the Food Industry, Online Publications, Pinner, UK. pp. 97-116.

Turner, A. P. F., Hendry, S and Cardosi, M. F. (1987b). Tetrathiafulvalene: a new mediator for amperometric biosensors. In: The World Biotech Report 87 1(3), Online Publications, Pinner, UK. pp. 125-138.

Turner, A. P. F., Hendry, S. P., Cardosi, M. F. and Neuse, E. W. (1987c). Polyferrocenes as mediators in amperometric biosensors for glucose. Diab.Med. (in press).

Turner, A. P. F, Karube, I. and Wilson, G. S. (Eds.). (1987a). Biosensors: Fundamentals and Applications. Oxford University Press, Oxford, UK. p 780.

Turner, A. P. F. and Pickup, J. C. (1985). Diabetes mellitus: biosensors for research and management. Biosensors 1, 85-115.

Wilson, G. S. (1987). Fundamentals of amperometric sensors. In: Biosensors: Fundamentals and Applications (Eds A. P. F. Turner, I. Karube and G. S. Wilson). Oxford University Press, Oxford, UK. pp. 165-179.

UNI-, BI- AND TRI-ENZYME ELECTRODES FOR ANALYSIS

J.D.R.Thomas
Applied Chemistry Department
Redwood Building
UWIST
PO Box 13
Cardiff CF1 3XF
Wales, United Kingdom

ABSTRACT. Amperometric enzyme electrodes with immobilized enzymes have been modelled on glucose oxidase immobilized on nylon mesh and placed over a platinum electrode for monitoring substrate in solution by the amperometric response to hydrogen peroxide produced during the enzyme catalyzed oxidation. The uni-enzyme glucose oxidase electrode, tested in a flow injection analysis assembly, exhibited good linearity (0.1 to 3.0 mM glucose), short response times (< 45 s), good capacity (24 h with a continuous flow of 2.5 mM glucose) and long lifetimes (4 months, with storage when not in use at 4°C).

A cholesterol oxidase electrode was linear between 0.01 and 0.7 mM, but the need for surfactant to solubilize cholesterol standards leaves some problems unresolved.

The bi-enzyme electrodes discussed are based on peroxidase with the oxidase of the substrate being analyzed. Thus, the glucose oxidase/peroxidase electrode has been optimized for analyzing glucose in blood sera, and the xanthine oxidase/peroxidase electrode for analyzing hypoxanthine in fish meats. Illustrative data are given in each case and correlated with data obtained by established procedures.

A tri-enzyme electrode based on invertase, mutarotase and glucose oxidase for sucrose analysis is described. The optimum coimmobilization enzyme composition used for nylon net was 2000:1000:200 IUs of the respective enzymes. The electrode showed good linearity over 0.001 to 1 mM sucrose, and was free of interference from several sugars and organic acids, except for ascorbic acid.

A thin membrane development of the glucose oxidase electrode is based on covalently attaching the enzyme to an activated platinum wire surface. The improved diffusion features give fast response times (≈25 s) with short wash times (≈ 30 s) between samples, and the electrode has a lifetime of 9 h during continuous exposure to 2.5 mM and 10 mM glucose and 10 days for normal use. The shorter lifetime compared with the nylon mesh membrane electrodes is attributed to the weakness of the Pt-O bonds used to form the link to the enzyme through the agency of silanization and glutaraldehyde.

G. G. Guilbault and M. Mascini (eds.),
Analytical Uses of Immobilized Biological Compounds for Detection, Medical and Industrial Uses, 141–152.

1. INTRODUCTION

An enzyme electrode consists essentially of an immobilized enzyme layer covering an electrode which is able to monitor the consumption of a reactant or the appearance of a product of an enzyme-catalyzed reaction. For example, in the glucose oxidase catalyzed oxidation of glucose,

$$\text{Glucose} + O_2 \xrightarrow[\text{oxidase}]{\text{Glucose}} \text{Gluconic acid} + H_2O_2 \qquad (1)$$

either oxygen or hydrogen peroxide is monitored.

Hitherto, enzyme electrodes could be summarized[1] as having a lifetime of two to three weeks, with a range of 0.1 to 10 mM and response times of 1 to 5 min. With improved methods of immobilization, lifetimes are now considerably improved, while use of enzyme electrodes under the dynamic approach of flow injection analysis ensures much faster sample throughput than is implied by response times of minutes. It was to test these hypotheses that the enzyme electrode studies described here were embarked upon.

The initial approach was to explore the rôle of poly-(vinyl chloride) as an enzyme support matrix for glucose oxidase[2]. The resulting enzyme membrane was used in association with a potentiometric iodide ion-selective electrode screening of the iodide ion consumed on oxidation by hydrogen peroxide enzyme reaction product which was itself catalyzed by molybdate or peroxidase:

$$H_2O_2 + 2I^- + 2H^+ \xrightarrow[\text{or peroxidase}]{\text{Mo(VI) catalyst}} I_2 + 2H_2O \qquad (2)$$

Glucose oxidase was immobilized in plasticised PVC, either alone or with peroxidase, to form an enzyme based membrane for covering the sensing face of the iodide electrode[2]. The data were compared with those obtained for an iodide/glucose oxidase electrode where the enzyme was chemically immobilized on bovine serum albumin[2]. The chemically immobilized enzyme membranes may be prevented from cracking by incorporating a small proportion of high molecular mass polyethoxylate humectant[3]. Also, chemical immobilizations on nylon mesh of glucose oxidase and peroxidase together have been explored[4].

The simplicity of the PVC entrapment gave enzyme electrodes of good response slopes (72 to 75 mV decade^{-1}) towards glucose in the millimolar and submillimolar range to compensate for their limited lifetimes (≈ 7 days) compared with the longer lives (>14 days) of electrodes with chemically immobilized enzyme on bovine albumin. However, the constraint of monitoring reaction (2) is not conducive to routine use. Therefore, the main investigations now reviewed are those based on voltammetric/

amperometric sensing set up in a dynamic flow injection analysis mode and with spontaneous wash stages by the carrier buffer stream in between each sample injection[5]. Also, it is to be noted that the enzyme-catalyzed reaction is sufficiently fast for the hydrogen peroxide product to be screened without the need for stopped-flow regimen.

2. AMPEROMETRIC NYLON MESH UNI-ENZYME ELECTRODES

2.1 Glucose Oxidase Electrode

Here, glucose oxidase was chemically immobilized[5,6] on nylon mesh with lysine spacer and glutaraldehyde following treatment with dimethylsulphate and sodium hydroxide[6,7].

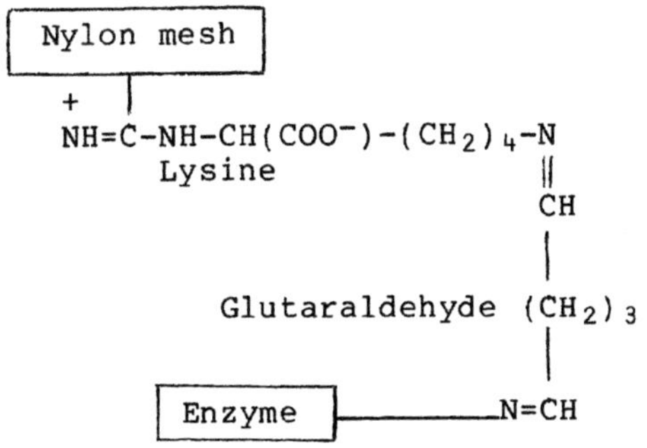

The resulting thin enzyme membrane was held with an 'O' ring over a platinum electrode and set up in a micro-cell in a flow-injection analysis arrangement[5] where glucose was monitored from the hydrogen peroxide produced by reaction (1).

This system, following optimization for flow rate, applied potential (+600 mV versus the silver/silver chloride electrode was chosen) and enzymatic reaction conditions gave long range and good linearity to glucose (0.01 to 3 mM) with short response times (<45 s) and wash times (≈45 s). The electrode possessed good capacity (24 h with continuous flow of 2.5 mM glucose) and long lifetimes of more than 4 months for storage at 4°C.

The arrangement was shown to be an effective system for analysing glucose in various foodstuffs, such as molasses, ice cream and syrups, and the data agreed well with those obtained by soluble enzyme kit and Yellow Springs Instrument glucose analyzer methods[5] (Table 1).

TABLE I Glucose in foodstuffs by enzyme electrode and soluble enzyme test kit (data from Ref.5)

Sample	Sample Pretreatment	Test Kit /g%	Enzyme Electrode /g%
Strawberry ice cream	Carrez solutions I and II	4.4	4.5
Vanilla ice cream	do	4.7	4.6
Glucose syrup	Dissolve in phosphate buffer (0.1M, pH 7) gentle warming at 35°C	14.8[a]	13.9
Glucose powder	Dissolve in buffer with gentle warming at 35°C	88.0	88.7

a Yellow Springs Instrument Glucose Analyzer

2.2 Cholesterol Oxidase Electrode

Cholesterol oxidase was immobilized[8] on nylon mesh as described above for glucose oxidase. The electrode may be used in the normal mode for flow injection analysis. However, a major difficulty for this electrode system is the lack of availability of suitable cholesterol standards for calibration. Commercial aqueous cholesterol standards contained an unspecified surfactant and sodium azide (0.1%) preservative, but gave sharp negative peaks thus masking the cholesterol electrode response. Hence, laboratory standards based on cholesterol (400 mg) dissolved in 50 cm^3 of Triton-X-100 (20%) in phosphate buffer (0.1M, pH 7.0), diluted to 100 cm^3 with the buffer were prepared. These standards of concentration up to 0.7 mM were used for calibrating the cholesterol electrode by injecting 500 mm^3 into a carrier stream of phosphate buffer (0.1M, pH 7.0) containing Triton-X-100 (1.0% v/v) at a flow rate of 2.3 cm^3 min^{-1}. The response[8] was linear between 0.01 and 0.7 mm.

The reproducibility of this electrode was only 5% compared with better than 2% for the glucose enzyme electrode. This may be attributed to the relatively poor quality of the enzyme as indicated by multiple peaks obtained for gel exclusion chromatography[8]. Also, the electrode was active for only six weeks compared with four months for the glucose electrode similarly stored at 4oC. The poor stability and reproducibility may also be attributed to the low activity of the enzyme used (25 to 50 IU

mg^{-1}) compared with that for glucose oxidase (100 IU mg^{-1}). In addition, the presence of surfactant may have a harmful effect on the cholesterol oxidase electrode[8].

3. AMPEROMETRIC ENZYME ELECTRODES IN THE REDOX MEDIATED MODE AND BI-ENZYME ELECTRODES

3.1 Glucose Determinations

At the high platinum electrode potentials employed in the above systems (+600 mV *versus* the silver/silver chloride electrode), other electroactive species, such as ascorbic acid and uric acid present in physiological samples, will be oxidized. Therefore, the glucose oxidase electrode system was assessed in the redox mediated mode[9].

The use of hexacyanoferrate(II) mediator permitted a low applied potential (-100 mV *versus* the silver/silver chloride electrode) enabling the determination of glucose in blood serum without difficult sample pretreatment[9]. Peroxidase was used to catalyse the hexacyanoferrate(II) reaction with hydrogen peroxide:

$$H_2O_2 + 2Fe(CN)_6^{4-} + 2H^+ \xrightarrow{\text{Peroxidase}} 2H_2O + 3Fe(CN)_6^{3-} \qquad (3)$$

The generated hexacyanoferrate(III) is reduced at a platinum electrode at a low applied potential, that is, -100 mV. The peroxidase was co-immobilized with glucose oxidase on nylon mesh,[9] when the optimum glucose oxidase: peroxidase rate was 2:1.

The blood serum samples (3.8 mM to 19.2 mM in glucose) were diluted 10-fold in phosphate buffer (100 mM at pH 7.0) containing potassium hexacyanoferrate(II) (2.5 mM). 500 mm^3 samples were used for flow injection analysis of glucose using the glucose oxidase/peroxidase electrode, and the results showed good correlation[9] with analyses carried out by a colorimetric soluble enzyme kit method according to

$$[\text{Glucose}]_{\text{Electrode}} = 1.00[\text{Glucose}]_{\text{Soluble enzyme test kit}} - 0.219 \qquad (4)$$

with a correlation coefficient of 0.997.

The glucose oxidase/peroxidase electrode was linear over the range 0.02 to 3 mM glucose[9] where

$$\log(\text{current}/A) = 1.09 \log([\text{glucose}]/M) - 2.91 \qquad (5)$$

with a correlation coefficient of 0.999.

3.2 Hypoxanthine Analysis

The above studies on glucose served as a model for studying xanthine-hypoxanthine analysis, which demanded a lower applied potential because of the electroactivity of xanthine and hypoxanthine at the 600 to 700 mV required for direct hydrogen peroxide monitoring[9]. Also, the uric acid product of enzymic xanthine oxidation interferes at high applied potentials.

Here, again, the bienzyme electrode approach was used[9] whereby the enzymes (2:1::xanthine oxidase:peroxidase) were attached to the nylon by dipping the activated nylon mesh in a mixture of the enzymes for 2 h at room temperature and then overnight at 4°C. The bi-enzyme electrode was assembled by stretching the membrane over a smooth platinum electrode[5].

The electrode was readily calibrated for xanthine, its primary substrate, and for hypoxanthine[9]. A typical chart recorder output is shown in Fig.1 for the duplicate injection of xanthine standards[9]. The calibration is

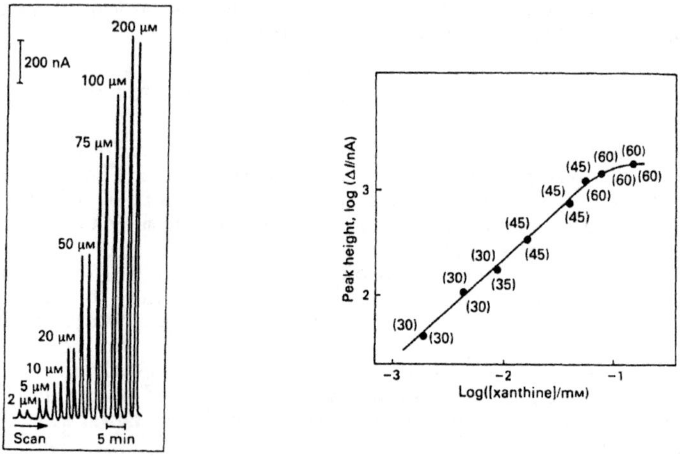

Fig.1 Typical chart recorder output for hexacyanoferrate(II) mediated xanthine calibration of the xanthine oxidase-peroxidase bi-enzyme electrode (from Ref.9).

linear over the range 2 to 100 μM according to (correlation coefficient is 0.999)

$$\text{Log}(\text{current}/\text{A}) = 0.93([\text{xanthine}]/\text{M}) - 2.08 \qquad (6)$$

with a correlation coefficient of 0.999. There is a similar calibration profile for hypoxanthine according to[9] (correlation coefficient is 0.999)

$$\text{Log(current/A)} = 0.895 \log([\text{hypoxanthine}]/\text{M}) - 2.62 \quad (7)$$

An application of the xanthine oxidase electrode is to the determination of the deterioration of fish meat quality, as reflected by increases in hypoxanthine levels during storage[10]. The electrode described here was tested on four different kinds of fish meat[9] and the results compared with those of the spectrophotometric based method recommended by the Analytical Methods Committee[11] (Table II). Thus, for the combined data relating to fresh fish and stored fish, the line

$$[\text{Hypoxanthine}]_{\text{Electrode}} = 1.01[\text{hypoxanthine}]_{\text{AMC}} + 4.0 \times 10^{-3} \quad (8)$$

with a correlation coefficient of 0.998 was obtained[9].

TABLE II. Hypoxanthine analysis in fish meat, using the bienzyme electrode (BE) of xanthine oxidase and peroxidase and AMC spectrophotometric approach (data from Ref.9)

Fish meat type	Hypoxanthine/μmol g^{-1}			
	Fresh fish		After 20 h storage/20°C	
	BE	AMC	BE	AMC
Rainbow trout	0.32	0.32	0.42	0.39
Herring	0.74	0.88	1.87	1.84
Hake	0.59	0.63	2.19	2.40
Plaice	0.95	0.91	1.89	1.76

4. A TRI-ENZYME ELECTRODE FOR THE DETERMINATION OF SUCROSE

Various approaches have been described of the use of enzyme electrodes for the analysis of sucrose (see Ref.12). These generally involve the coimmobilization of enzymes in one way or another. The simplest approach is the random coimmobilization on a single membrane. Thus, coimmobilization of invertase, mutarotase and glucose oxidase on nylon mesh placed over a platinum electrode used in the amperometric mode in a modified three-electrode Stelte cell gives a very effective arrangement for the flow injection analysis of sucrose[12].

As previously, the nylon mesh was activated with dimethyl sulphate, reacted with lysine spacer and glutaraldehyde, and then treated with the triple enzyme solution for 2 hours at room temperature followed by overnight contact at 4°C. The resultant sucrose-sensing membrane was then firmly fitted over the platinum indicating electrode of the Stelte cell with an 'O' ring. Varying proportions of invertase, mutarotase and glucose oxidase, based on enzyme units (Table III), were taken and diluted in phosphate buffer (0.1 mM, pH 7 at 25°C) to give a total protein concentration of about 3 mg cm^{-3}.

TABLE III. Trienzyme electrode response (/nA) to glucose and sucrose at pH 6.5, 1 mM of each sugar, flow rate 4.2 cm^3 min^{-1}, and 500 mm^3 sample volume (data from Ref.13).

Membrane type	A	B	C	D	E
IU ratio/200 INV:MUT:GO[a]	10:10:10	10:10:1	10:5:1	5:10:1	20:10:1
Enzyme/buffer for [Enzyme] of 3mg cm^{-3}/cm^3	7.5	1.9	1.85	1.25	3.25
I_G/nA	3613	1195	1373	1283	1328
I_S/nA	200	160	923	103	145
I_G:I_S ratio	18.1	7.5	1.5	12.5	9.2

a: INV=Invertase; MUT=mutarotase; GO=glucose oxidase:

Except for glucose (as expected), sucrose response is not affected by the presence of fructose, arabinose, galactose, raffinose, maltose, lactose, sorbose[12]. Also, there is no interference from gluconic, citric, lactic and acetic acids, but ascorbic acid inflates the readings.

The optimum enzyme composition was found to be 2000:1000:200 IUs of invertase:mutarotase:glucose oxidase, respectively, that is, that corresponding to membrane C (Table III). The corresponding electrode[12] exhibited good linearity range (0.001 to 1 mM sucrose) short response times (10 s to 20 s), long lifetime (just 6% reduction in signals after >14 h continuous flow of 1 mM sucrose) and good storage stability for intermittent use (38 d when stored in 0.1M phosphate buffer of pH 7 at 4°C). Furthermore, the electrode showed good resistance towards radiation (2.1 Mrad of ^{60}Co-γ source), thus offering the prospect for sterilization and of use in on-line monitoring of sucrose in clinical and food processed samples[12].

5. THIN MEMBRANES OF ENZYMES ON MODIFIED ELECTRODES

Although immobilization of enzyme on nylon mesh gives effective enzyme electrodes, permitting a glucose analysis every 90 s or so, the membrane is still relatively thick. Therefore, glucose oxidase has been immobilized on to silanized anodized platinum wire to give a micro enzyme electrode[13].

PtO[3-aminopropyltriethoxysilane)-glutaraldehyde-enzyme]

The approach is similar to a procedure[14] of silanizing a 0.5 cm^2 platinum plate, followed by enzyme immobilization of a mixture of glutaraldehyde, bovine albumin and enzyme coated onto the silanized platinum, except that here[13] the agency of bovine albumin was not used.

The microenzyme wire electrode was assembled in a micro flow-through cell (Fig.2) and gave fast and sensitive response (Fig.3). For a continuous exposure to glucose, achieved by pumping glucose (10 and 2.5 mM) over the electrode at 3.5 cm^3 min^{-1}, the electrode functioned well[13] for 9 h. This is a shorter lifetime than the 24 h observed for

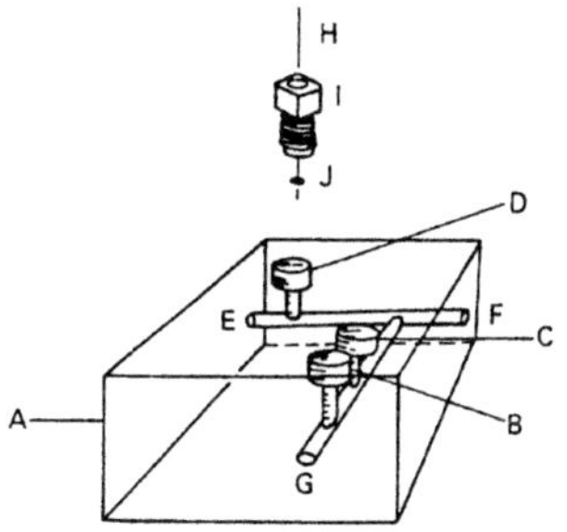

Fig.2 Amperometric flow-through cell with a micro enzyme electrode (H) (from Ref.13). A: Perspex block; B: reference electrode; C: auxiliary electrode; D: enzyme electrode chamber; E: sample outlet; G: from saturated potassium chloride reservoir; H: platinum wire enzyme electrode; I: electrode connector; J: silicone rubber seal.

an electrode with glucose oxidase immobilized on nylon mesh[5] (loading 22 nmol cm^{-2} min^{-1}) and may relate to the smaller enzyme loading on the wire (5-10 nmol cm^{-2} min^{-1}), small surface area and/or weakness of the nature of the Pt-O bonding. Linearity was

$$\log(\text{current}/A) = 0.992 \log([\text{glucose}]/M) - 3.94 \qquad (9)$$

with a correlation coefficient of 0.999. Lifetime in normal use was 10 days[13].

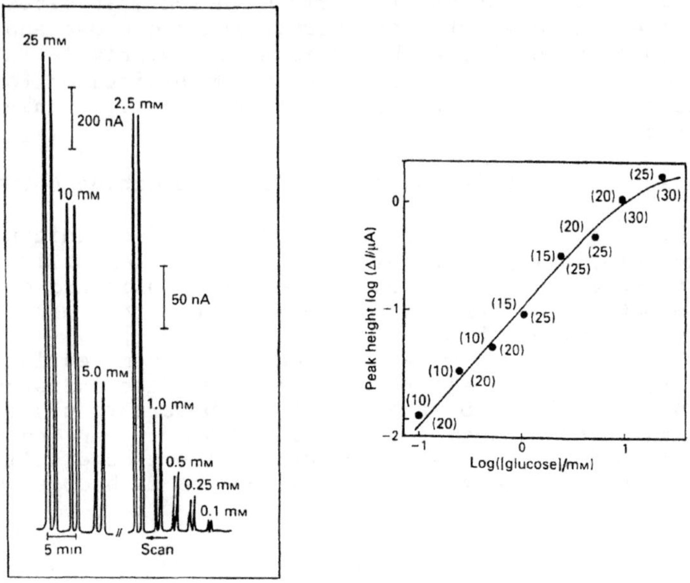

Fig.3 Chart recorder output (left) for glucose calibration of 3 day old platinum wire micro enzyme electrode and calibration graph (right) showing response times (s) above the line and wash times (s) below the line (from Ref.3).

6. CONCLUSION

Clearly, attention to immobilization of enzymes yields enzyme electrodes for direct analysis of substrates in flowing streams. These can be based on multi-enzyme membranes and are adaptable for a wide range of applications.

The immobilization of glucose oxidase on to platinum wire to give microenzyme electrodes prompts the possibility of in vivo determination of glucose. In this respect, it is noted that in vivo needle glucose electrodes have been described[15]. These are essentially based on a platinum wire encased on Teflon insulation inside a stainless steel tube; the platinum sensor tip is in contact with a permselective membrane followed by a glucose oxidase matrix and polyurethane outer membrane[15]. The direct immobilization of glucose oxidase on silanized anodised platinum could be an improvement on this but, of course, regard must be had of the possibility of mediated systems to overcome possible interferences and of the electrode functioning independently of oxygen[16,17] for substrate oxidation. Nevertheless, the multienzyme electrode approach can be extended to embrace these extensions.

ACKNOWLEDGEMENT

The UK Department of Trade & Industry (Laboratory of the Government Chemist) is thanked for financial support and the co-workers named in the various references are thanked for their dedicated cooperation.

REFERENCES

1. Guilbault GG Ion-Selective electrodes applied to enzyme systems. Ion-Selective Electrode Revs., (1982) **3** , 187-231.

2. Al-Hitti IK, Moody GJ and Thomas JDR Glucose oxidase membrane systems based on poly(vinyl chloride) matrices for glucose determination with an iodide ion-selective electrode. Analyst (1984) **109** , 1205-1208.

3. Louie KK Studies on enzyme membranes for use with ISEs. B.Sc Project, Applied Chemistry Department, UWIST (1984).

4. Gonji AN Studies on enzyme electrodes in chemical analysis. B.Sc Project, Applied Chemistry, UWIST (1986).

5. Moody GJ, Sanghera GS and Thomas JDR Amperometric enzyme electrode system for the flow injection analysis of glucose. Analyst, (1986) **111** , 605-609.

6. Hornby WE and Morris DL In HH Weetall (Editor) Immobilized enzymes, antigens, antibodies and peptides. Marcel Dekker, New York (1975).

7. Mascini M, Ianello M and Palleschi G Enzyme electrodes with improved mechanical and analytical characteristics obtained by binding enzymes to nylon nets. Anal.Chim.Acta, (1986) **146** , 135-148.

8. Moody GJ, Sanghera, GS and Thomas JDRT A cholesterol oxidase enzyme electrode. To be published.

9. Moody GJ, Sanghera GS and Thomas JDRT Chemically immobilized bi-enzyme electrodes in the redox mediated mode for the flow injection analysis of glucose and hypoxanthine. Analyst, (1987) **112** , 65-70.

10. Karube I and Suzuki S Amperometric and related determinations with immobilized enzymes and micro-organisms. Anal.Proc., (1983) **20** , 556-559.

11. Analytical Methods Committee (Fish products Sub-Committee) Recommended General Methods for the Examination of Fish and Fish Products. Analyst, (1979) **104** , 434-450.

12. Abdul-Hamid J, Moody GJ and Thomas JDRT A chemically immobilized tri-enzyme electrode for the flow injection analysis of sucrose. Analyst, (1987) In the press.

13. Moody GJ, Sanghera GS and Thomas JDR Modified platinum wire glucose oxidase amperometric electrode. Analyst, (1986) **111** , 1235-1238.

14. Yao T A chemically-modified enzyme membrane electrode as an amperometric sensor. Anal.Chim.Acta, (1983) **148** , 27-33.

15. Churchouse JJ, Mullen WH, Keedy KH, Battersby CM and Vadgama PM Studies on needle glucose electrodes. Anal.Proc., (1986) **23** , 146-148.

16. Cass AE, Davies G, Francis GD, Hill HAO, Aston WJ, Higgins IJ, Plotkin EV, Scott LDL and Turner APF Ferrocene-mediated enzyme electrode for amperometric determination of glucose. Anal.Chem., (1984) **56** , 667-671.

17. Hill HAO Ferrocene-based electrodes. Anal.Proc., (1985) **22** , 201-202.

ENZYME ELECTROCHEMICAL SENSORS FOR "IN VIVO" ANALYSIS WITH ARTIFICIAL PANCREAS

M. MASCINI
Istituto Chimica Analitica
Universita' di Firenze
Via Gino Capponi 9,
50121 Firenze, Italy

Abstract.
Lactate and pyruvate electrochemical biosensors based on oxygen or hydrogen peroxide sensors and oxidase enzymes have been developed and coupled to an artificial pancreas like Biostator from Miles (USA) or Betalike from Esacontrol (Italy) in experiments with extracorporeal circulation. Procedures to develop sensors are detailed and results obtained with them are presented. These new sensors used in line with the extracorporeal circulation should be able to clarify the fate of the intermediary metabolites of glucose in diabetic patients receiving infusion of insulin or drugs.

INTRODUCTION

The concept of using extracorporeal circulation and devices in the treatment of diseases or physiological defaults dates back to the 19th century (1). Today over 100,000 cardiopulmonary bypasses are performed each year with the aid of heart-lung bypass equipment, while over 5 millions of treatments of renal diseases only in the USA (32,000 patients) are performed with the use of "artificial" kidneys.
In respect to these examples, artificial pancreas is in a preliminary stage for the routine application, while the research to develop it is very active at the present (2-5).
The mass transfer process is of critical importance in the use of extracorporeal devices designed to maintain the metabolic functions or for the treatment of a specific disease.
The value of concentration of a specific biochemical, can drive mechanisms to infuse drugs or physiological solutions or can be the alarm for changing filters, varying specific rates etc. realizing, of course, a better control of the precedure.
Therefore the need of biochemical sensors - miniaturized, disposable,

G. G. Guilbault and M. Mascini (eds.),
Analytical Uses of Immobilized Biological Compounds for Detection, Medical and Industrial Uses, 153–167.

sterile-is urgent.
Our interest in the last few years was focused to the development of biosensors for use in conjunction with an artificial pancreas, such as Biostator from Miles or Betalike from Ansaldo, which are glucose controlled insulin infusion system for diabetic patients, or so-called closed-loop automatic insulin delivery systems.
The blood is taken from the patient, diluted with a physiological solution and passes in a flow cell where glucose enzyme sensor gives continuously the concentration of perypheric venous blood glucose. This value drives, in a closed-loop system, the delivery of insulin with a feedback mechanism.
However the infusion of insulin provokes changes in the concentration of intermediate metabolites, including L-lactate, pyruvate, alanine and ketone bodies.
Then although these instruments represent a great improvement in therapy, they still do not completely normalize the altered concentrations of intermediary metabolites, in part because perypheral blood glucose is unsuitable as the only signal on which to base the control algorithms of the artificial pancreas.
Once adequate systems for continuous monitoring of metabolites have been realized, algorithms based on multiple signals presumably will be developed.
Therefore, we thought it necessary to combine the blood glucose signal with other metabolic signals. To this purpose, electrodes for the continuous monitoring of lactate and pyruvate, in line with the blood glucose electrodes of the artificial endocrine pancreas, have been developed.
Here we report the development of L-lactate and pyruvate sensors obtained by coupling oxidase enzymes on oxygen and hydrogen peroxide electrochemical sensors.
We were able to place these sensors in the flow stream of blood of diabetic patients submitted to the treatment of artificial pancreas, in order to monitor continuously in vivo the lactate and pyruvate concentrations as well the glucose values.
The reaction with lactate oxidase is

Lactate + Oxygen = Pyruvate + Hydrogen Peroxide

The reaction with pyruvate oxidase is

$$\text{Pyruvate} + H_2PO_4^- + O_2 = \text{Acetylphosphate} + H_2O_2 + CO_2$$

Therefore an oxygen sensor or a hydrogen peroxide sensor could be

coupled with such reactions.
To obtain an interference free lactate sensor for clinical determination an oxygen sensor has been used which is separated from chemicals in solution by the gas permeable membrane excluding any non-gaseous compounds.
However to obtain a pyruvate sensor for clinical determination a hydrogen peroxide sensor must be used because the normal range of pyruvate in serum (0.04-012 mmol/L) is too low to be determined with an oxygen sensor.

LACTATE SENSOR

Materials

L-lactate oxidase (activity 18 kU/g) from Pediococcus sp. was obtained from Toyo Yozo Co., Shizuoka, Japan. The nylon net used (A. Bozzone, Appiano Gentile, Italy) has a mesh of 120/cm , is 100 µm thick, and has 35% free surface area.
The oxygen probe was a Clark electrode (Instrumentation Laboratory, Milano, Italy) consisting of a 0.1 mol/l KCl internal solution, a 12 µ-thick Teflon membrane , a platinum cathode, and a silver anode. To measure the current we used a Model 213 ammeter (Instrumentation Laboratory) with an Omniscribe recorder (Houston Instruments).

Procedures

We immobilized the enzyme on nylon net as described previously (2), then secured the nylon net to the oxygen probe with the Teflon membrane and fixed the probe in a flow-through cell with an estimated volume of 40 µL.
Before determining L-lactate in serum samples, we diluted them 10- to 20- fold with the commercially available buffer used in the Betalike (an isotonic solution of acetate, gluconate and sodium chloride, pH 7.4 called Normosol) or in the Biostator (phosphate isotonic buffer).
In a series of preliminary experiments, we had checked these conditions by using standard solutions, reconstituted sera obtained from the hospital, and whole blood.
Before and after each sample, the buffer was pumped at the same flow rate.

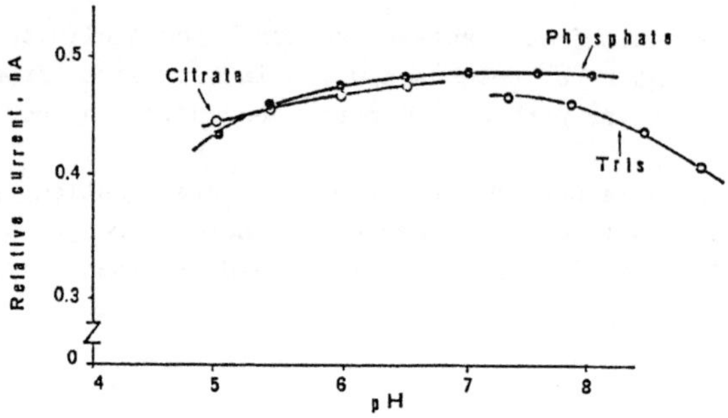

Fig 1 pH and buffer effect on the lactate sensor with enzyme from Pediococcus sp.

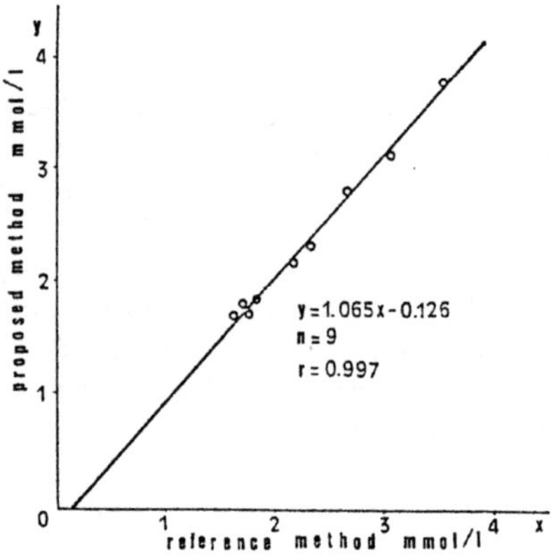

Fig. 2 Correlation of lactate retermination with the lactate sensor and a reference method (Sigma No 726-UV) with 9 reconstituted human sera.

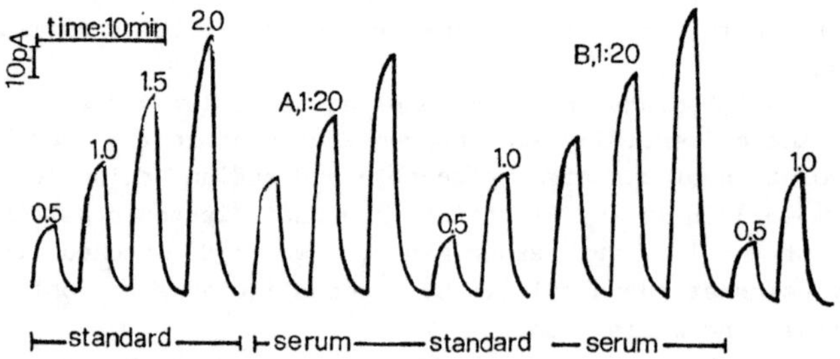

Fig. 3 Experimental recording of lactate standard solutions (values in mmol/l), serum diluted and spiked with standard.

Characteristics of the Sensor

The features of this L-lactate sensor are very similar, in terms of activity of the immobilized enzyme, the range of the calibration curve, and useful lifetime- to results previously published for a similar probe obtained by immobilizing L-lactate oxidase obtained from *Mycobacterium smegmatis*(6). The major difference is in the buffer and in the pH effect. The results (Fig. 1) clearly show the lack of the inhibitory effect that was seen for the previous enzyme.

Application to serum samples was therefore straightforward, and a correlation of 0.985 was found for nine samples of reconstituted commercially available sera as analyzed spectrophotometrically and with the sensor (Fig. 2).

Fig. 3 reports an experimental recording of L-lactate standard solutions, human reconstituted serum samples and the same samples spiked with L-lactate standards. Each sample was manually changed every two minutes.

Experiments with Biostator

Figure 4 reports the results of an "in vivo" determination of L-lactate in heparinized blood from a diabetic being treated with the artificial pancreas (the Biostator) in a "glucose-clamped" experiments. Blood was taken from the patient via a double-lumen catheter, heparinized, diluted with buffer, and pumped through the L-lactate sensor that had been placed in series with it. The value for glucose read .by the computer drives via a complex algorithm, an infusion pump for insulin and a second pump for glucose.

We calibrated the L-lactate sensor by using the device used for glucose calibration in the Biostator system, substituting for the glucose standard provided with the Biostator an L-lactate standard prepared in the same buffer.

During the first hour we recorded the glucose and the L-lactate concentration and calibrated the L-lactate sensor.

At the time indicated by "ins" in Fig.4, insulin was infused; the glucose concentration soon returned to normal, but the L-lactate concentration suddenly increased.

At the second hour, at "glu" in Fig. 4, glucose was introduced into the bloodstream to "clamp" the glucose concentration at a predetermined value. To this we ascribe the changes in concentration recorded for the L-lactate as a result of the glycolytic pathway.

At "pos" in Fig. 4 the position of the patient was changed such that the catheter briefly dropped from its initial position and no more

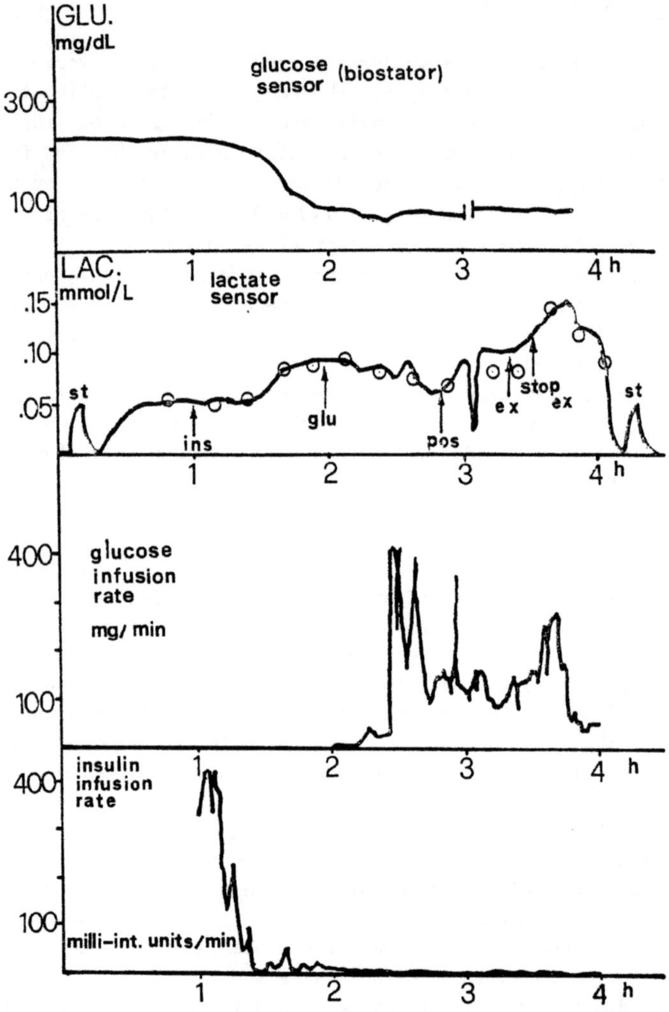

Fig. 4 Glucose and L-lactate concentration during a "glucose clamp" experiment with the Biostator. The top panel shows the glucose recording obtained by the Biostator, the secon panel shows the lactate conc. See text.

blood was aspirated. When this happened, the Biostator stopped recording glucose values, an alarm rang, and the printer discarded the fault values (shown in Figure 4 by a break in the top line).
At "ex" in Figure 4 the patient was requested to do a simple physical exercise and at "stopex" he was told to stop.
At the fourth hour the experiment was stopped and the sensor was recalibrated against the same standard. During this period L-lactate was continuously determined without delay for the patient by using the same sample of the glucose monitor without any reagent or any pretreatment, a measurement in real time on a specimen that ordinarily is wasted.
The comparison of calibration at the beginning and at the end of the experiment shows that the sensitivity of the probe did not vary during the experiment. Moreover, every 15 min blood was sampled from the patient for an indipendent analysis of the L-lactate, done the next day. The dots on the second panel of Figure 4 show the results of these separate L-lactate determinations and illustrate how comparable the results by both methods are.
The third and fourth panels of Figure 4 show the infusion rates for insulin and glucose, obtained by the computer. These illustrate the "delay" between the infusion of insulin and the response in the blood of glucose and of L-lactate or between the glucose infusion and the L-lactate response in the blood. The experiment shows also the large variations in L-lactate concentration that take place within a few minutes.

PYRUVATE SENSOR

Materials

Pyruvate oxidase (POD) E.C. 1.2.3.3., from Pediococcus sp. was obtained from Toyo Yozo Co. Shizuoka, Japan (21 U/mg).
Cellulose acetate (53 % acetyl) and polyvinyl acetate with high molecular weight was obtained from Farmitalia Carlo Erba.
The Biodyne membrane, nylon 6,6 , porosity 0.2 μ, with carboxylic groups on the surface was obtained from Pall Filtration Corporation, Glen Cove, N.Y. (USA).
Dialysis membrane thickness 0.001 inch with molecular cutoff 12,000 from A.H. Thomas Co. Philadelphia (Pa) USA.

Procedures

Casting the Cellulose Acetate Membrane

For protecting the anode from interfering chemicals which could be oxidized, a cellulose acetate membrane was prepared (7).
The dried membrane has a thickness of 20 μ.

Immobilization of Pyruvate Oxidase

Immobilization has been performed by chemical bonding and by physical entrapment on acetylcellulose membrane.
We tried four main procedures:
First we tried the immobilization on nylon net. This procedure proved to be effective with a number of oxidase enzymes but gave very low yields with POD and a very short lifetime.
The second was a physical entrapment on cellulose acetate filter (Millipore HA 0.45 μ) (8).
In the third we tried a new developed and commercially available nylon 6,6 filter membrane with carboxylic groups on the surface obtained from Pall Corp. Company.
The membrane (0.8 cm in diameter) was soaked with a solution of 0.1 mol/l 1-ethyl 3-(3-dimethylaminopropyl) carbodiimmide, in buffer phosphate 0.5 mol/l pH 4.8 for 40 min at room temperature with continuous stirring. Then the membrane was washed with a buffer bicarbonate pH 7.0, 0.4 mol/l and on the surface 1-2 mg of enzyme were mixed with 10 μl of buffer pH 7.0 and stand in humid atmosphere at 4 °C.
Finally a new prepolymer, polyazetidine (PAP) recently described for the immobilization of Escherichia Coli (9) was used. On a dialysis membrane (0.8 cm of diameter) we spread 10 μl of the prepolymer solution, as obtained, and 1 mg of enzyme mixing uniformly . The membrane was left for 24 hours at 4 °C.
The polyazetidine reacts with various functional groups, carboxyl, hydroxyl, amine, mercaptan and free amine on another PAP chain.

Assembly of the Sensor

On the platinum surface a first membrane of cellulose acetate was placed to eliminate intereferences from electroactive chemicals (ascorbic acid etc.). A second membrane with immobilized enzyme was placed on the first and finally a third dialysis membrane to prevent microbial attack of the enzyme and the leaching of the enzyme from the second membrane.
The three layers were blocked with an O rubber ring.
The sensor was placed in a flow cell where, by a peristaltic pump, pyruvate standard solutions were aspirated.

Features of the Sensor

The pyruvate oxidase can be coupled to oxygen sensor or to hydrogen peroxide sensor. Because of the low concentration range expected in serum samples (0.04-0.12 mmol/l) we coupled the enzyme with a hydrogen peroxide sensor.

The reaction proceeds in the presence of several cofactors which optimize the reaction. In all immobilization procedures we observe a bell shape curve of the activity of the POD in function of the phosphate concentration and of the calcium and it was estimated a critical value of these parameters. This behaviour exclude the use of a phosphate buffer because of the low activity of the enzyme at the phosphate concentration higher than 1 mmol/l. TPP (thiaminepyrophosphate) is another cofactor necessary for all immobilization procedures. Its concentration should be at least 0.1 mmol/l in the sample solutions. Fig. 5 reports the pH and buffer effect in the useful range of 6.5-7.5.

We can observe the low activity of POD in presence of citrate; higher activity is, in all procedures, displayed by Tris buffer.

From our experiments we concluded that the cofactor concentrations in the samples for pyruvate determination should be 0.1 mmol/l of TPP, 0.5 mmol/l of phosphate and 2.5 mmol/l of calcium chloride in the buffer Tris 0.04 mol/l at pH 7.0. Fig.6 reports the behaviour as function of time, which is the most interesting feature of a pyruvate sensor.

The nylon net procedure failed after few days and it seems that POD is inactivated by all procedures using glutaraldheyde as a final step.

By the physical adsorption on acetylcellulose the activity of POD decreases as expected for a physical immobilization. With a carboxylic membrane, the decrease of the signal is only 13% after 30 days, while with a polyazetidine procedure the decrease after the same period is 64%.

In Fig. 7 the correlation between the amperometric sensor and the spectrophotometric measurement of a series of human blood serum obtained from the hospital, are reported.

A System for Continuous Monitoring of Lactate and Pyruvate

We have previously described the connection of the lactate sensor to the artificial pancreas Biostator (Life Science Division, Miles,

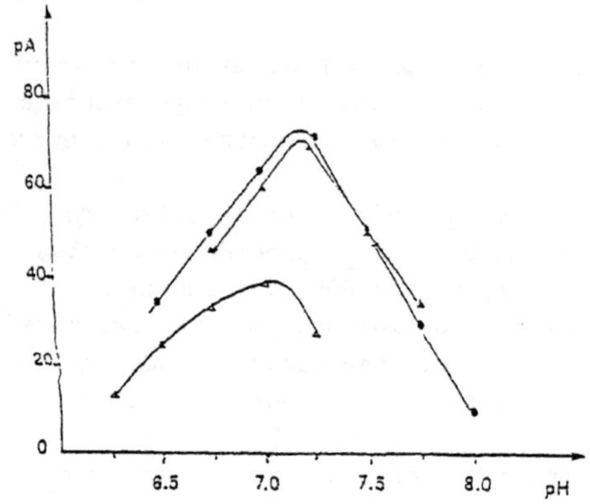

Fig. 5 pH and buffer effect on the pyruvate sensor.
Buffers: (●)Tris;(▲) HCO_3^- (△) citrate.

Fig. 6 Lifetime of pyruvate sensors prepared with different procedures:(○) carboxylic; (●) polyazetidine;(▲) acetylcellulose; (■) nylon.

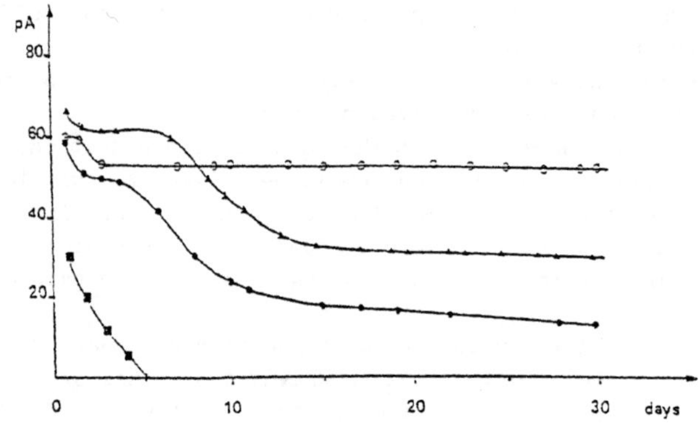

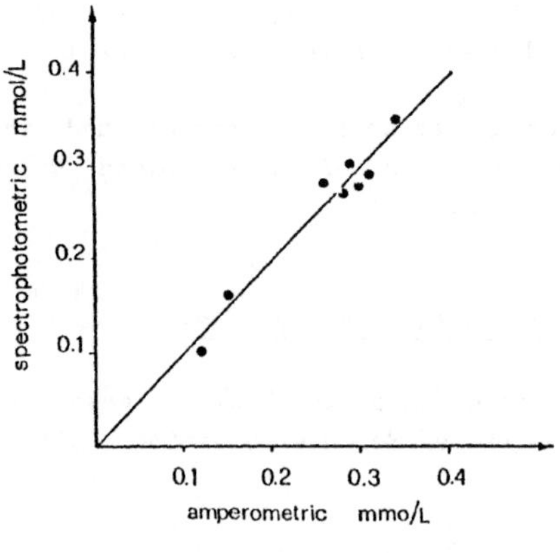

Fig. 7 Correlation between spectrophotometric measurements and with the pyruvate amperometric sensor of blood serum.

Elkhart, IN. USA) (4).
Recently we turned our attention towards the newly developed artificial pancreas Betalike (Esacontrol S.p.A., Genova, Italy) which represents a marked improvement with respect to the devices previously available.
One of the most interesting features is a miniaturization hollow fibre hemofiltrate cartridge (filtration surface: 50 cm; membrane cutoff: about 35,000 daltons) which allows only the hemofiltrate to reach the sensors while the blood cells and proteins are reinfused into the patient bloodstream. It results in a more stable signal, a longer life of the membranes and avoids the blood loss. Other technical improvements include a system for the control and adjustement of the blood dilution, a program for automatic selection and periodic calibration of the membranes and a compact disposable system for blood handling and insulin and glucose infusion. The features of the device greatly reduce the assistance required for its running and make it suitable for clinical use.
The blood taken from the patient via a double lumen catheter (6 ml/h) is diluted (1:9) with a buffer solution (Normosol, pH 7.4, Abbott, Italy) with the addition of 3 units/ml of heparin. The diluted blood is then dialyzed by the miniaturized hemofilter, and the dialysate passed through the glucose sensor. The flow cell (dead space : 40 μl) containing the lactate and pyruvate was placed downstream after the glucose sensor (Fig. 8).
A concentrated solution of cofactors is added to the dialysate before entering the cell for pyruvate measurements at a flow rate of 0.05 ml/min to obtain the optimal concentration of TPP, phosphate and calcium chloride within the cell. An auxiliary circuit has been added and connected to the system with a mechanical valve for the calibration of the sensors with baseline and standard solutions (Fig 8).

Experiments with the Betalike Artificial Pancreas

Figure 9 shows the results of an "in vivo" determination of glucose, L-lactate, and pyruvate in heparinized blood from a normal subject (a volunteer nondiabetic, 23 year-old man) being treated with the artificial pancreas (Betalike).
Every 15-20 min. we collected blood samples from the patient and assayed lactate and pyruvate spectrophotometrically within the next few days. These results, represented as dots in Figure 9, illustrate how comparable the results by both methods are.
The good correlation between the results by continous monitoring and by spectrophotometry excludes the possibility of reciprocal

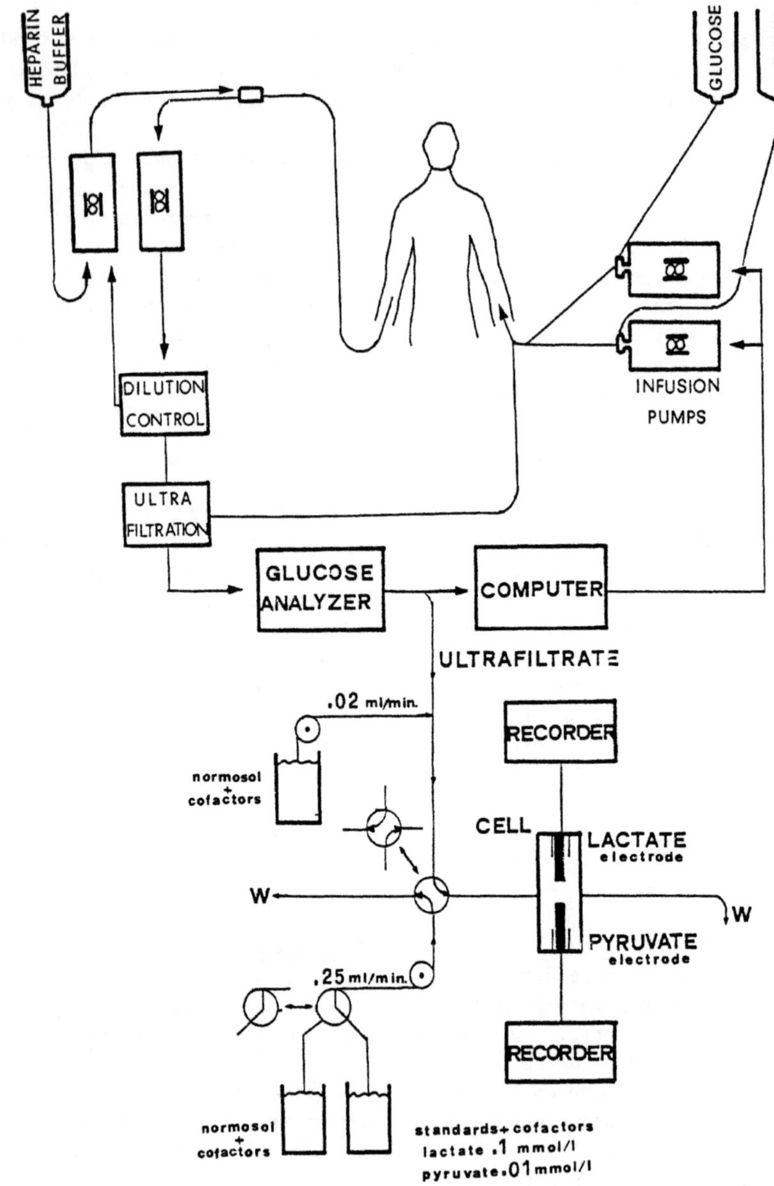

Fig. 8 General scheme for the calibration and measurements of lactate and pyruvate sensors coupled to the "Betalike"

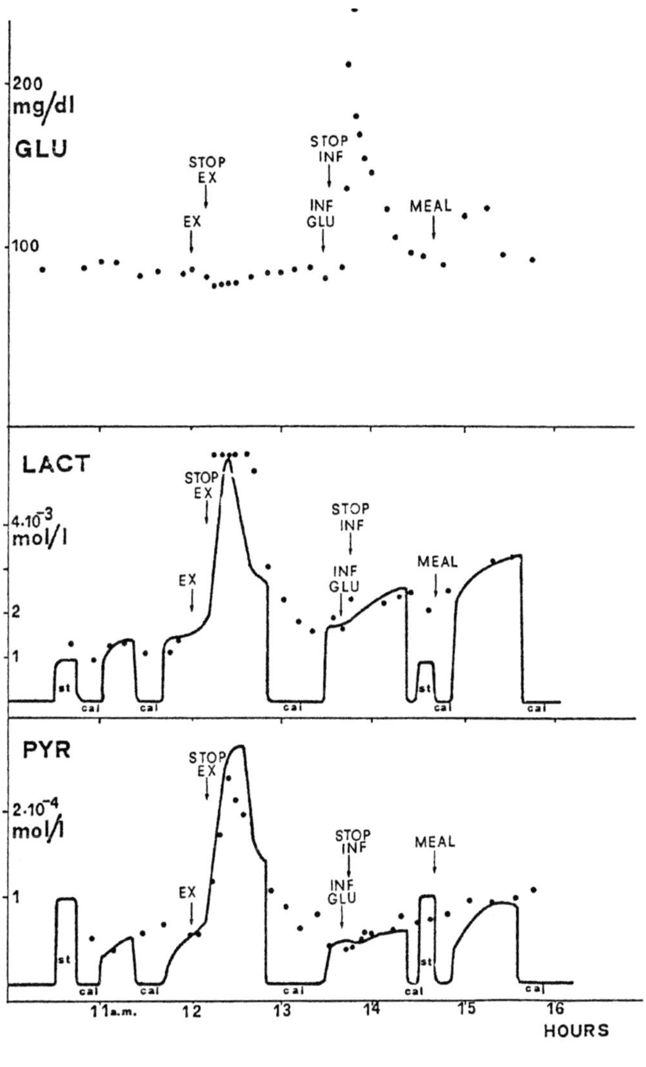

Fig. 9 Continuous monitoring of glucose, lactate and pyruvate in vivo during an experiment with the artificial pancreas "Betalike". See text for explanations and abbreviations.

interference between different sensors. The possibility that the oxygen consumed at the glucose sensor might interfere with the performance of the lactate or pyruvate sensors was excluded in preliminary experiments, in which standard solutions of lactate and pyruvate, with or without glucose, gave the same signals. Moreover, one feature of an enzyme sensor is that the enzymatic reaction occurs only at the electrode surface in a very limited area, the bulk concentration of the sample being unaffected by such measurement. The limited extent of the reaction means that the sample composition is not altered by the act of measurement. Such is the case here. The consumption of oxygen, calculated from the nanoamperes measured at the glucose electrode, shows a possible decrease of oxygen of a few nanomoles per liter, i.e. very negligible oxygen concentration.

As Figure 9 illustrates, at the time marked ST, a standard solution of lactate and pyruvate was passed through the cell to calibrate the sensors. At the time marked CAL, the glucose sensor of Betalike was calibrated. The blood flow was disconnected from the sensors during such periods. At the time marked EX, the patient was requested to do a short physical exercise, which was stopped at the time marked STOPEX. At the time INFGLU, a load of glucose was infused (50 g) within a short period. MEAL, indicates the time at which the patient ate a normal meal.

It is interesting to note the large effect of physical exercise on the concentration of lactate and, even more, pyruvate, followed by the parallel return of both analytes to "normal" concentrations. As expected the glucose load had a large effect on the glucose concentation, but caused only small variations in the concentrations of lactate and pyruvate. After the meal, all three analytes showed large and continuous variations.

Numerous factors influence fuel consumption; each can produce different variations in metabolic substrates. The lack of insulin that characterizes diabetes causes these variations to assume pathological values.

The correction of these variations, to complete normalization, will be possible only if the insulin replacement therapy is regulated on the basis of the complex relationships among the intermediate metabolites (2-5).

The opportunity to create algorithms that induce optimal therapeutic results is strictly related to the amount of information available. The reliability of the immobilization procedures described here provides diabetological studies with a new possibility for regulating the insulin therapy, as well interesting speculations regarding a "cure" for diabetes.

REFERENCES

1. Malchesky PS and Nose' Y (1978) in: Theory, Design and Biomedical Application of Solid State Chemical Sensor p.269, ed Cheung PW, Fleming DG, Neuman MR, Ko WH, CRC Press, Florida.
2. Mascini M, Fortunati S, Moscone D, Palleschi G, Massi-Benedetti M, Fabietti P (1982) Clin Chem, 31 451.
3. Second Assisi International Symposium on Advanced Models for the Therapy of Insulin Dependent Therapy, Assisi, Italy, 20-23 April, 1986, Session IV.
4. Mascini M, Mazzei F, Moscone D, Calabrese G, Massi-Benedetti M (1987) Clin Chem, 33 591.
5. Fogt EJ, Dodd LM, Clemens AH (1978) Clin Chem, 24 1366.
6. Mascini M, Moscone D, Palleschi G (1984) Anal Chim Acta, 157 45.
7. Mizutani F, Tsuda K, Karube I, Suzuchi S, Matsumoto K Anal Chim Acta 118 51.
8. Tsuchida T, Yoda K (1983) Clin Chem, 29 51.
9. Wood LL, Calton GJ (1984) Biotechnology Dec 1081.

BIOSENSORS IN CLINICAL AND PHARMACEUTICAL ANALYSIS

L. Campanella and M. Tomassetti
Department of Chemistry, University of Rome
"La Sapienza"
Piazzale A. Moro, 5

ABSTRACT.
The field of the sensors applied to biological and pharmaceutical analysis is summarely reviewed with particular reference to the researches of the authors in this area and to the enzyme immobilization aspects.

1. INTRODUCTION

Membrane electrodes are continuously increasing their importance both for the analytical difficulties they can concurr to overcome and for their operating simplicity and feasibility. Also the commercial aspects are not omittable: in the still developing membrane market, sensors accupy a meaning position (about ten million of dollars for year).

2. CLASSIFICATION

The traditional classification of the membrane electrodes is the following one (1,2,3) (Table I).

TABLE I

Solid membrane electrodes	- Homogeneous - Heterogeneous
Liquid membrane electrodes	- (Organic lipophilic solvents, polymer supports, PVC, ...)
Sensibilized electrodes	- Gas electrodes - Enzyme electrodes - Bacteria electrodes - Tissue electrodes
Immuno electrodes (antigen-antibody)	- Indirect measurements - Amplified detection - Direct measurement

G. G. Guilbault and M. Mascini (eds.),
Analytical Uses of Immobilized Biological Compounds for Detection, Medical and Industrial Uses, 169–176.

3. LIQUID MEMBRANE ELECTRODES

In the field of the applications to pharmacological and clinical matrices the first important results were obtained with the liquid membrane electrodes (2, 4, 5-12).

TABLE II

Liquid membrane electrodes selective to pharmacological active substances.

Detected drug	Counter ion
Codeine, morphine, dionin	dipicrylaminate
vitamin B_6	dipicrylaminate
novocaine	tetraphenylborate, dipicrylaminate
codeine	dipicrylaminate
propanol	dodecylnaphtalensulphonate
strichnine, papaverine	picrate
quinine, cocaine	picrate
papaverine, procaine	tetraphenylborate
pyridoxine, ethacrine	hexadecylsulphate
quinidine, amylocaine	octadecylsuphate
chloropromazine	1,4-azimino-dibenzensulphonate
ephedrine, emetine	1,4-azimino-dibenzensulphonate
thyamine, amphetamine	1,4-azimino-dibenzensulphonate
phenobarbital	aliquat 336S

4. BIOSENSORS

But the real jump of quality was performed by the enzyme sensors, containing in the membrane an immobilized enzyme (3, 13-16), or assimilated systems such as bacteria (17), plant tissues, animal tissues (4). Compared to isolated enzymes, these biocatalytic systems often provide an enhanced stability of the biosensor. An other advantage is their cheap and generally simple preparation; the applicability of sensors, assembled with such systems, can be widely expanded by additional coupling of immobilized isolated enzymes, thus forming hybrid electrodes, characterized by a bacterial strain and a separate enzyme catalysts jointly immobilized at the electrode surface.

4.1. Enzyme sensor using Clark oxygen electrode

The most common case in the chemical and pharmaceutical analysis by biosensors is the couplement between an enzymatic membrane and an amperometric Clark oxygen electrode as indicating (Fig. 1).

$$\text{Ascorbic acid} + O_2 \xrightarrow[\text{oxidase}]{\text{ascorbate}} \text{dehydroascorbic acid} + H_2O_2$$

$$\text{Alcohol} + O_2 \xrightarrow[\text{oxidase}]{\text{alcohol}} \text{aldheyde} + H_2O_2$$

$$\text{Galactose} + O_2 \xrightarrow[\text{oxidase}]{\text{galactose}} \text{galattoesodialdose} + H_2O_2$$

$$\text{Cholesterol} + O_2 \xrightarrow[\text{oxidase}]{\text{cholesterol}} \Delta^4\text{-cholestenon} + H_2O_2$$

$$\text{Uric acid} + O_2 + 2H_2O \xrightarrow{\text{uricase}} \text{allanthoin} + H_2O_2 + CO_2$$

$$\text{L-lactate} + O_2 \xrightarrow[\text{oxidase}]{\text{lactate}} \text{acetate} + CO_2 + H_2O$$

Figure 1. Some classic enzymatic reactions utilized.

4.2. Analysis of drugs and some biological fluids by enzyme sensors

Basing on the same kind of assembly we determined (Fig. 2) phospholipids (18-20) and cholanic acids (21-22), the latter ones compounds which recently became of great interest for their pharmaceutical activity, especially against gallstones; correspondingly their determination is of continuously increasing interest. The results of the choline containing phospholipid, in drugs and amniotic fluids are shown in fig.s 3 and 4 and in table III.

$$7\alpha\text{-(or } 3\alpha\text{-) hydroxysteroid} + NAD^+ \xrightarrow{7\alpha\text{-(or } 3\alpha\text{-)HSD}} \text{7-(or-3-)chetosteroid} + H^+ + NADH$$

$$NADH + H^+ + 1/2\ O_2 \xrightarrow{\text{peroxidase, } Mn^{2+}} NAD^+ + H_2O$$

$$\text{Phospholipid (Lecithin)} + H_2O \xrightarrow{\text{phospholipase D}} \text{choline} + \text{phosphatidic acid}$$

$$\text{Choline} + 2O_2 + H_2O \xrightarrow{\text{choline oxidase}} \text{betaine} + 2H_2O_2$$

Figure 2. Enzymatic-amperometric determination developed by us.

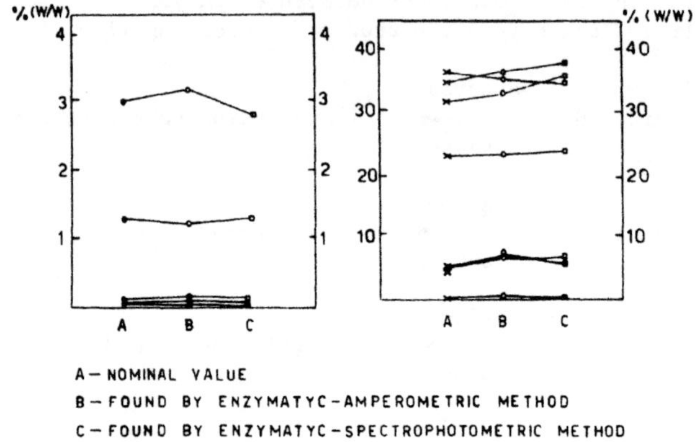

Figure 3. Correlation between enzymatic-amperometric and enzymatic-spectrophotometric methods in the analysis of drugs containing choline (enzyme physically entrapped in cellulose triacetate membranes).

Figure 4. Correlation between enzymatic-amperometric and enzymatic spectrofotometric methods, in the analysis of drugs containing lecithin (enzyme chemically immobilized on nylon net).

TABLE III
Results of choline containing phospholipids analysis, obtained by enzyme electrode, for twelve human amniotic fluid samples. Values are in mmol/L. Each value is the mean of at least three determinations.

Amniotic fluid sample number n.	Pretreatment	Phospholipids founds by enzyme electrode (a)	choline found by enzyme electrode
1	centrifugated	0.06	0*
2	"	0.12	0*
3	"	0.14	
4	"	0.13	
5	uncentrifugated	0.20	
6	"	0.22	0*
7	"	0.21	0*
8	"	0.18	
9	"	0.27	
10	"	0.31	
11	"	0.31	
12	"	0.38	0.01

(*) < 1.5 . 10^{-5} mol/L.

5. ENZYME IMMOBILIZATION

Probably the key to open the door of the success, to an enzyme sensor is the adopted immobilization method that must be such that, on one hand enzyme is not too free to diffuse away, othervise the lifetime of the electrode is too short and the efficiency poor and on the other one, it is not too much protected, to be practically unable to contact the system to be analysed. Chemical and physical immobilization methods are known (3,23,24). Chemical ones need generally longer and more laborious procedures as shown in previous work (25).

5.1. Development in the physical method of enzyme immobilization.

We like here to call the reader's attention to physical methods. The classical scheme is the following one:

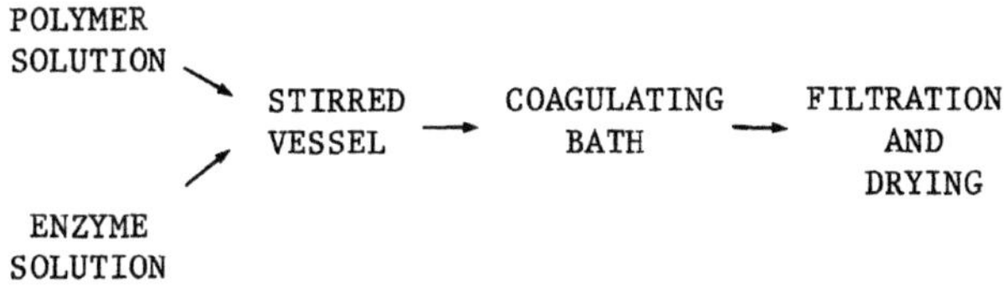

Figure 6. Usual scheme of physical immobilization procedure.

It suffers the great disadvantage that the enzyme is contacted with organic solvents with great danger of it denaturation.
We replaced this scheme with a new one, able to avoid this possible damage, as by it enzyme solution is not mixed with the organic solution of the support polymer, but an already preformed membrane is dipped into the solution of the enzyme, which is let to diffuse inside (26).

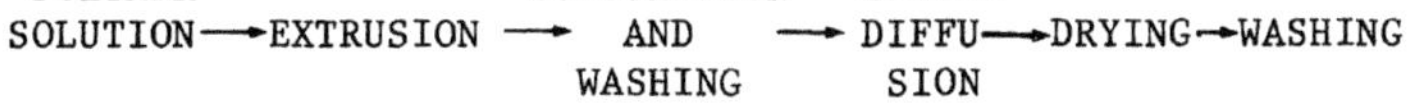

Figure 7. Our new scheme of physical immobilization procedure.

Also by this method we prepared enzymatic sensors for many common compounds to be determined. A list of the sensors we prepared and of their main applications is reported (In table V).

TABLE V

Developed enzyme sensors, with the enzymes physically entrapped in cellulose triacetate membranes.

Sensor	Method	Determined species	Nature of the sample
choline electrode	enzymatic-amperometric	choline	- drugs - biological fluids
		acethylcholine	- applications in course
		lecithin (choline containing phospholipids)	- drugs, foods - human bile - control sera - amniotic fluids
glucose electrode	enzymatic amperometric	glucose	- control sera
saccarose electrode	enzymatic-amperometric	saccarose	- applications in course

6. PERSPECTIVES

The last part of our presentation is devoted to discuss some perspectives. The greatest limits of the enzyme sensors are generally two: the unadequate detection limits to the aim of many kinds of analysis, the too great dimensions.

A contribution to solve both them can be given firstly by the combination of an enzyme membrane with a MOS (metal oxide semiconductor) or FET (field effect transistor) and by partial or total substitution of the support polymer with a conductor polymer. Many references in literature are already available in the first direction; in this case the more dramatic aspect is the combination of the "dead nature" of the semiconductor, or of the transistor, with the "live nature" of the enzyme: so all the studies concern the set up of a sort of chain, the first ring of which is the inorganic surface and the last one is the enzyme. Many compounds were tested in this direction such as polyvinylalcohol, polyazetidine (fig.s 8 and 9), polyethyleneimine and other ones, with the double exigence of continuing the chain and of crosslinking, in order yield stability to the membrane. Some examples are also reported of glueing the enzymatic membrane on the electronic system, or of direct immobilisation of the enzyme on it.

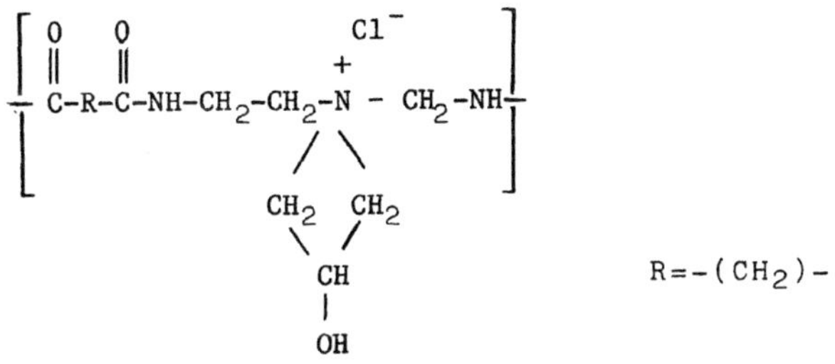

Figure 8. Polyazetidine prepolymer (P.A.P.)

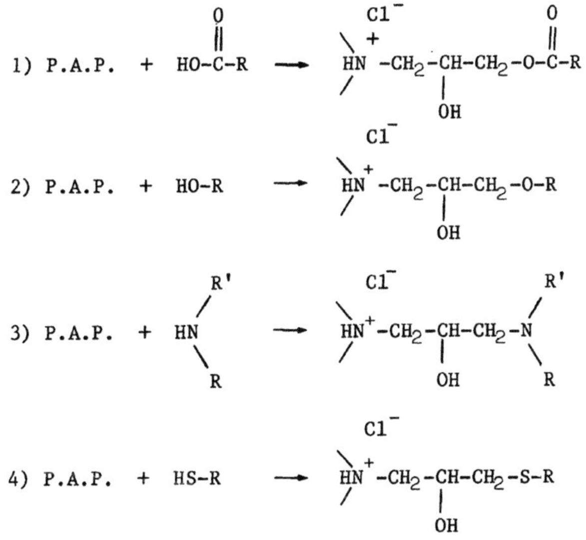

Figure 9. Reactions of polyazetidine prepolymer with compounds containing carbonilic, hydroxilic, aminic, thiolic groups.

For the second line we have immobilized different enzymes in membranes of polyphenylacetylene or polybenzylpropargylamine. The most meaning results of the presence of these polymers in the membrane composition are: faster response time (less then 15 sec), lowering of detection limit by one-two decades, greater reproducibility (27).

7. REFERENCES

1. Moody G.J., Thomas J.D.R., Editors, Ion-selective Electrodes, Merrow Technical Library, Watford (1971).
2. Baiulescu G.E., Cosofret V.V., Editors, Application of ion-selective membrane electrodes in organic analysis, Ellis Harwood Li-

mited, Coll House Press, Chichester, England (1977).
3. Guilbault G.G., Analytical uses of Immobilized Enzymes, Marcel Dekker, Inc, New York (1984).
4. Solsky R.L., Ion-selective electrodes in biomedical analysis, C.R.C. Press, 14, 1 (1982).
5. Campanella L., De Angelis G., Ferri T., Gozzi D., Analyst., 102 723 (1977).
6. Gozzi D., Campanella L., Ferri T., Electrochim. Acta, 24, 817 (1979).
7. Amoroso P., Campanella L., De Angelis G., Ferri T., Morabito R., J. Membr. Sci., 16, 259 (1983).
8. Campanella L., Sorrentino L., Tomassetti M., Anal. Lett., 15, 1515 (1982).
9. Campanella L., Sorrentino L., Tomassetti M., Analyst, 108, 1490 (1983).
10. Campanella L., Sorrentino L., Tomassetti M., Ann. Chim., 74, 483 (1984).
11. Campanella L., Tomassetti M., Cordatore M., J. Pharm. Biomed. Anal., 4, 155 (1986).
12. Campanella L., Tomassetti M., Sbrilli R., Ann. Chim. (in press).
13. Guilbault G.G., Ion. Select. Elect. Rev., 4, 187 (1982).
14. Senn D.R., Carr P.W., Klatt L.N., Anal. Chem., 48, 954 (1976).
15. Luzzana M., Dossi G., Mosca A., Granelli A., Berger D., Rovida E., Ripamonti M., Musetti A., Rossi Bernardi L., Clin. Chem., 29, 80 (1983).
16. Mascini M., Palleschi G. Anal. Chim. Acta, 136, 69 (1982).
17. Mascini M., Memoli A., Anal. Chim. Acta, 182, 113 (1986).
18. Mascini M., Tomassetti M., Iannello M., Clin. Chim. Acta, 132, 7 (1983).
19. Campanella L., Mascini M., Palleschi G., Tomassetti M., Clin. Chim. Acta, 151, 71 (1985).
20. Campanella L., Tomassetti M., Bruni M.R., Mascini M., Palleschi G., Food Add. Cont. Anal., 3, 277 (1986).
21. Campanella L., Bartoli F., Morabito R., Tomassetti M., in: Membranes and Membrane Processes, Drioli E. and Nakagaki M., Editors, Plenum Pub. Corp., 543 (1986).
22. Campanella L., D'Ascenzo G., De Angelis G., Ferri T., Mascini M., Palleschi G., Tomassetti M., in: La bile: aspetti chimici, farmacologici e fisiologici, Roda A., Roda E., D'Ascenzo G., Barbara L., Editors, Bologna, 97 (1985).
23. Zaborsky O., Immobilized Enzymes, CRC press, Boca Raton, Fla. (1972).
24. Hornby W.E., Morris D.L., in: Immobilized enzymes, antigens, antibodies and peptides, Weetal H.H. Editor, New York, M. Dekker Inc. (1975).
25. Campanella L., Tomassetti M., Rappuoli B., Bruni M.R., Int. Prod. Rev., Sept./Oct., 44 (1985).
26. Campanella L., Tomassetti M., Sammartino M.P., in: Proceedings of 2 International Meeting on Chemical Sensors (Bordeaux) 580 (1986).
27. Campanella L., Salvi A.M., Sammartino M.P., Tomassetti M., Chim. Ind., 68, 41 (1986).

GRAFT COPOLYMERS AS SUPPORTS FOR THE IMMOBILIZATION OF BIOLOGICAL COMPOUNDS

M. Alves da Silva & M. H. Gil
University of Coimbra
Department of Chemistry
3000 Coimbra
Portugal

ABSTRACT. On a search for suitable supports on which to covalently bind biocatalysts and other biological compounds, the development of graft copolymers based on hydrophilic natural polymers (cellulose and agar) was undertaken. The prefered graft was hydroxyethyl methacrylate (HEMA), sometimes partially hydrolyzed. The suitability of the graft copolymers prepared was assessed by the amount of enzyme which could be coupled with the substrate by different methods of activation of support groups and by the retention of activity of the bound enzyme in comparison with that of the free enzyme. Good results were obtained with papain (460 mg bound enzyme per g support, 130 mg remaining active) and trypsin (400 mg bound enzyme per g support, 25 remaining active). Glucose oxidase and urease showed modest values. Some of the systems developed can be of use in the food industry and in Medicine.

1. INTRODUCTION

The different methods of immobilizing enzymes and biological cells have been dealt with previously in this ARW and have been extensively reviewed in the literature (1,2,3). Here we are concerned with binding biological compounds, and particulary enzymes, onto graft copolymers. The value of using these supports has been reported by some authors (4,5). Graft copolymers offer several advantages as supports for the immobilization of biocatalysts, since they can be prepared in an enormous variety of forms, allowing the choice of the microenvironment for the enzyme on the support, and the control of the hydrophobicity and mechanical properties of the copolymer. In addition it is possible to use a great variety of reactive groups through which bind the biological compounds onto the polymer base.

Among the earliest reported studies on the immobilization of enzymes and proteins on graft copolymers are those of Hoffman _et al_ (6), who grafted hydroxyethyl methacrylate and N-vinylpirrolidone on silicone rubber films, and used them to couple human serum albumin, heparin and streptokinase. Garnett _et al_ (7,8) used the system polypropylene-g-co--p-nitrostyrene to immobilize trypsin.

G. G. Guilbault and M. Mascini (eds.),
Analytical Uses of Immobilized Biological Compounds for Detection, Medical and Industrial Uses, 177–185.

More recently other workers have prepared grafts on different supports for the immobilization of biological compounds. Beddows et al have grafted polyacrylamide onto nylon and onto cellulose (5), polyfluorostyrene and polynitrofluorostyrene onto nylon (4), poly(acrylic acid) onto polyethylene (9,10), poly(maleic anhydride/styrene) and poly(maleic anhydride/vinyl acetate) onto polyethylene (11) and onto cellulose, poly(galacturonic acid) and starch, respectively (12), and poly(hydroxyethyl methacrylate) onto cellulose and pectin (13). The afore-mentioned authors utilized these supports to immobilize pepsin, β-galactosidase, bovine serum albumin, trypsin, acid phosphatase, alkaline phosphatase and glucose oxidase by covalent binding.

Muller-Schulte et al (14,15) grafted different monomers carrying carboxyl, hydroxyl and amino groups onto polyethylene foil. They use these graft copolymers to couple human serum albumin and antibodies against thyrokine. More recently, Raghunath et al (16) have immobilized urease onto collagen-poly(glycidyl methacrylate) graft copolymers, while Simionescu et al (17) have immobilized α-amilase on cellulose acetate membranes grafted with poly(acrylic acid). Lately Hoffman and coworkers (18) have fixed L-asparaginase and Schistosoma antibody onto polyethylene and polypropylene based graft copolymers.

A different technique has been used by Angiuro and coworkers (19, 20) for the immobilization of enzymes onto graft copolymers. These authors grafted the vinylated enzyme onto a non-activated polymeric matrix, either by chemical initiation or by UV irradiation. In each case the monomer was previously attached to the enzyme by covalent binding.

In our laboratory, we have used polyethylene based graft copolymers to immobilize acid phosphatase, alkaline phosphatase, glucose oxidase and trypsin (21). We have observed that the coupling yield and the activity retention of the enzyme increased with the hydrophilic character of the graft copolymer. In consequence we have turned our attention to the use of cellulose, agar and poly(vinyl alcohol) based graft copolymers, as supports for the immobilization of glucose oxidase, trypsin, urease, invertase and bovine serum albumin (22,23). We believe these supports to be especially appropriate for the immobilization of biological compounds, and are currently studying the possible use of such immobilized systems in the food industry and in Medicine.

In the present paper we wish to report some results as regards the immobilization of trypsin, papain, urease and glucose oxidase onto cellulose and agar based graft copolymers. The grafting has been done with hydroxyethyl methacrylate on either natural polymer to protect it from rottening by microorganisms, to increase its hydrophilic character and the number and kind of reactive groups available for the binding of the enzyme.

2. EXPERIMENTAL

2.1. Materials

Trypsin, papain, urease and glucose oxidase were obtained from Sigma Biochemicals Ltd, Missouri, U.S.A. Hydroxyethyl methacrylate and all other reagents (analytical grade) were obtained from BDH Ltd, Poole, Dorset, U.K. The hydroxyethyl methacrylate (HEMA) was used as supplied. Wood pulp cellulose supplied by Celbi SA, Figueira da Foz, Portugal was boiled in distilled water for 6 hours, filtered, washed with distilled water and dried to constant weight under vacuum at 40ºC. Agar was supplied by Sicomol, Sociedade Industrial de Coloides do Mondego, SA, Figueira da Foz, Portugal, and used without further purification.

2.2. Preparation of Graft Copolymers

2.2.1. Cellulose graft copolymers. Dry cellulose (1g portions) was dispersed in aqueous solutions (30 cm^3) of hydroxyethyl methacrylate of varying concentration, containing 0,3g concentrated nitric acid (65%) and 0,3g cerium ammonium nitrate, and the suspensions heated at 50ºC for one hour. Monomer concentrations ranged from 2 to 5% (w/v). Each suspension was then poured into an excess of methanol, filtered, washed thoroughly methanol to remove homopolymer and dried under vacuum at 40ºC.

2.2.2. Agar graft copolymers. Agar (5g) was treated with 100 cm^3 of an aqueous solution of ferrous ammonium sulfate hexahydrate (6.4×10^{-7} M) at 25ºC for 30 minutes. The sample was then filtered, washed with water and dried. Samples (1g) of the Fe(II) treated agar was dispersed in aqueous solution (8% w/v) of hydroxyethyl methacrylate (30 cm^3) containing hydrogen peroxide (0,029M) and heated at 40ºC for one hour. The mixture was then poured into an excess of methanol, filtered, and the solid washed thoroughly with methanol and dried under vacuum at 40ºC.

2.3. Immobilization of Enzymes

The prepared graft copolymers were activated with p-benzoquinone (p-BQ) (24), and p-toluene sulphonyl chloride (p-Ts) (25). The activated copolymer (100 mg) was introduced into the solution of the enzyme (40 mg) in an appropriate buffer, and shaken at 4ºC for 18 hours. The solids were then filtered off, washed with distilled water and with the buffer solution. The filtrate and washings were made up to 100 cm^3 and the protein content determined by the Lowry method (26). The amount of protein bound onto the graft copolymer was determined by the difference. Portions (10 mg) of the solid with the bound enzyme were assayed for enzymic activity.

In another experiment, portions of the graft copolymers (100 mg) were refluxed with NaOH 0.5 M for 18 hours, thereby the ethyl methacrylate groups being partially hydrolyzed. The solids were then filtered off, washed with distilled water, 0,1 M hydrochloric acid solution, again with distilled water and dried under vacuum at 25ºC. Portions of the hydrolyzed graft copolymer (50 mg) were added to an aqueous solution (30 cm^3) of the enzyme (40 mg) and 1-cyclohexyl-3-(2-morpholinoethyl)-carbodiimide metho-p-toluenosulphonate (CMC) (40 mg), and the suspension shaken at 4ºC for 18 hours. The product was filtered off and the protein was determined as indicate above.

2.4. Assay of Enzymatic Activity

Glucose oxidase was assayed using o-dianisidine (27). In the case of trypsin, p-nitrophenol phosphate was used (28). For the assay of papain, the method of Anson (29), using haemoglobin was followed, and urease was assayed by measuring the NH_3 liberated with the Nessler reagent (30).

3. RESULTS AND DISCUSSION

3.1. Immobilization onto Cellulose Graft Copolymers

3.1.1. Papain. The enzyme was immobilized on the prepared cellulose graft copolymers, using different activation reagents. Table I shows the results obtained when coupling the enzyme onto cellulose-g-co-HEMA (42% graft, on the initial weight).

TABLE I. Immobilization of papain onto cellulose-g-co-HEMA (42% graft), with different activating reagents, at 4ºC for 18 hours.

Activation	Coupling medium	Bound enzyme ($mg\ g^{-1}$)	Active bound enzyme ($mg\ g^{-1}$)
p-Bq	$NaHCO_3$ 0.1 M	190	9
p-Bq	Phosphate buffer 0.1 M (pH=7)	90	7
p-Ts	$NaHCO_3$ 0.1 M	104	1

The results show that a very good yield of coupling of the enzyme on the support was obtained. However, the enzyme activity was severely impaired on binding. To improve the activity retention the coupling procedure was repeated, but with the enzyme in presence of casein, which acts as substrate to papain. It was observed that the bound active enzyme in the first case of Table I, rose from 9 to 12 mg per g support, which was not considered to be satisfactory.

In consequence it was decided to modify the graft copolymer by partial hydrolysis, as described in 2.3., introducing on the support a number of carboxyl groups, which renders the substrate more hydrophilic and allows the use of other activation methods. A partially hydrolyzed cellulose-g-co-HEMA, containig 24% hydroxyethyl methacrylate and 18% methacrylic acid (18% hydrolysis) was used.

TableIIshows that the presence of carboxyl groups improved the retention of activity by the enzyme on binding, the best result being obtained when the coupling was carried out in $NaHCO_3$ 0.1 M.

Another technique, which has been used to improve the retention of activity by the bound enzyme, consists in the introduction of a "spacer arm" between the support and the enzyme, aiming at reducing the interaction between them. Possibly the "spacer arm" diminishes the distor-

TABLE II. Immobilization of papain onto cellulose-g-co-HEMA (42% graft, 18% hydrolyzed), activated with p-benzoquinone, at different pH values, at 4°C for 18 hours.

Coupling medium	Bound enzyme ($mg\ g^{-1}$)	Active bound enzyme ($mg\ g^{-1}$)
$NaHCO_3$ 0.1 M	140	18
Phosphate buffer 0.1 M (pH 7)	200	10
Acetate buffer 0.1 M (pH 5)	150	5

tion of the enzyme on binding, which is one of the possible causes of the loss of activity. We have tried the use of lysin and aspartic acid as "spacer arms".

TABLE III. Immobilization of papain onto cellulose -g-co-HEMA (42% graft, 18% hydrolysis) coupled with "spacer arm", using different activation reagents, at 4°C for 18 hours.

"spacer arm"	Activation	Coupling medium	Bound enzyme ($mg\ g^{-1}$)	Active bound enzyme ($mg\ g^{-1}$)
Lysin	glutaraldehyde	$NaHCO_3$ 0.1 M	560	6
Lysin	"	Phosphate buffer 0.1 M (pH 7)	190	108
Lysin	"	Acetate buffer 0.1 M (pH 5)	274	66
Aspartic acid	CMC	Acetate buffer 0.1 M (pH 5)	460	130

The results obtained are collected in Table III. They show that the insertion of an appropriate "spacer arm" between support and enzyme increases strikingly the coupling yield and also the activity retention, except in the first case of the table, possibly due to an excessive amount of bound protein on the support. These results as regards papain are very satisfactory, and we consider that they provide a basis for the development of an immobilized biocatalytic system for industrial application. We are currently studing the kinetics of the immobilized enzyme in comparison with the free enzyme, but we can already say that the Km for the bound enzyme is of higher order than the Km for the free enzyme

3.1.2. Trypsin. In a preliminary experiment we observed that trypsin, like papain did not show appreciable coupling yield when the unhydrolyzed graft copolymer was used. We have thus studied in detail the coupling of the enzyme onto the partially hydrolyzed copolymer, cellulose-g-co-HEMA (100% graft, 17% hydrolyzed). The groups activated for the binding of the enzyme were either the carboxyl groups with CMC, or the hydroxyl groups with p-BQ or p-Ts. The results obtained are shown in Table IV.

TABLE IV. Immobilization of trypsin onto cellulose-g-co-HEMA (100% graft, 17% hydrolyzed), using different activating reagents, at 4ºC for 18 hours.

Activation	Coupling medium	Bound enzyme ($mg\ g^{-1}$)	Active bound enzyme ($mg\ g^{-1}$)
CMC	Phosphate buffer 0.1 M (pH 7)	38	11
CMC	$NaHCO_3$ 0.1 M	18	18
p-BQ	$NaHCO_3$ 0.1 M	400	25
p-Ts	$NaHCO_3$ 0.1 M	n.d.	96

Table IV shows that there is some difficulty in binding the enzyme to the support, except when p-BQ is used as activating agent. However, the retention of activity by the bound enzyme is good, the best activating reagent in this respect being p-Ts.

We have thus studied the influence of the concentration of this activating agent on the coupling process. Table V shows the results obtained with concentrations of p-Ts ranging from 0.01 M to 0.05 M, the

TABLE V. Influence of the p-Ts concentration on the amount of bound active enzyme in the immobilization of trypsin onto cellulose-g-co-HEMA (100% graft, 17% hydrolyzed), in $NaHCO_3$ 0.1 M, at 4ºC for 10 hours

p-Ts concentration (M)	Bound p-Ts ($mmol\ g^{-1}$)	Active bound enzyme ($mg\ g^{-1}$)
0.01	0.26	66
0.02	0.26	69
0.03	0.77	96
0.04	0.12	47
0.05	0.14	49

optimal concentration for activation and for coupling being 0.03 M. We are currently studying the kinetics of bound enzyme, which has a higher Km value than the free enzyme, like papain.

3.2. Immobilization onto Agar graft Copolymers

Previously (23) we have grafted agar with hydroxyethyl methacrylate, and have used this graft copolymer to couple various enzymes. We have observed that the yield of coupling and the activity retention were very low. In view of the results obtained with cellulose-g-co-HEMA partially hydrolyzed, we proceeded to carry out the same hydrolylitic treatment with the agar graft copolymers. We have tried to immobilize papain, trypsin, glucose oxidase and urease onto these partially hydrolyzed agar graft copolymers. The results obtained are shown in Table VI.

TABLE VI. Immobilization of papain, trypsin, glucose oxidase and urease onto agar-g-co-HEMA (42% graft, 15% hydrolyzed), using different activating agents, at 4ºC for 18 hours.

Enzyme	Activation	Coupling medium	Bound enzyme ($mg\ g^{-1}$)	Active bound enzyme ($mg\ g^{-1}$)
papain	HMD + glutaraldehyde (*)	Phosphate buffer 0.1 M (pH 8)	134	13
trypsin	p-Bq	$NaHCO_3$ 0.1 M	64	37
trypsin	p-Ts	$NaHCO_3$ 0.1 M	107	77
trypsin	CMC	Acetate buffer 0.1 M (pH 5)	8	8
glucose oxidase	p-Bq	$NaHCO_3$ 0.1 M	19	0.04
urease	CMC	Acetate buffer 0.1 M (pH 5)	n.d.	8

(*) A "spacer arm" of hexamethylene diamine (HMD) was previously introduced with p-Bq activation, and the coupling of the enzyme carried out after activation of the support with glutaraldehyde (31).

These results show the agar graft copolymers to be also good supports for the immobilization of papain and trypsin. With glucose oxidase and urease, however, the results were not satisfactory. We are trying at present to modify these supports and using other activation methods in an attempt to improve substantially the yield of binding and the activity retention of these enzymes. We are using agar and cellulose graft copolymers both in powder and in film form.

4. CONCLUSIONS

Hydroxhethyl methacrylate grafted cellulose and agar can be used as satisfactory supports for the immobilization of papain and trypsin if the hydrophilic character of the support is regulated through partial hydrolysis of the ester groups of the graft copolymer, and if an appropriate activation method for the reactive groups is used. The activity retention by the immobilized enzyme can be improved through the insertion of a "spacer arm" between support and enzyme.

The same supports did not give satisfactory results with glucose oxidase and urease, probably because the appropriate activation method for the binding groups has not been used. To a better understanding of the effect of immobilization on the activity of the enzyme is necessary to study the structure of the surface of the support and the microenvironment it provides to the immobilized enzyme.

5. ACKNOWLEDGMENTS

We wish to acknowledge the help of our students Manuela Moreira and Isabel Andrade with the experimental work and the financial support provided by Junta Nacional de Investigação Científica e Tecnológica - - Lisbon.

6. REFERENCES

1. O.R. Zaborsky, Immobilized Enzymes, C.R.C. Press, Cleveland, Ohio, U.S.A., 1973
2. I. Chibata, Immobilized Enzymes, Kodansha, Tokyo, J. Wiley & Sons, London, New York, 1978
3. J. Woodward, Immobilized Cells and Enzymes. A Pratical Approach, IRL Press, Oxford, 1985
4. C.G. Beddows, R.A. Mirauer, J.T. Guthrie, F.I. Abdel-Hay and C.E. Morrish, Polym. Bull., 1, 749, (1979)
5. F.I. Abdel-Hay, C.G. Beddows and J.T. Guthrie, Makromol. Chem., 182, 717, (1981)
6. A.S. Hoffman, G. Schmer, C. Harris and W.G. Kraft, Trans. Amer. Soc. Artif. Int. Organs, 18, 10, (1972)
7. J.L. Garnett, R.S. Kenyon and M.J. Liddy, J. Chem. Soc. Chem. Comm., 735 (1974)
8. M.J. Liddy, J.L. Garnett and R.S. Kenyon, J. Polym. Sci. Polym. Symp., 49, 109 (1975)
9. C.G. Beddows, M.H. Gil and J.T. Guthrie, Polym. Bull., 3, 645 (1980)
10. C.G. Beddows, M.H. Gil and J.T. Guthrie, Biotechnol. Bioeng., 34, 1371 (1982)
11. C.G. Beddows, M.H. Gil and J.T. Guthrie, Biotechnol. Bioeng., 27, 579 (1985)
12. C.G. Beddows, M.H. Gil and J.T. Guthrie, Biotechnol. Bioeng., 28, 51 (1986)
13. C.G. Beddows, M.H. Gil and J.T. Guthrie, Polym. Bull., 11, 1 (1984)
14. D. Muller-Schulte and F.A. Horster, Polym. Bull., 7, 77 (1982)
15. D. Muller-Schulte and F.A. Horster, Polym. Bull., 7, 395 (1982)
16. K. Raghunath, K. Panduranga Rao and K. Thomas Joseph, Biotechnol. Bioeng., 26, 104 (1984)

17.C.I. Simionescu, S. Branisteam and S. Dumitriu, Acta Polymerica, 35, 383 (1984)
18.A.S. Hoffman, W.G. Gombotz, S. Uenoyama, L.C. Dong and G. Schmer, Radiat. Phys. Chem., 27, 265 (1986)
19.L. D'Angiuro and P. Cremonesi, J. Appl. Biochem., 3, 459 (1981)
20.L. D'Angiuro and P. Cremonesi, Biotechnol. Bioeng., 25, 735 (1983)
21.M. Alves da Silva, C.G. Beddows, M.H. Gil and J.T. Guthrie, Actas IX Simpósio Iberoamericano de Catálise, Lisboa 1984, pg. 358
22.M. Alves da Silva, M.H. Gil, M.L. Lopes, P. Fernandes and P. Piedade, Actas II Encontro Nacional, Sociedade Portuguesa de Materiais, Porto, 1985, pg.1
23.M. Alves da Silva, M.H. Gil and P. Fernandes, Actas X Simpósio Ibero-americano de Catálise, Mérida, Venezuela, 1986, pg. 825
24.J. Brandt, L.O. Andersson and J. Porath, Biochim. Biophys. Acta, 386, 196 (1975)
25.K. Nilsson, O. Norrlow and K. Mosbach, Acta Chem. Scand., B35, 19, (1981)
26.O.H. Lowry, N.J. Rosebrough, A.L. Farr and R.J. Randall, J. Biol. Chem., 193, 265 (1951)
27.A.G. Huggett and D.A. Nixon, Lancet, 2, 368 (1957)
28.B.F. Erlanger, N. Kokowsky and W. Cohen, Arch. Biochem. Biophys., 95, 271 (1961)
29.M.L. Anson, J. Gen. Physiol., 22, 79 (1939)
30.A. Vogel, Textbook of Quantitative Inorganic Analysis, Longman, London and New York, 731 1978
31.A.C. Johansson and K. Mosbach, Biochem. Biophys. Acta, 370, 348 (1974)

USE OF PROTEIN COATINGS ON PIEZOELECTRIC CRYSTALS FOR ASSAY OF GASEOUS POLLUTANTS

G. G. Guilbault and J. Ngeh-Ngwainbi
Chemistry Department
University of New Orleans
New Orleans, Louisiana 70148

ABSTRACT. Preliminary studies utilizing biological substrates, such as enzymes and antibodies, as coatings for a piezoelectric crystal detector have proven successful. Immobilization of cholinesterase and parathion antibodies allows for the detection of organophosphorous pesticides at the ppb level with the cholinesterase coatings exhibiting a more specific response to parathion. Excellent reproducibilities, coating lifetimes, response times and selectivities are observed. This research is the first use of proteins as coatings for the direct assay of gaseous compounds, thus making this method an attractive alternative to some conventional techniques currently in use.

1. INTRODUCTION.

Piezoelectricity was first discovered in 1880 by the brothers Pierre and Jacques Curie (1). They observed that when pressure is applied to the surface of a quartz crystal, an electrical potential forms between the deformed surfaces, and also that the application of a voltage to a quartz crystal effected physical distortions. The crystal, when activated by a concise electrical charge, will vibrate mechanically for a short period until physical equilibrium is established.

Alpha quartz is most often used for piezoelectric crystal detectors. The properties which make it most suitable for this application are its water insolubility, and resistance to temperatures up to 579°C with no loss of piezoelectric properties. The resonant frequency of the quartz crystal depends on the physical dimensions of the quartz plate and the thickness of the electrode deposited. For use as a piezoelectric detector, only AT- or BT-cut quartz plates are useful (2). AT and BT refer to the orientation of the plate with respect to the crystal structure. The temperature coefficient is a critical function of the angle of cut. The AT-cut crystal is more stable than most other piezoelectric cuts and has a temperature coefficient of about one ppm per degree centigrade over a temperature

G. G. Guilbault and M. Mascini (eds.),
Analytical Uses of Immobilized Biological Compounds for Detection, Medical and Industrial Uses, 187–194.

range of 10-50°C. The crystals most frequently used are 9MHz quartz of 10 to 16 mm disks, squares and/or rectangles that are approximately 0.15 mm thick. The metal electrodes are 3000 to 10,000 A° thick, 3 to 8 mm in diameter, and can be made of gold, silver, aluminum or nickel. The resonant frequency of a plated crystal can be obtained using a frequency counter attached to the output of an oscillator circuit.

The theoretical foundation for the use of quartz crystals as a microbalance or detector can be traced to Lord Rayleigh, who showed that small changes in the inertia of a mechanically vibrating system perturbed the resonant frequency of the system. One first considered the problem of a quartz crystal coated with a surface film in 1957. In that same year, Sauerbrey suggested the use of the piezoelectric crystal as a sensing device for the measurement of the thickness of thin films. He showed that the shift in resonant frequency of the crystal was proportional to the mass deposited to within ± 2%.

In 1959, Sauerbrey developed an empirical equation for AT-cut crystals vibrating in the thickness shear mode that describes the relationship between the mass of thin metal films deposited on quartz crystals, and the corresponding change in resonant frequency of the crystal (3,4):

$$\Delta f = -2.3 \times 10^6 \; F^2 \Delta m / \; A \qquad (1)$$

where Δf is the change in frequency due to a coating (Hz), F is the resonant frequency of the piezoelectric crystal before coating (MHz), m is the mass of the deposited coating (g), and A is the area coated (cm^2).

This equation predicts that a commercially available 9MHz piezoelectric crystal would have a mass sensitivity of about 400 Hz/μg or a 15 MHz crystal a sensitivity of 2600 Hz/μg. From this, it can be seen that a vibrating quartz crystal can be an extremely sensitive weight indicator. The limit of detection is estimated to be about 10^{-12} gram (5).

Metals and a number of other solids have been found to change the resonant frequency of a crystal when deposited without affecting the crystals' ability to oscillate. However, when liquids are deposited on the crystal's surface, the ability to oscillate is often impared due to the dissipation of energy in the liquid from the vibrating crystal. When a gas is allowed to absorb onto a suitable coating, the amplitude of vibration of the crystal changes in proportion to the concentration of the gas.

2. GENERAL EXPERIMENTAL APPARATUS.

A typical experimental set-up used for the evaluation of piezoelectric detectors is shown schematically in Fig. 1. The cell is designed with optimum sensitivity for use in a flow system. The piezoelectric crystals are usually AT-cut mounted in HC 25/u holders with a 9MHz resonant frequency. The electrodes are usually constructed of silver or gold. Instrumentation consists of a low frequency OX oscillator powered by a regulated power supply set at 9 volts. The frequency output from the oscillator is monitored with a frequency counter, coupled with a digital to analog frequency converter so that the frequency changes may be recorded. The frequency can be read from either the counter or recorder.

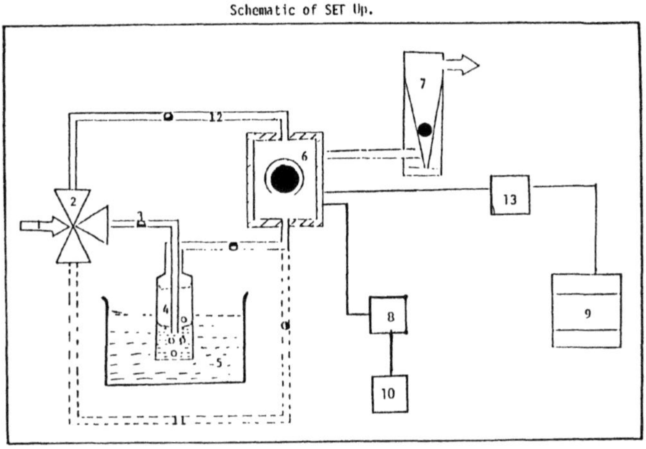

Figure 1. Schematic of experimental setup: (1) carrier gas, (2) 4-port valve, (3) one-way valve, (4) liquid sample, (5) constant-temperature bath, (6) antibody-coated crystal, (7) flow meter, (8) oscillator, (9) digital read out, (10) power supply, (11) diluent gas stream, (12) background gas stream, and (13) frequency counter.

The carrier gas can be supplied from compressed cylinders or, for air, by a vibrating diaphragm pump. The optimum flow rate must be determined for each coating and application and it is usually around 100 mL/min.

The 9 MHz crystal is very small, about a U.S. nickel. The crystal is mounted in a ceramic holder. A reference crystal is used to correct for temperature fluctuations and differences in pressure or air flow.

For field measurements, this type of apparatus is inappropriate. A self-contained, small, portable detector was designed using the basic components, but with 6V battery, a mini pump, and digital readout. The inside diagram of such a unit is simple. Using clean-up columns to purge the environment of residual gases, a base line is set. Then the unit is switched to sampling of the presence of contaminants. The unit can operate for 8 hours without recharging, has 8 gates for sensitivity control, and a pump adjustable for 10-500 ml/min sampling volume. It is sold commercially by Universal Sensors, New Orleans.

3. PROTEIN COATINGS.

The piezoelectric effect was first described by Pierre and Jacques Curie in 1880. Cady built the first oscillator in 1911, but the first analytical papers were not published until 1964, by King, who proposed the detector for gas chromatography, and by Guilbault and co-workers, who used it as a detector for organophosphorus compounds in air. King's detector was marketed first by Esso Research, and now Dupont, as a moisture analyzer, and also by a now defunct Florida company, as a GC instrument utilizing the PZ detector. Over a hundred papers have been published, on the use of the PZ detector for assay of atmospheric pollutants (6-8).

Preliminary studies utilizing biological substrates, such as enzymes and antibodies, have proven successful as coatings for a piezoelectric crystal detector. Immobilization of cholinesterase and parathion antibodies allows for the detection of organophosphorus pesticides at the ppb level with the cholinesterase coatings exhibiting a general response to all organophosphorus pesticides and the antibody coatings exhibiting a more specific response to parathion. Excellent reproducibilities, coating lifetimes, response times and selectivities are observed. This research is the first use of proteins as coatings for the direct assay of gaseous compounds, thus making this method an attractive alternative to some conventional techniques currently in use.

Before measurements of the response of a coated crystal to a particular pesticide, were made, the gas delivery system was purged for at least 24 hours with the substrate saturated carrier gas. This ensured that only the pesticide of interest was in the delivery system when measurements were recorded. The gas permeation tubes allow entry of a fixed and known concentration of pesticide into the detection chamber.

In each case, the electrode portion of the crystals were coated by first applying equal amounts of protein to both sides of the crystal using a microliter syringe. The protein could then be immobilized using appropriate ratios of glutaraldehyde and bovine serum albumin, applied to the crystal in the same manner as the protein. Excess glutaraldehyde was washed off using a 0.1 M phosphate buffer. The crystal was placed in a desiccator for at least one hour before testing to allow the coating to dry.

The coated crystals were then placed in the test cell. The pure carrier gas was introduced to the cell using a four-way valve and the subsequent decrease in frequency was recorded as the response of that coating to the pesticide being tested. Generally, four measurements of this type were made in succession in order to evaluate the reproducibility of each response, as for a cholinesterase coating with pesticide. This was continued throughout the lifetime of each coated crystal.

The PZ detector was mounted for analytical work. The inlet from the atmosphere is split into a T junction, to allow gas to pass on both sides of the coated crystal. A set frequency, usually 9 or 14 MHz, is applied, and the frequency of oscillation, which indicates the amount of pollutant present, is recorded using a frequency counter and recorder.

In order to generate a known concentration of the test gas at very low part-per-billion to part-per-trillion concentrations, a gas permeation tube is used. Pure carrier gas mixes with the diffused test gas to generate a known concentration.

3.1. Cholinesterase Coatings (9)

The reaction sequence for organophosphorus detection using an enzyme coating is shown below:

$$\underset{I}{\text{Malathion}} + \underset{E}{\text{Acetylcholinesterase}} \rightleftharpoons \underset{\Delta F \text{ (reversible response)}}{\text{E-I}}$$

Typical response times were on the order to 30-60 seconds with complete recovery in 5-10 minutes.

A plot of the response of PZ crystals coated with various enzymes (horse serum, eel and bovine cholinesterases) is shown in Fig 2. All enzymes used here were soluble, and coated on the crystal from water solutions. Similar plots are obtained with immobilized enzymes. All enzymes appear to have good reactivities and follow a linear relationship. Immobilization was via Glutaraldehyde + Enzyme + Albumin as described before.

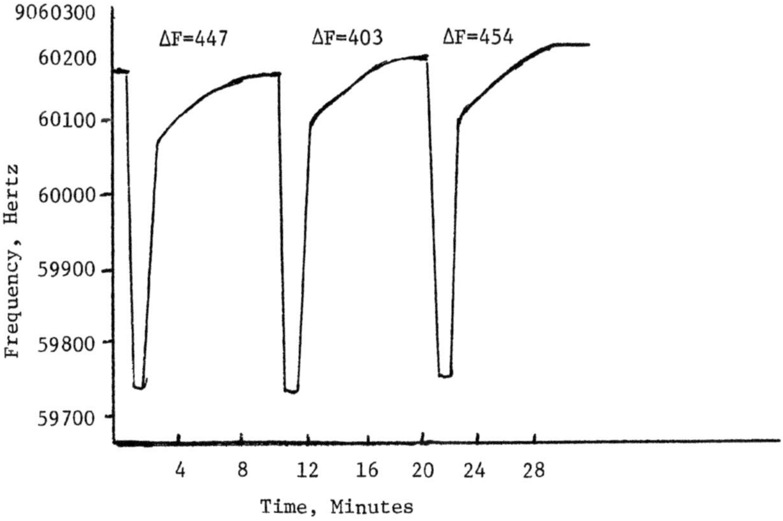

Figure. 2. Response to repeated exposures to Malathion.

The responses of various pesticides to the cholinesterase coated crystals are good. The order of reactivity follows that observed in the solution phase, giving evidence that a gas phase enzyme-inhibition reaction is indeed being monitored.

The effect of relative humidity on the response of the cholinesterase coated crystals was tested. As expected, the higher the R.H., the greater the response. Only a little water is required for the gas phase reaction to proceed. Dry lab air, which has about 5 ppm of H_2O, is sufficient for a good response to be observed. The instrument is calibrated before every use to the relative humidity present - then response curves can generally be recorded for several hours without need for recalibration.

Finally, some studies of the effect of various atmospheric gases on the response were made. Essentially no interferences were observed.

3.2. Antibody Coatings(10)

A typical response curve for the antibody coating to parathion is shown in Fig. 3. Note the fast response and good reproducibility. Responses were complete in 2-3 minutes, while recovery time was on the order of 1-2 minutes.

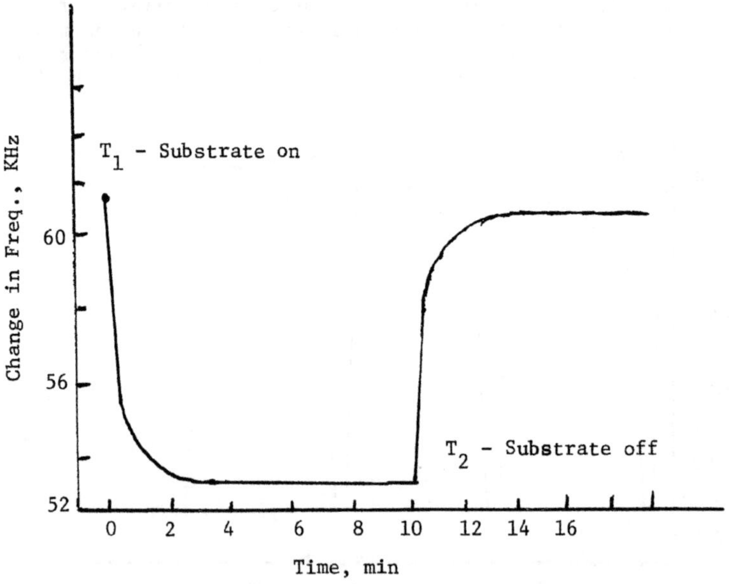

Fig. 3. Response of Detection to 35 ppb of Parathion.

The response of the antibody coating in the gas phase to other similarly structured pesticides is shown in Table 1. Comparison is made to the response of the antibody to these same compounds in solution. The similarity in response shows that indeed an antigen-antibody reaction is being monitored in the gas phase.

Table I. Comparison of Antibody Specificity in the Gas and Liquid Phases

pesticide	amt (μg) in solution[a]	amt (ppb) in gas phase[b]
parathion	50	36
malathion	70 000	106
methyl parathion	15 500	158
p-nitrophenol	10 000	680
disulfoton		560
ethion		

[a] Amount of pesticidal material (μg) causing 50% inhibition of parathion binding to antiserum in solution (Ercegovich et al J. Agric. Food Chem. **1981**, 29, 559-563). [b]Amount (ppb) needed to produce a 400-Hz change in the gas phase.

A reproducibility of better than 4% for all concentrations of soluble antibody tested was achieved, and the calibration plot for parathion is linear in the part-per-billion to part-per-million range. A calibration plot using immobilized antibodies is linear on a wide range.

No interferences from humidity were observed as long as it remained constant. A larger response was recorded at higher humidities than at relatively low humidities, hence calibration curves at different humidities are first required.

The enzyme coated crystals had a lifetime of approximately 40 days when stored in a desiccator at room temperature while the antibody coatings had a lifetime of approximately 18 days when stored in the same manner. The stability of the antibody coated crysal is quite good.

These preliminary results make the future use of piezoelectric crystals as immunochemical sensors very promising. This technique is also potentially useful and convenient in the area of environmental analysis, and an alternative to radioimmunoassay. Use of this technique for the assay of substrates, such as formaldehyde in air with formaldehyde dehydrogenase, for detection of contraband drugs with cocaine or morphine antibodies, or explosives in airport security with antibodies specific for those compounds, are only a few of the applications we have developed. The detectors have the sensitivity of a dog's nose, and the same high selectivity and fast response. We believe the future for this technology to be bright.

4. REFERENCES.

1. F. W. Maarsen, M. C. Smit, and J. Matze, Recueil, 76, (1957) 713.
2. R. A. Heising, Quartz Crystal for Electric Circuits, Van Nostrand, NY, 1946, p. 24.
3. G. Z. Sauerbrey, Z. Phys., 155 (1959) 206.
4. G. Z. Sauerbrey, Z. Phys., 178 (1964) 457.
5. A. W. Warner and C. D. Stockbridge, in K. Behrndt (Ed.), Vacuum Microbalance Techniques, Vol. 3, Plenus Press, NY, 1963, p. 55.
6. G. G. Guilbault, M. Ho., and B. Rietz, Anal. Chem., 52 (1980) 1489.
7. W. H. King, Jr., Anal. Chem., 36 (1964) 1735.
8. G. G. Guilbault and J. Jordan, A Review of Piezoelectric Crystals in Analytical Chemistry, CRC Press, Cleveland, 1987.
9. G. G. Guilbault, J. Ngeh-Ngwainbi, P. Foley and J. Jordan, in Electrochemistry, Sensors and Analysis. Proceedings of the International Conference "Electroanalysis na h'Eireann", M. R. Smyth and J. G. Vos (Eds.), Elsevier Publishing Co., Dublin, Ireland, June 10-12 1986.
10. J. Ngeh-Ngwainbi, P. H. Foley, Shia S. Kuan and G. G. Guilbault, Parathion Antibodies on Piezoelectric Crystals, Journal of the American Chememical Society, 108 (1986) 5444.

USE OF IMMOBILIZED ENZYMES IN FET-DETECTORS

Bart H. van der Schoot and Piet Bergveld
Dept. of Electrical Engineering
Twente University
P.O. Box 217, 7500 AE Enschede
The Netherlands

ABSTRACT. This paper presents an overview of the various enzyme-FET's that have been developed up to now. The advantages of ISFET's over the use of conventional glass membrane electrodes are; a planar construction, small dimensions, low impedance, fast response, large scale production, ease of multi sensor realization and direct applicability after dry storage. A major drawback of enzyme sensors based on the measurement of pH, including the ENFET's, is the dependence of the response on the buffer capacity of the sample solution. To overcome this problem, we present the integration of an urea sensitive FET with a coulometric pH-actuator electrode. It is thus possible to control the pH in the enzyme membrane and keep this pH equal to that in the sample solution. The generating current needed to obtain this situation is directly proportional to the substrate concentration.The resulting sensor has a linear ouput that is independent of the buffer capacity of the sample solution and has an expanded detection range.

INTRODUCTION

Potentiometrically operating enzyme sensors suffer from a number of problems, partly due to the properties of immobilized enzymes and partly resulting from the use of the particular ion sensitive electrode. Focussing on enzyme electrodes in which the conversion of a substrate causes a change in the local pH, these problems are:

- pH-dependence of response, detection limit and dynamic range
- response dependency on the analyte buffer capacity
- dependence on immobilisation technology
- individual membrane casting and control of membrane thickness

During the 3-day workshop on the theory, design and biomedical application of solid-state chemical sensors held at Case Western Reserve University in Cleveland, Ohio on March 28-30, 1977, Professor Guilbault made a request to the audience, saying that he would welcome as much help as possible in the area of enzyme electrodes from workers in the

G. G. Guilbault and M. Mascini (eds.),
Analytical Uses of Immobilized Biological Compounds for Detection, Medical and Industrial Uses, 195–206.

field of ion-sensitive field effect transistors [1]. He probably expected that at least some of the problems could possibly be solved by the application of ISFET's.

It will be investigated in this paper whether this expectation is realistic or not. We therefore have to list the properties of ISFET's and conventional electrodes first of all in order to be able to compare the performances and operational mechanism.

GLASS ELECTRODE VS. ISFET

pH-glass electrodes are based on the properties of a glass membrane which is contacted on both sides by an electrolyte solution. In practice, the glass membrane is formed at the end of a glass tube which is filled with a reference solution. At both sides of the membrane a diffusion potential is formed from which the difference is measured by a reversible electrode, such as an Ag/AgCl electrode, in the inner solution and a reference electrode, e.g. a saturated calomel electrode, in the analyte [2].

The ISFET belongs to the class of Chemically Sensitive Electronic Devices. This means that the device is in the first place a transistor and can be operated also as such. The so-called threshold voltage is however a function of the pH of the analyte which contains a reference electrode contacted to the common of the specially developed ISFET amplifier. The operational mechanism of the ISFET originates from the pH sensitivity of the inorganic gate oxide, such as SiO_2, Al_2O_3, Si_3N_4 or Ta_2O_5 [3].

It should be mentioned that ISFET's are essentially chips which are of a planar construction, are very small and can be manufactured with IC-technology, which means that mass production of completely identical chip sensors can easily be performed. The advantages and drawbacks of both types of pH sensors can be deduced from table I.

TABLE I

	glass electrode	ISFET
appearance	bulb type surface	flat surface
dimensions	> 100 mm^2	< 1 mm^2
impedance	very high	very low
sensitivity	59 mV/dec	< 55 mV/dec
drift	< 1 mV / week	0.1 - 1 mV / hour
multi sensor	individual types	integrated
production	individual	large scale
time constant	seconds	milliseconds
conditioning	24 hours	not required

Table 1. A comparison of the essential features of glass membrane electrodes and ISFET's as pH sensors.

From this table it can be concluded that ISFET's have in comparison with glass membrane electrodes the advantages of: planar construction, small dimensions, low impedance, fast response, large scale production, ease of multi sensor (differential) realization and directly applicable after dry storage. A drawback is the larger drift rate and the necessary stringent encapsulation of the chip edges and bonding leads.

It will be obvious that membrane casting onto flat surfaces gives a better control of membrane thickness and improved sealing possibilities than with the use of curvilinear surfaces from bulb type electrodes. Furthermore the deposition of membrane material is no longer individual per sensor, but in the ISFET processing the membranes can be deposited on a whole wafer containing hundreds of ISFET's. In this way very small quantities of enzymes are necessary per sensor, which is advantageous especially for expensive enzymes.

In order to compensate for the pH of the analyte, it is very easy to carry out a differential measurement between an enzyme modified ISFET and a bare ISFET, because two ISFET's can easily be integrated in one chip without increasing the price.

The advantages mentioned above look very promising and we will now first of all summarize the literature concerning ISFET-based enzyme electrodes in order to be able to discus the real significance of the relevant ISFET features.

REPORTED RESULTS ON ISFET-BASED ENZYME SENSORS

Although Janata and Moss mentioned already in 1976 the possibility of designing an enzyme-modified ISFET [4], the first preliminary results were not published until 1980. In that year Caras and Janata [5] described a penicillin-responsive device, which comprised a dual pH sensitive ISFET, one ISFET having a membrane on the gate of cross-linked albumin-penicillinase and the other having a membrane of only cross-linked albumin. The device was called an ENFET. When penicillin was present in the analyte, the penicillinase present in the active gate membrane catalysed the hydrolysis of penicillin to penicilloic acid. The released protons caused a local decrease in the pH near the ISFET gate, resulting in an ISFET output signal. The second ISFET remained however unaffected, because its membrane did not carry the enzyme penicillinase, so this ISFET can be operated as a reference device. A schematic drawing of their measurement setup can be seen in figure 1. Response times of ca. 25 seconds and lifetimes of two months with intermittent usage were reported.

It was mentioned however that the buffer capacity of the analyte had a profound influence on the sensor sensitivity, its linearity as well as on the concentration range. A decrease in analyte buffer capacity suppressed less the internal membrane pH, enabling greater device sensitivity, whereas an increase in buffer capacity caused the reversed.

It can be concluded that the first reported results of ENFET's show that these devices have indeed some advantages, especially with respect to the dimensions, the membrane fixation and the dual sensor design for easy differential measurements, but that various problems already known

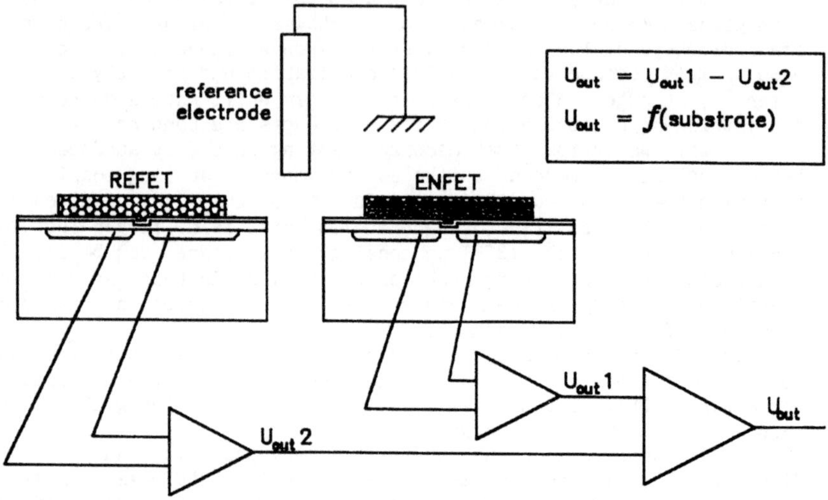

Figure 1. Experimental setup for ENFET's using one ISFET with an enzyme loaded membrane as the sensor and one ISFET with an unloaded membrane as a reference.

from convential enzyme electrodes, such as buffer dependence of sensor response and pH dependence of enzyme activity, are not solved. This is of course not surprising, because the ISFET application does not influence these typical enzyme electrode related properties.

The proceedings of the first international meeting on chemical sensors, held in september 1983 in Fukuoka, Japan, contain two contributions on ENFET's. [6,7]. The experiments of Miyahara et al [6] confirm the results mentioned above. They describe an urea and an acetylcholine ENFET, measured with respect to a reference FET on which glycine was immobilized instead of respectively urease and acetylcholinesterase.

Hanazato et al [7] report the results of a glucose oxidase ENFET, measured with respect to a bare ISFET. This paper explicitly describes the use of the differential measurement through the application of a platinum pseudo reference electrode. This is an important advantage because this electrode is much easier to fabricate and miniaturize than the Ag/AgCl/saturated KCl reference used by others.

More papers on ENFET's appear in the open literature in 1985, concerning the theoretical basis as well as technology and application. Both Caras and co-workers [8-11] and Eddowes and co-workers [12-14] independently publish extensive theoretical models. The main factors, as indicated in figure 2, on which the device response depends are; the relative rates of diffusional mass transport of all the species involved, the enzyme kinetics (which are pH dependent!) and the buffer capacity of the sample solution. This was confirmed by the corresponding

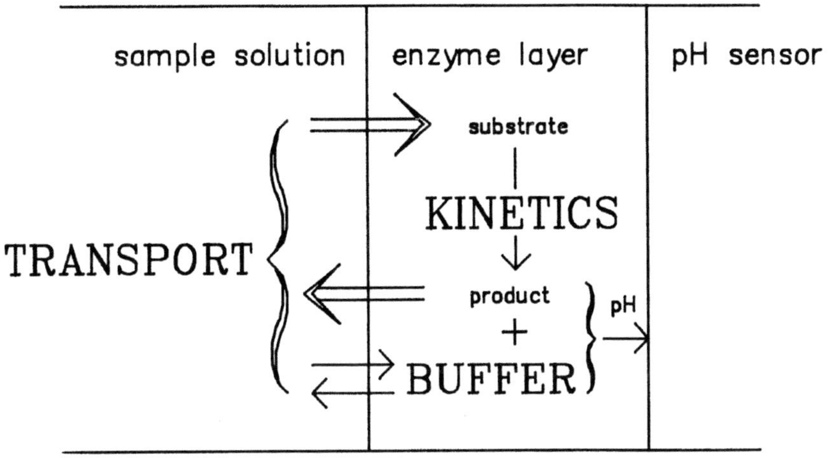

Figure 2. The response of pH-based enzyme sensors is governed by the transport rates of the chemical species, the enzyme kinetics and the buffer capacity of the sample.

experimental verification on glucose oxidase-modified ENFET's and penicillinase-modified ENFET's.

For the fabrication of the ISFET-part of the corresponding enzyme sensors, Miyahara et al [15,16] and Kimura et al [17-19] use the silicon on sapphire (SOS) technology. This means that on the wafer all ISFET's are electrically insulated. Furthermore the chips are surrounded by sapphire and surface insulation layers of SiO_2 or Si_3N_4, so that the edges are insulated from measurement solutions too. Accordingly, after many ISFET's are formed on the wafer, when a sensor is taken by cutting the wafer, there is no need for insulating the cut edge, so that an actual sensor can be fabricated with ease and reliable encapsulation. Only the bonding wires should be insulated carefully.

Both research groups mentioned above with respect to the SOS technology also introduce IC-technology compatible methods for the deposition of enzyme membranes on ISFET's. This is an important advantage over the previously described individual membrane casting techniques, because only then the full potential of microfabricated sensors can be exploited. Moriizumi and Miyahara [16] use photosensitive PVA for selectively depositing membranes that contain urease. The polymer films are applied by spin coating and then patterned photolithographically. The lifetime of the sensors was reported to be 20 days.
Kuriyama et al [18] show a different approach. They deposit an acetylcellulose membrane over an entire wafer by spin coating and succesively immobilize urease in this membrane using glutardialdehyde. Finally the enzyme is locally inactivated by ultraviolet light through the use of a photomask. An attractive feature of this technique may be that ENFET/REFET pairs that are created this way, behave identical in every

respect, electrically as well as chemically, except for the substrate sensitivity of the ENFET.

A lift-off method for selectively depositing membranes over the pH-sensitive gate area of ISFET's, using membrane materials that are not photosensitive themselves, was described by Kuriyama et al in 1986 [20]. First a layer of photoresist is deposited over the entire wafer and selectively removed from the gate areas. After silanization of the surface, a solution containing albumin, the enzyme and an amount of glutardialdehyde for crosslinking is spincoated on the wafer. After the crosslinking is completed, the enzyme membrane other than over the gate areas is removed by dissolving the underlying photoresist in acetone.

A further advantage of the use of ISFET's for enzyme sensors could be the construction of multi-sensors. The possibility of fabricating several transistors on one chip surely surpasses the glass electrode as a detector. In the literature, a number of multi-enzyme FET's is reported, however, up to now these devices are always fabricated by individual casting of the different membranes.

Kimura et al [17-19] use membranes of enzyme crosslinked with albumin, deposited in micropools over the ISFET gates. A sensor that simultaneously measures glucose and urea is described and the cross-talk between the different sensors located at a distance of 300 micrometers is investigated [19].

Another construction of a multi-enzyme sensor is described by Hanazato et al for the case of a combined urea, glucose, pH and pseudo reference electrode [21]. Here separate chips are used, assembled together on an epoxy laminate. The particular enzyme solutions were dropped over the gate areas and spun out to relatively thin layers without mixing. The experimental results are again similar to those already described above.

A fabrication method for multi-enzyme sensors that overcomes the problem of individaul manual membrane casting is proposed by Kuriyama et al [20]. On the wafer, micro pools are formed around the gate areas using a photo-polymer. In these micro pools, minute drops of enzyme solution are injected by means of an ink jet nozzle. The wafer is mounted on a dirigible X-Y stage, so that mass production should be possible. Unfortunately, no experimental details are given yet.

Urea sensors are again described by Anzai et al [22] and by Karube et al [23]. Both authors describe the influence of pH on the sensor response.

The investigations decribed up to now deal mainly with the urea and glucose sensors. The positive and negative experiences are however not caused by the enzymes chosen for these feasibility studies. Gotoh et al report similar experience with an ATP modified ISFET [24]. Hanazato et al describe an ENFET for neutral lipid determination [25,26]. Here photosensitive PVP is used for the membrane and the authors claim that the composition of this material is much easier to optimize than that of the previously used PVA.

COMMENTS ON REPORTED RESULTS

This paper started with an inventory of the problems that are related to the fabrication and application of pH sensor based enzyme electrodes. When the reported ENFET results as reviewed above are analysed, it can be concluded that so far the most important improvements from the use of ISFET's are in the methods of fabrication. ENFET's have the potential of being mass fabricated and in the light of the number of papers that arise from industrial research laboratories, it is to be expected that these devices will become commercially availble in the not to distant future.

Up to this moment, ENFET's are completely analogue to sensors based on pH glass electrodes, albeit fabricated in a more sophisticated manner. Problems with regard to the pH and buffer dependence of the response have not been solved. However, a number of ENFET-papers have made meaningful contributions to the theoretical understanding of these devices. It is important to recognize that enzyme sensors actually consist of two distinctive functional parts as was already shown in figure 2. First there is the immobilized enzyme membrane that takes care of the selective recognition of the substrate. In this selector part of the biosensor, protolytes are generated through the catalytic conversion of substrate and thus a pH change is created in the membrane. The second part of the biosensor is the transducer part, i.e., the FET that converts the selectively induced pH change into an electrical signal. Through this bipartition it is very easy to see that the adjective "Nernstian" that is used by some authors to describe the response of enzyme sensors has no meaning. The term Nernstian should be reserved to describe the potential that arises from the thermodynamical equilibrium condition that exists in an electrochemical cell. The response from enzyme electrodes may in some cases be equal to a change of 1 pH per decade of substrate concentration but this is only so under very specific conditions.

The response of ENFET's depends on the pH and buffer capacity of the sample and thus for a practical application, these variables have to be rigidly controlled. So far, this problem has not been brought any nearer to a solution by the application of an ISFET instead of a pH glass electrode and thus the general applicability of ENFET's is still limited. A further improvement in the operation of enzyme sensors has to be found in a compensation method for the buffer capacity and in stabilizing the operation of the enzyme in the membrane. In the next section we will present a possible solution for this problem.

COMPENSATION METHODS FOR THE BUFFER CAPACITY OF THE SAMPLE

The sensitivity of pH based enzyme sensors is strongly dependent on the buffer capacity and pH of the sample solution. Furthermore, the response is non-linear for two reasons. First, the buffer capacity of a solution is pH dependent. A second, and perhaps even more important, cause for the non-linear response are the pH dependent enzyme kinetics. As a consequence of this pH dependent activity, the dynamic range of enzyme electrodes is limited as was already described by Caras and Janata in their first ENFET paper [5]. In low capacity buffers, the sensitivity is high but the upper limit of the linear range is restricted. For buffers with a high capacity the opposite is true. For practical use of enzyme electrodes it is thus required that pH and buffer capacity of the solutions used for calibration and of the sample solutions are equal.

As a potential solution for the problem of buffer dependency we have investigated the possibilities of a coulometric pH control system. This system consists of an ISFET with an integrated noble-metal electrode around its pH-sensitive gate. This electrode can either be used as an anode or cathode to produce hydroxyl ions or protons through the electrolysis of water. Through the generation of these ions, pH changes are created in the vicinity of the ISFET, changes that can be used to determine the buffer capacity. Previously, we have shown the practical value of such a system for rapid acid-base titrations [27] and in a new type of carbon dioxide sensor with an excellent long-term stability [28]. The application of this technique for the improvement of enzyme electrodes has also been proposed by Chandler and Eddowes [29].

In principle, there is a great similarity between an enzyme-FET and an ISFET with a pH-actuator electrode. In both cases, local pH changes are created, either through an enzymatic reaction or through the conversion of electric charge into an equivalent amount of ions. The electroanalytical technique that uses the equivalence between an electrical charge (in Coulombs) and an amount of ions is called coulometry.

To create an enzyme electrode that is independent of the buffer capacity of the sample solution, we have integrated the ENFET with a pH-actuator electrode. In this configuration the coulometric system can be used to control the pH inside the immobilized enzyme layer. The sensor is operated in a control system in which the output of the ENFET is compared with that of a reference-ISFET. The reference-FET is continuously measuring the pH in the bulk of the sample solution. When a substrate is offered and the pH in the enzyme layer is increasing, this is automatically compensated through the generation of H^+ ions at the actuator electrode. Thus, the pH in the enzyme layer is kept equal to that in the bulk of the sample solution. The enzyme operates at a constant pH and therefore we propose to call this sensor a <u>chemostatic enzyme sensor</u>.

Experimental

To investigate the possibilities of the coulometric buffer compensation system we have made use of urea-sensitve ISFET's. The enzymatic membrane consists of a cross-linked mixture of albumine and urease and was prepared according to the method described by Kimura et al [17-19].

Of course, these ENFET's can be operated in the usual way and their characteristics are very similar to those described by the previously reviewed authors. 90% response times are in the order of 10-20 seconds. The response of these sensors is, of course, also dependent on the buffer capacity of the sample solution as is clearly demonstrated in figure 3.

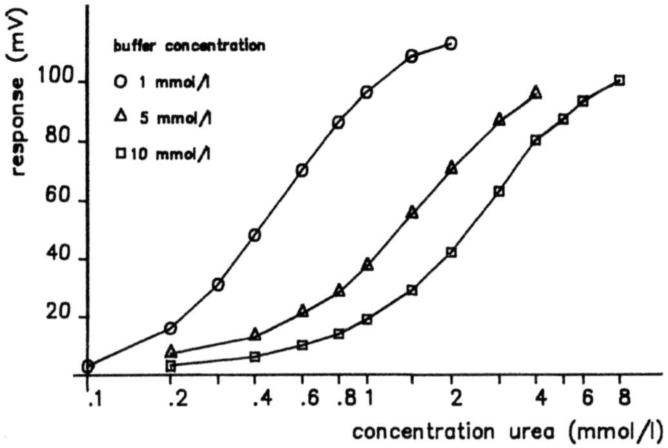

Figure 3. The sensitivity and dynamic range of the urea-ENFET depend on the buffer capacity of the sample.

The output of the chemostatic enzyme sensor is buffer independent. Because the feedback takes place directly in the enzyme layer, changes in the buffer capacity by the products of the enzymatic reaction are also automatically compensated. Figure 4 shows the relationship between generating current and substrate concentration for two different buffer solutions. A second important advantage of this system is that the enzyme, because it operates at a constant pH, has a constant activity. This means that the dynamic range is expanded as can be seen by comparing the responses in a 1 mmol/l buffer in figures 3 and 4.

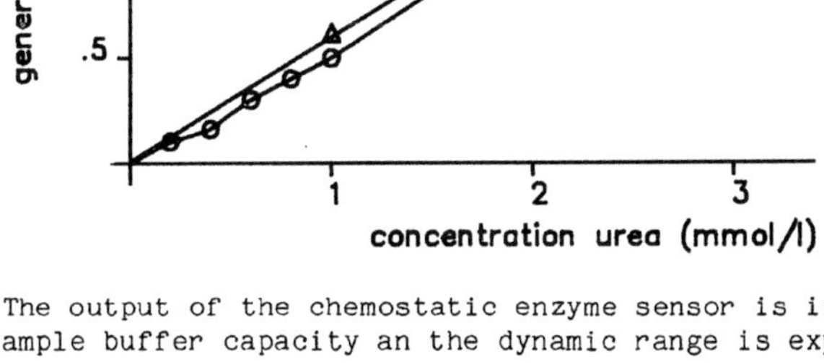

Figure 4. The output of the chemostatic enzyme sensor is independent from the sample buffer capacity an the dynamic range is expanded.

CONCLUSION

The integration of an enzymatic membrane and a pH-actuator offers the unique possibility to control the internal membrane pH. Of course, the method is not limited to keeping the membrane pH equal to that of the sample solution. As we have seen in a previous section, each enzyme has its own optimal pH range. If the sample pH deviates from this optimal pH, by this coulometric method it should also be possible to shift the pH in the membrane to a more favourable value. In this way it should be possible to decrease the lower detection limit of the electrode and to improve the stability of the control system.

The first results with the chemostatic enzyme sensor are very promising. Through the application of a direct feedback the sensor becomes independent of the buffer capacity of the sample solution, the response is strongly linear and the dynamic range can be expanded. The general applicability of pH-based enzyme sensors will surely be enlarged by this new method of measurement.

Part of this work has been financed in the Dutch Biosensor Stimulation Programme by the Centre for Micro-Electronics Twente.

REFERENCES

1 G. G Guilbault, in P. W. Chueng (ed.), Theory, Design and Biomedical Applications of Solid-state Chemical Sensors, CRC Press, 1978, pp 193-205.

2 R. G. Bates, Determination of pH, Theory and Practice, 2nd Ed., Wiley, New York, 1973, Ch. 11.

3 A. Sibbald, Recent advances in field-effect chemical sensors, J. Molecular Electronics, 2 (1986) 51-83.

4 J. Janata and S. D. Moss, Chemically sensitive field effect transistors, Biomed. Eng., July (1986) 241-245.

5 S. Caras and J. Janata, Field effect transistor sensitive to penicillin, Anal. Chem., 52 (1980) 1935-1937.

6 Y. Miyahara, F. Matsu, T. Moriizumi, H. Matsuoka, I. Karube and S. Suzuki, Micro enzyme sensors using semiconductor and enzyme-immobilization techniques, Proc. International Meeting on Chemical Sensors, Fukuoka, Japan, September 19-22, 1983, Analytical Chemistry Symposia Series, Vol. 17, Elsevier, (1983) 501-506.

7 Y. Hanazato and S. Shiono, Bioelectrode using two hydrogen ion sensitive field effect transistors and a platinum wire pseudo reference electrode, Proc. International Meeting on Chemical Sensors, Fukuoka, Japan, September 19-22, 1983, Analytical Chemistry Symposia Series, Vol. 17, Elsevier, (1983) 513-518.

8 S. D. Caras and J. Janata, pH-based enzyme field effect transistors, Proc. 3rd International Conference on Solid-State Sensors and Actuators (Transducers '85), Philadelphia, PA, U.S.A., June 11-14,1985, (1985) 158-161.

9 S. D. Caras, J. Janata, D. Saupe and K. Schmidt, pH-based enzyme potentiometric sensors. Part 1. Theory, Anal. Chem., 57 (1985) 1917-1920.

10 S. D. Caras, D. Petelenz and J. Janata, pH-based enzyme potentiometric sensors. Part 2. Glucose-sensitive field effect transistor, Anal. Chem., 57 (1985) 1920-1923.

11 S. D. Caras and J. Janata, pH-based enzyme potentiometric sensors. Part 3. Penicillin-sensitive field effect transistor, Anal. Chem., 57 (1985) 1923-1925.

12 M. J. Eddowes, Response of an enzyme-modified pH-sensitive ion selective device; consideration of the influence of the buffering capacity of the analyte solution, Sensors and Actuators, 7 (1985) 97-115.

13 M. J. Eddowes, D. G. Pedley and B. C. Webb, Response of an enzyme-modified pH-sensitive ion-selective device; experimental study of a glucose oxidase-modified ion-sensitive field effect transistorin buffered and unbuffered aqueous solution, Sensors and Actuators, 7 (1986) 233-244.

14 M. J. Eddowes, D. G. Pedley and B. C. Webb, Corrigendum, Sensors and Actuators, 8 (1985) 179-180.

15 Y. Miyahara, T. Moriizumi and K. Ichimura, Integrated enzyme FETs for simultaneous detections of urea and glucose, Sensors and Actuators, 7 (1985) 1-10.

16 T. Moriizumi and Y. Miyahara, Monolithic multi-function ENFET biosensors, Proc. 3rd International Conference on Solid-State Sensors and Actuators (Transducers '85), Philadelphia, PA, U.S.A., June 11-14,1985, (1985) 148-151.

17 J. Kimura, T. Kuriyama and Y. Kawana, An integrated SOS/FET multibiosensor and its application to medical use, Proc. 3rd International Conference on Solid-State Sensors and Actuators (Transducers '85), Philadelphia, PA, U.S.A., June 11-14,1985, (1985) 152-155.

18 T. Kuriyama, J. Kimura and Y. Kawana, Development of biosensors with immobilized enzyme, Chemical Economy & Engineering Review, 17 (1985) 22-27.

19 J. Kimura, T. Kuriyama and Y. Kawana, An integrated SOS/FET Multi biosensor, Sensors and Actuators, 9 (1986) 373-387.

20 T. Kuriyama, S. Nakamoto, Y. Kawana and J. Kimura, New fabrication methods of enzyme immobilized membrane for ENFET, Proc. 2nd International Meeting on Chemical Sensors, Bordeaux, France, July 7-10, 1986, (1986) 568-571.

21 Y. Hanazato, M. Nakako and S. Shiono, Multi-enzyme electrode using hydrogen-ion-sensitive field-effect transistors, IEEE Trans. Electron Devives, ED-33 (1986) 47-51.

22 J. Anzai, Y. Okhi, T. Osa, H. Nakajima and T. Matsuo, Urea sensor based on an ion sensitive field effect transistor. II. Effects of buffer concentration and pH on the potentiometric response, Chem. Pharm. Bull., 33 (1985) 2556-2559.

23 I. Karube, E. Tamiya, J. M. Dicks and M. Gotoh, A microsensor for urea based on an ion-selective field effect transistor, Anal. Chim. Acta, 185 (1986) 195-200.

24 M. Gotoh, E. Tamiya, I. Karube and Y. Kagawa, A microsensor for adenosine-5'-triphosphate pH-sensitive field effect transistor, Anal. Chim. Acta, 187 (1986) 287-291.
25 Y. Hanazato, M. Nakako, M. Maeda and S. Shiono, Application of water-soluble photocrosslinkable polymer to enzyme membrane for FET-biosensor, Proc. 2nd International Meeting on Chemical Sensors, Bordeaux, France, July 7-10, 1986, (1986) 576-579.
26 M. Nakako, Y. Hanazato, M. Maeda and S. Shiono, Neutral lipid enzyme sensors based on ion-sensitive field effect transistor, Anal. Chim. Acta, 185 (1986) 179-185.
27 B. H. van der Schoot and P. Bergveld, An ISFET based microliter titratrator: integration of a chemical sensor-actuator system, Sensors and Actuators, 8 (1985) 11-22.
28 B. H. van der Schoot and P. Bergveld, The coulometric carbon dioxide sensor, a new type of chemical transducer, Proc. 2nd International Meeting on Chemical Sensors, Bordeaux, France, July 7-10, 1986, (1986) 665-668
29 G. K. Chandler and M. J. Eddowes, Enzyme-mediated pH sensitive FET devices; problems of the non linear response, Proc. 2nd International Meeting on Chemical Sensors, Bordeaux, France, July 7-10, 1986, (1986) 531-533.

MICRO-BIOSENSOR

Isao Karube and Izumi Kubo
Research Laboratory of Resources Utilization
Tokyo Institute of Technology
Nagatsuta, Midoriku, Yokohama
227 JAPAN

ABSTRACT. ISFET is utilized as a pH sensitive transducer, so that the enzyme or micro-organisms, which catalyze the reaction that involves pH changes, were coupled with the ISFET to construct a micro-biosensor. A micro-electrode, such as, an O_2-electrode or H_2O_2 electrode based on amperometric determination, was developed. Therefore, combining with the oxidase, a micro-biosensor based on an amperometric determination was constructed.

In this chapter, micro-biosensors for urea, ATP (adenosine-5'-triphosphate), ethanol, glucose, and glutamate constructed from micro-transducers and immobilized enzyme thin membranes, are detailed and their characteristics are discussed.

1. INTRODUCTION

Methods for the selective determination of organic compounds in materials, such as biological fluids or waste water, are very important. Most analyses of organic compounds can be performed by spectrophotometric methods based on specific enzyme-catalyzed reactions. However, these methods often require a long reaction time and complicated procedures. On the other hand, electrochemical sensors employing immobilized biocatalysts have definite advantages. Namely, an enzyme sensor possesses excellent selectivity for biological substrates and can directly determine single compounds in a complicated mixture without need for a prior separation step [1,2].

The application of the biosensor for clinical analyses, has been extended to bedside monitoring or to direct implanting into an organ. This has been achieved with the use of an ISFET or a micro-electrode as a transducer, to miniaturize the sensor. These micro-devices are inexpensive because they are manufactured by the silicon fabrication technology.

G. G. Guilbault and M. Mascini (eds.),
Analytical Uses of Immobilized Biological Compounds for Detection, Medical and Industrial Uses, 207–218.

The miniatualized biosensor has been developed by combining immobilized enzymes or micro-organisms with the micro devices.

2. MICROBIOSENSORS BASED ON ISFET

The ISFET was first demonstrated by Bergveld [3], and used a silicon dioxide layer to impart sensitivity to hydrogen ions upon an insulated gate field effect transistor (IGFET). The principle of the ISFET is similar to the potentiometric electrode. The determination of analyte is performed by the measurement of the gate potential change caused by ion absorption. Matsuo and Wise [4] improved the ISFET properties by utilizing silicon nitride (Si_3N_4) as the gate insulator, reporting its use as a pH sensor. In 1980, Caras and Janata demonstrated that an immobilized pencillinase layer over the gate insulator of the ISFET could be used as a pencillin sensor [5]. We have also reported an enzyme-FET sensor [6].

Fabrication of the ISFET uses basically the same procedures employed for the metal-insulator-semiconductor FET(MISFET), as reported by Matsuo and Esashi [7]. The structure of the ISFET is shown in Fig. 1.

The gate insulator of the ISFET is composed of two layers; the lower is thermally grown silicon dioxide (SiO_2), the upper being silicon nitride (Si_3N_4), which is sensitive to H^+ ions and also has a barrier effect on ion penetration. The thickness of the SiO_2 and Si_3N_4 layers are approximately 0.1 μm. The gate of the ISFET is modified with immobilized biocatalysts membrane to produce highly selective micro-biosensors.

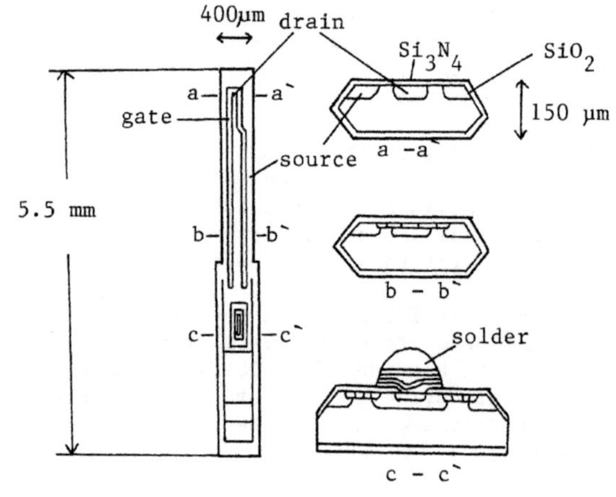

Figure 1. Structure of ISFET used in this study.

2.1. Micro-urea sensor

The assay of urea in biological fluids is a very important diagnositic test to evaluate kidney function. Besides, the qualitative analysis of urea is necessary in the well-planned artificial dialysis of blood. Conventional assay methods for urea are based on spectrophotometry, but involve complicated and delicate procedures. Therefore, the development of an inexpensive and miniaturized sensor that is highly selective and sensitive, yet easy to use, is extremely desirable. Realization of these goals can be achieved using the ISFET transducer.

The sensor system consists of two ISFETs; one ISFET is covered with an immobilized urease membrane and the other is covered with the same membrane, but without enzyme. The technology and materials for enzyme immobilization is almost similar to the conventional one. In the micro urea sensor reported by Miyahara et. al [6], urease was immobilized onto a cellulose triacetate membrane by reaction with glutaraldehyde and triamine 1,8-diamino-4-aminomethyloctane. The characteristics of the sensor were strongly dependent on the immobilized enzyme membrane and especially on the membrane thickness, which affected the response time because of its effect on diffusion. One of the major properties required for enzyme immobilization is good adhesion to the Si_3N_4 insulated gate surface.

Gotoh et al proposed the method utilizing polyvinylbutyral which is widely used as the photoresistive resin for integrated circuit fabrication [8].

The polyvinylbutyral membrane was spread onto the gate insulator of the ISFETs by a dropping method. Approximately 0.1 g of polyvinylbutyral resin and 1 ml of 1,8-diamino-4-aminomethyloctane were dissolved in 10 ml of dichloromethane. This polymer solution was dropped onto the gate insulator of the two ISFETs and was then immersed in a 5% glutaraldehyde solution at room temperature for approximately one day to advance the cross linking reaction. Urease was immobilization on the ENFET (enzyme FET) by immersing it in a 5 $mg.ml^{-1}$ urease solution at 4^oC for approximately one day.

Measurements of urea concentration were performed in a differential mode, by comparing the difference in gate output voltage of the urea-sensing gate and the reference gate. A schematic diagram of the circuit is shown in Fig. 2. An Ag/AgCl reference electrode was placed directly in solution with the ENFET and the REFFET (Reference FET), and a gate voltage was applied between the Ag/AgCl reference electrode and the source of the ENFET and the REFFET. A change in solution pH affects the gate insulator surface potential, with a concomitant proportional change in the gate output voltage. 100 μl aliquots of urea were injected into a solution of 5 mM Tris-HCl buffer at $37^oC \pm 1^oC$, and the differential gate output voltage change recorded for 10-20 min. The differential gate output voltage reached a steady state approximately two minutes after injection of urea. The initial rate of change of the differential gate output voltage after

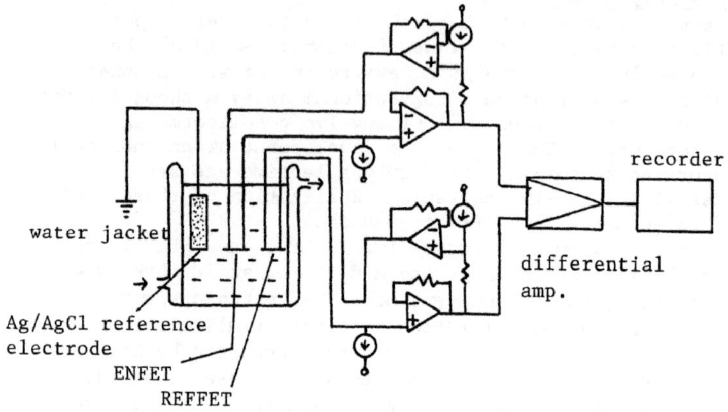

Figure 2. Circuit diagram of measuring system using ENFET. ENFET; enzyme FET, REFFET; referential FET.

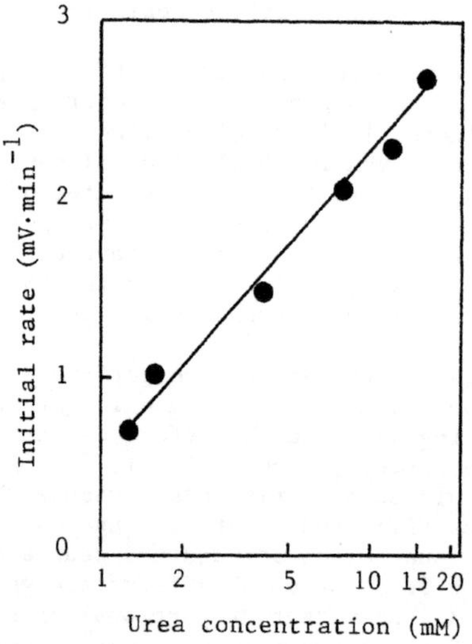

Figure 3. Calibration curve for the urea sensor.

injection was plotted aainst the logarithm of the urea concentration. Fig. 3 shows a calibration curve of the urea sensor system. A linear relationship was obtained between the initial rate of voltage change and the logarithm of urea concentration over the range 1.3 to 16.7 mM ura. An examination of the selectivity of the urea sensor system showed that it did not respond to 6.3 mM glucose, 10 mM creatinine, and 3.6 M albumin.

The stability of the urea sensor system was also examined. The ENFET was stored at 4°C between measurements and exhibited a response to 16.7 mM urea for at least two weeks.

2.2. Micro ATP sensor

The determination of ATP (adenosine triphosphate) is important in fermentation processes and for clinical analysis. Conventional methods of ATP assay are based on spectrophotometric and bioluminescence measurements. These methods, however, require such complicated and delicate procedures that a simple and more inexpensive assay is desirable.

H^+-ATPase(EC 3.6.1.3.) in cellular membranes catalyses the production or hydrolysis of ATP. Furthermore, the enzyme has many functions, such as proton transport, which could be utilized for a bio-molecular device. Several studies on the properties and function of the enzyme in biological membranes have been reported by Kaqawa and co-workers.

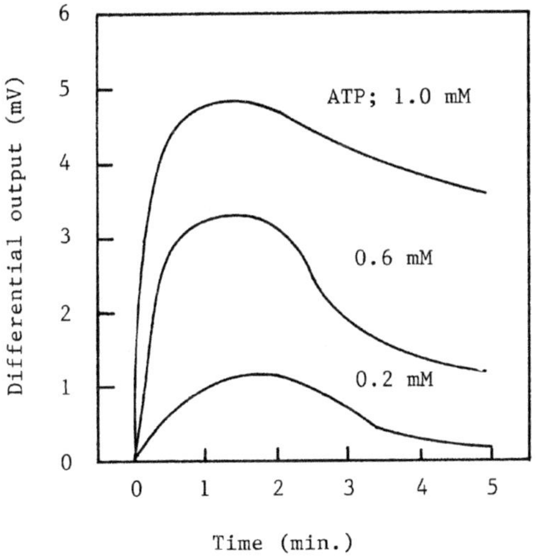

Figure 4. Sensor responce to ATP.
At time zero, ATP was injected.

The ATP sensor combining H^+-ATPase and ISFETs was described by Gotoh et al [9]. H^+-ATPase was prepared from thermophilic baterium PS3 and is classified as thermophilic $F_1(TF_1)$ ATP. The procedures employed in constructing the ATP sensor and measurements of gate voltage were identical to those of the urea senor, described above. For the measurement of ATP, 50 mM Tris malate buffer was used at $40^oC \pm 1^oC$.

After injection of the sample solution containing ATP, a rapid initial increase in the differential output was observed. The initial rate of change of the differential gate output voltage after injection of ATP was plotted against the logarithm of the ATP concentration. Fig. 4 shows a calibration curve of the ATP sensor system. A linear relationship was obtained between the initial rate of the voltage change and the logarithm of ATP concentration over the range 0.2 to 1.0 mM ATP. No detectable response was observed below 0.2 mM ATP, and above 1.0 mM the response did not increase appreciably.

Other chemicals (1 mM glucose, urea or creatinine) exhibited only very slight responses when the measurements were made in the differential mode. The reponse of the system to 1 mM ATP retained 90% of the initial value after 18 days.

2.3 Micro alcohol sensor

The determination of alcohol concentration is one of the most important and routinely performed tests in the field of fermentation processing, as well as in the clinical laboratory. For the on-line monitoring of processes, a small and selective ethanol sensor is important. Kitagawa et al developed a micro alcohol sensor consisting of immobilized acetic acid bacteria, a gas permeable membrane and an ISFET [10]. This sensor was utilized for the determination of ethanol.

Acetic acid bacteria have been widely used, to make vinegar, and ascorbic acid by a oxidative fermentation. It is in acetic acid fermentation, that both alcohol dehydrogenase (ADH) and aldehyde dehydrogenase (ALDH) are involved in the oxidation of ethanol to acetic acid via acetaldehyde (Fig. 5). When acetic acid bacteria are immobilized on an ISFET, ethanol is oxidized to acetic acid by those bacteria, causing a pH change to occur at the surface of an ISFET. Therefore, the ethanol concentration can be measured by combining an ISFET and acetic acid bacteria.

Acetobactor aceti IAM 1802 was used for the micro alcohol sensor. The microorganisms were immobilied on the gate surface of an ISFET, in calcium alginate gel. The ISFET-immobilized micro-organisms and wire-like Ag-AgCl electrode were placed in a small shell (6x3x22 mm), that had a gas permeable membrane (a porous Teflon membrane, 0.5 m pore size) fitted to the side. The inside of the shell was filled with inner buffer solution (5 mM tris-HCl buffer containing 0.1 M $CaCl_2$, pH 7.0).

Figure 6 shows a schematic diagram of the sensor system. The system consisted of a thermostatic circulating jacketed vessel (3 ml volume), the microbial-FET alcohol sensor, the circuit for

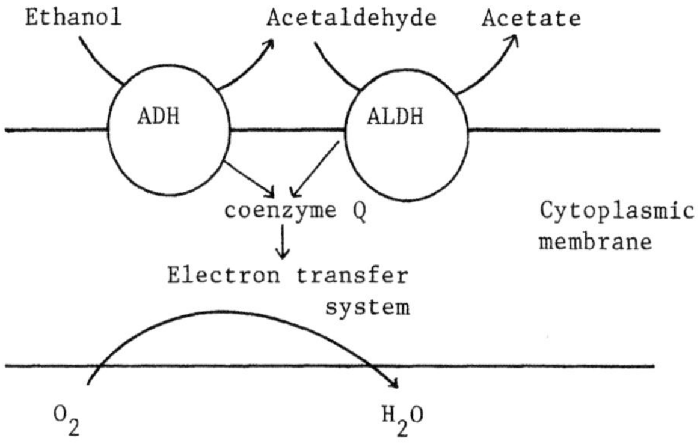

Figure 5. Proposed ethanol oxidation pathway in Acetobacter.

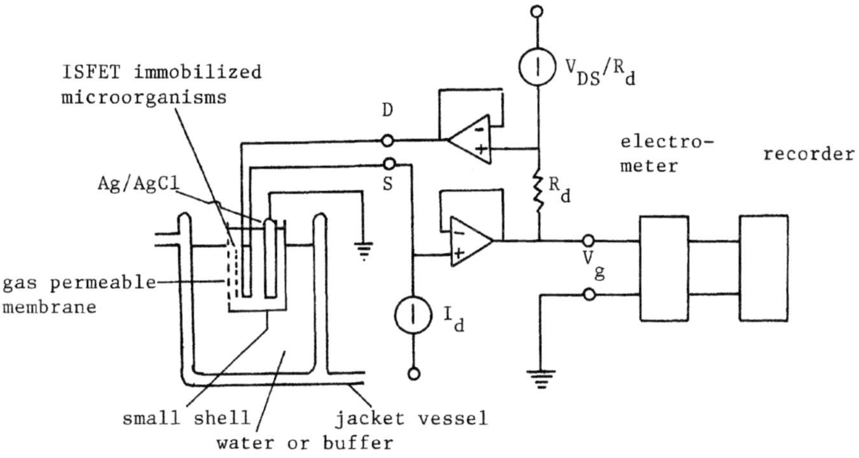

Figure.6 Schematic diagram of micro-alcohol sensor system.

measurement of the gate output voltage (Vg), an electrometer and a recorder.

The initial rate of change of gate voltage with time, dV_g/dt was plotted against the logarithmic value of the ethanol concentration. The minimum detectable response to ethanol was obtained at approximately 0.1 mM. A linear relationship was observed over the range between 3-70 mM.

The response of the micro-alcohol sensor was stable over a wide pH range (pH 2-12), however, at pH's lower than 6, the sensor responded to acetic acid. Volatile organic acids such as acetic acid do not dissociate in lower pH solutions, and hence penetrate through the gas permeable membrane. At pH's higher than 6, on the other hand, the volatile organic acids dissociate to the individual ions, and do not interfere with the measurement of ethanol by the sensor. The sensor was stable for 15 h, at 15°C. The sensor showed usefulness for the determination of ethanol, especially in terms of its good selectivity and potential for miniaturization, and it is expected to be integrated into a multi-functinal sensor capable of simultaneously determining substrates contained in a solution having a complex composition.

3. A MICRO BIOSENSOR BASED ON MICRO-ELECTRODE

3.1. Micro-glucose sensor

A glucose sensor based on a H_2O_2 electrode or an O_2 electrode has been practically used in diagnostic analyses, is rapid, simple, and possesses excellent selectivity. Therefore , its application to implantable sensing is desired. Glucose is easy to determine by amperometric devices but not potentiometirc devices such as ISFET. For this reason, a micro H_2O_2 sensor has been developed, utilizing integrated circuit technology.

The structure of the micro-H_2O_2 sensor is illustrated in Fig. 7. Two Au electrodes, 100 μm wide were deposited on the silicon nitride surface partially insulated by coating with Ta_2O_5. One of the electrodes is used as a cathod, the other is an anode. An oxidation potential, of 1.1V was applied to the anode vs. the cathode.

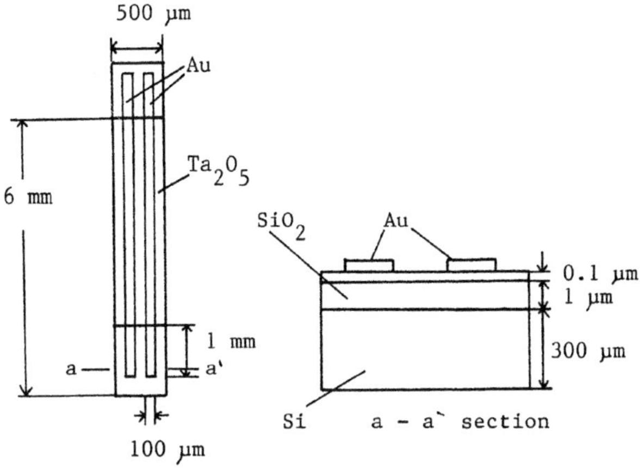

Figure 7. Schematic diagram of a micro-electrode.

When the sensor was placed in the sample solution containing H_2O_2, its output current immediately increased and reached a steady state within one minute. A linear relationship was observed between the H_2O_2 and the current. This electrode was then employed as the transducer in a micro-glucose sensor [11]. The procedure for glucose oxidase (GOD) immobilization onto the micro-electrode is as follows. Approximately 100 µl of γ-aminopropyl-triethoxysilane was vaporized at 80°C, 0.5 torr. for 30 min onto the electrode surface, followed by 100 µl of 50% glutaraldehyde vaporized under the same conditions. The modified micro-electrodes were then immersed in GOD solution containing BSA and glutaraldehyde, the GOD becoming chemically bound to the surface of the microelectrode by a Schiff linkage. Fig. 8 shows a typical response curve for the micro-glucose sensor. The output current increased after injection of a sample a solution, steady state being reached within 5 min.

A linear relationship was observed between the current increase (the difference between the initial and steady state currents) and glucose concentration in the range 0.1 to 10 $mg.dl^{-1}$ glucose. Examination of the selectivity of the glucose micro-sensor indicated no response to other compounds such as galactose, mannose, fructose and maltose. Therefore, the selectivity of this sensor for glucose is highly satisfactory. The optimum temperature

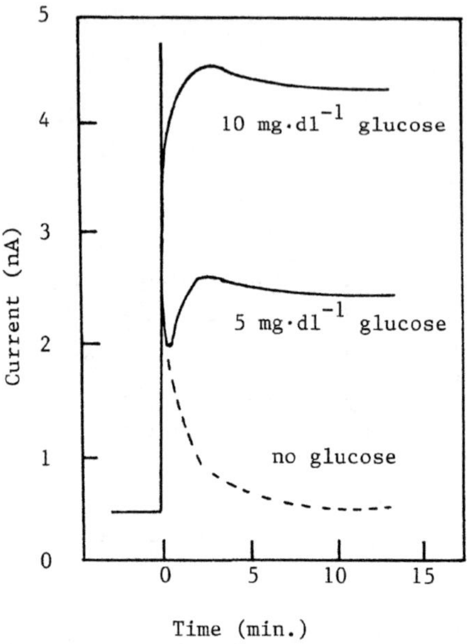

Figure 8. Time course of sensor responce to glucose. (pH 7.0, 37°C)

for the sensor was 55°C, but the stability of the sensor at this temperature is poor. Therefore, all experiments were performed at 37°C. Continuous operation of the sensor in 10 $mg.dl^{-1}$ glucose produced a constant output for more than 15 days and 150 assays. Therefore, this micro-glucose sensor possesses both selectivity and good stability, its potential for use as a micro-glucose sensor being very good.

3.2 Micro glutamate sensor based on a micro-O_2 electrode

The determination of L-glutamate(L-Glu) is very important in the food industry, because large amounts of L-Glu are produced by fermentation for use in food seasoning. Various glutamate sensors, consisting of immobilized enzyme and an electrochemical device, have been developed for the fermentation and food industries. Glutamate oxidase catalyses the oxidation of L-Glu, oxygen being consumed by the reaction. Therefore, an oxygen sensor can be employed as the transducer for a glutamate sensor. A micro-oxygen sensor was developed by modifying the micro-H_2O_2 electrode, and prepared by the method described in 3.1.

The structure of the micro-oxygen sensor is schematically shown in Fig. 9. A gas permeable teflon membrane covered a well filled with electrolyte (0.1 M KOH), and two micro-Au electrodes are placed at the bottom of the well. The characteristics of the micro- oxygen electrode were examined by cyclic voltammetry at various concentrations of dissolved oxygen (an oxygen and nitrogen mixture was passed through the sample solution). A peak current was observed due to the reduction of oxygen, when a voltage of approximately 1.1 V was applied to one of the Au electrodes vs. the other. A linear relationship was observed between the oxygen concentration and the peak current obtained from the cyclic voltammograms. These results indicate that the micro-oxygen electrode can be used for oxygen concentration determination. Therefore, the micro-oxygen electrode was employed as the transducer in a micro-glutamate sensor [11].

Glutamate oxidase was immobilized on a cellulose triacetate membrane containing glutaraldehyde and triamine (1,8-diamino-4-aminomethyloctane). The glutamate oxidase membrane was placed on the telfon membrane of a micro-oxygen sensor and covered with a nylon net. Application of a L-Glu sample solution to the sensor system produced a rapid drop in the current output to a steady state value, resulting from L-Glu oxidation.

When the current decrease at 5 min was used as the measure of activity, a linear relationship was observed between the current decrease and the L-Glu concentration in the range 5-50 mM. The optimum temperature for the sensor was approximately 40 ^{o}C, but gradual denaturation of the enzyme reduced the stability of the sensor. Therefore, all other experiments were performed at 30°C.

The selectivity of the sensor for L-Glu was found to be satisfactory and hence its application to fermentation process control and food analysis is very promising.

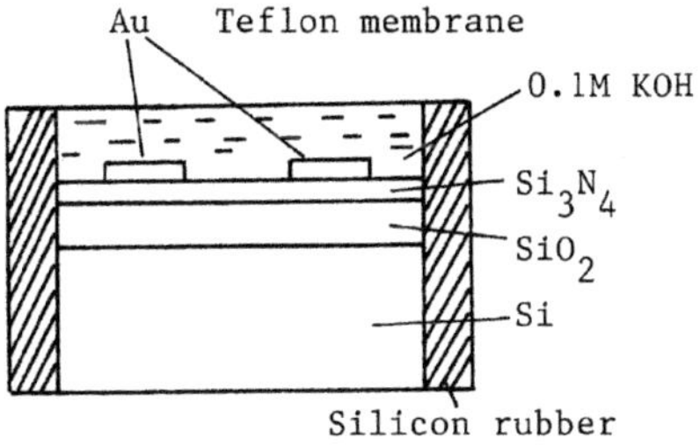

Figure 9. Schematic diagram of a micro-oxygen electrode.

4. REFERENCES

1. I. Karube and S. Suzuki. 'Amperometric and Potentiometric Determination with Immobilized Enzymes and Microorganism'. Ion Selective Electrode Review, 6, 15 (1984).
2. I. Karube and S. Suzuki, 'Immobilized Enyzmes for Clinical Analysis'. Enzyme and Immobilized Cells in Biotechnology (ed. A.I. Laskin), pp. 209 et seq., Benjamin/Cumming Publishing , London (1985).
3. P. Bergveld, 'Development of an Ion-sensitive Solid-state Device for Neurophysiological Measurements'. IEEE Trans. on BME, 17, 70 (1970).
4. T. Matsuo and K.D. Wise. 'An Integrated Field-effect ELectrode for Biopotential Recording'. IEEE Trans on BME. 21, 485 (1974).
5. S. Caras and J. Janata, 'Field Effect Transistor Sensitive to Pencillin'. Anal Chem. 52, 1935 (1980).
6. Y. Miyahara, F. Matsu, T. Moriizumi, H. Matsuoka, I. Karube and S. Suzuki Proc. Int. Meeting on Chem. Sensors, pp. 501-6, Kodansha, Tokyo, Elsevier, New York (1983).
7. T. Matsuo and M. Esashi. 'Methods of ISFET Fabrication'. Sensors and Actuators, 1, 77 (1981).
8. M. Gotoh, E. Tamiya and I. Karube. 'Polyvinylbutyral Resin Membrane for Enzyme Immobilization to an ISFT Microbiosensor'. J. Mol. Catalysis, 37, 133 (1986).
9. M. Gotoh, E. Tamiya, I. Karube and Y. Kagawa. 'A Microsensor for Adenosine-5'-triphosphate pH-sensitive Field Effect Transistor'. Anal. Chim. Acta., 187, 287 (1986).
10. Y. Kiagawa, E. Tamiya and I. Karube 'Microbial FET Alcohol Sensor'. Anal. Letters, 20, 81 (1987).
11. I. Karube, E. Tamiya, T. Murakami, M. Gotoh and Y. Kagawa. 'Microbiosensor Based on Silicon Fabrication Technology'. Enz. Eng., 8, 256 (1987).

THE DEVELOPMENT OF FIBRE-OPTIC SENSORS BY IMMOBILIZATION OF FLUORESCENT PROBES

Otto S. Wolfbeis
Analytical Division
Institute of Organic Chemistry
KF-University, 8010 Graz, Austria

Optical spectroscopy in combination with fibre-optics can offer a number of advantages over classical methods, in particular in the field of chemical and biomedical analysis. Since light is transported from the meter to the sample, fibre-optic sensors are most useful for remote and invasive sensing. Moreover, they are small and light, and provide a well-defined sampling area at the fibre end. One of the most attractive features, however, is the lack of reference cells, so that no problems are encountered as, for instance, in the case of electrochemical sensors with their rather sensitive liquid junctions.

Notwithstanding these advantages, interference by ambient light is a potential source of error. This disadvantage has to be overcome by either complete optical isolation of the sensor head, or by light modulation. Also, opto-electronic equipment is not ideal yet for sensing purposes. The lack of violet and ultraviolet LEDs, and the lack of inexpensive blue lasers is particularly annoying.

This contribution starts with a brief introduction into fibre-optic sensing, then a number of representative examples of sensors for fundamental analytes will be described, and finally a quite new approach for in-vivo determination of enzyme activities will be presented.

INTRODUCTION

Optical fibres are now produced in large quantities for use in telecommunication systems. They are available in high quality so that they can be used for optical spectroscopy. The path of light in an optical waveguide is shown in Fig. 1. Light is totally reflected only when (a) the index of refraction of the core is higher than that of the cladding and (b) light enters the light guide within the angle of acceptance. Both the cone angle and the angle of total reflection are governed by the refraction indices of core, clad, and sample.

This paper is part 17 in a series on Optical Sensors.

G. G. Guilbault and M. Mascini (eds.),
Analytical Uses of Immobilized Biological Compounds for Detection, Medical and Industrial Uses, 219–226.

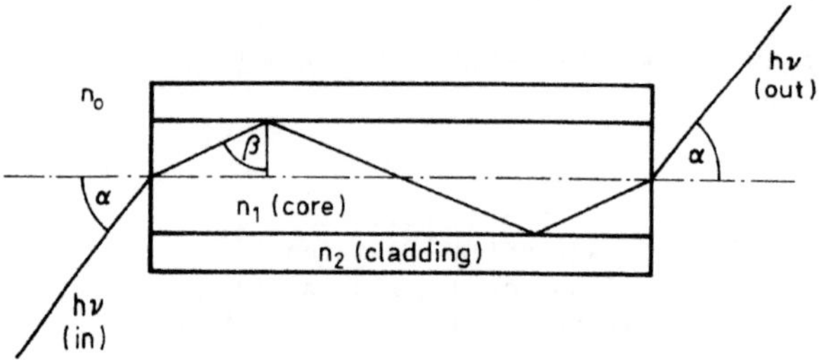

Fig. 1. Path of light in an optical waveguide. Angles and are determined by the indices of refraction of core, clad and sample.

An attractive feature of optical fibres is the flux of information that can be maintained in one fibre. For instance, blue and yellow light can be guided into one direction, and green and red light into the other. Moreover, light can be modulated by phase, intensity and, with certain limitations, by polarisation. Therefore, a number of different kinds of spectroscopy can be performed with one single fibre.

Fig. 2 shows how fluorescence is produced at the end of a fibre: Exciting light leaves the fibre, hits a sample to produce fluorescence, which then enters the same fibre and is guided back to an optical system that separates scattered light from fluorescence. The input beam may be considered as the question, and fluorescence is the encoded answer. The principle may, of course, also be applied to absorption, reflectance, or Raman spectrometry.

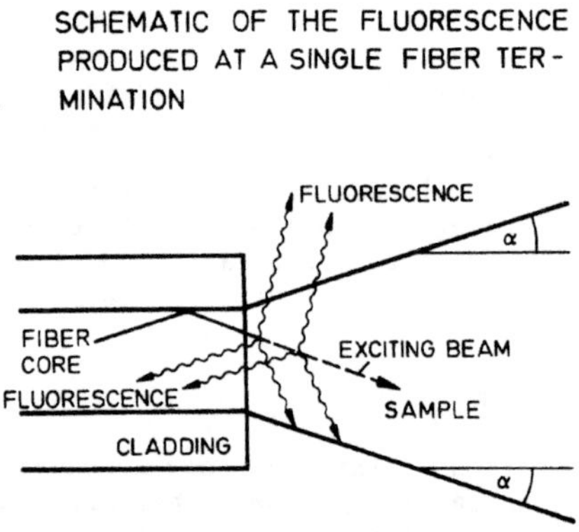

Fig. 2

Measurement of the intrinsic colour or fluorescence of analytes via fibres has been utilised for continous monitoring of groundwater contaminants such as uranium ions (1) and organic fluorescent pollutants (2), for drug assay in serum (3), and for continous monitoring of blood oxygen saturation (4), However, this approach is limited to analytes that have an intrinsic colour or fluorescence, and it is not a very specific method since any other species having similar spectral properties is likely to interfere.

SENSOR DEVELOPMENT

In 1979 we started a programme to develop optical sensors for the five chemical parameters that were considered to be of greatest importance in analytical biosciences: pH, oxygen, carbon dioxide, ammonia and hydrogen peroxide. All these are colourless and non-fluorescent, so that it became necessary to provide the light guide with an appropriate indicator chemistry that responds with high specificity to the analyte of interest. It was also decided to measure all parameters by fluorescence, because fluorimetry is selective, sensitive, and versatile. Versatility is a particularly promising aspect since fluorescence spectrometry not only allows measurements of light intensities, but also of lifetime, polarisation, energy transfer efficiency, and even combinations thereof.

The first sensor developed was for pH (5). It consists of a dye covalently immobilized onto a glass support. An indicator had to be found that has longwave absorption and emission maxima because most fibres have poor transmittance at wavelengths below 420 nm, and there are no in-expensive light sources available with high intensity at below 400 nm. Problems were encountered with the shift in pKa after immobilisation (the ideal pKa for blood pH measurement is, of course, 7.3 +/- 0.2), and the ionic-strength sensitivity of all optical indicators. These problems were overcome by proper chemical modification of both micro-environment and dye.

Oxygen is almost exclusively determined via its ability to quench the fluorescence of dyes such as polycyclic aromatic hydrocarbons (6). Again, longwave-absorbing dyes had to found and to be immobilized in order to prevent leaching. Unfortunately, all PAHs suffer a considerable decrease in quenching efficiency when immobilized, so that the slope of the response curve becomes smaller. Oxygen-sensitive dyes have been covalently immobilized (7), adsorbed on polymer beads (8), and also simply dissolved in silicone after being made completely hydrophobic (9) so that it is not at all washed out by the aqueous sample. Problems associated with oxygen sensors are mainly due to the suspected toxicity of the dyes, their photolability, and their tendency to crystallize after a while so that the signal drifts away. Also, all PAH-based oxygen sensors suffer from interferences by halothane which acts as a quencher too.

The observation of quenching by halothane prompted us to develop a fibre-sensor for this species (9). Since a specific indicator was not found, a two-sensor technique had to be applied, with one sensor being made specific for oxygen, and the other responding to both oxygen and halothane. The double sensor technique allows the determination of halothane and/or oxygen over the range of clinical interest. The method is, of course, applicable to other situations as well, when no specific indicator is found.

Carbon dioxide is usually measured by monitoring the change in the pH of a buffer solution when exposed to CO_2. In the optical approach, the change in pH is monitored either by absorption (10) or fluorescence (11,12). In one of the methods, an emulsion of an aqueous buffer solution containing a pH-sensitive dye is attached to the end of a fibre-optic system. When exposed to various tensions of CO_2, the pH is lowered, which is indicated by a change in the fluorescence of the dye, which in turn is seen by the fibre. A similar principle (with the pH change going into the other direction) was applied to continuously sense ammonia (13).

Hydrogen peroxide (HP) is of considerable interest in biomedical analysis because it is produced during the enzymatic action of all kinds of oxidases. Consequently, a fibre-optic sensor for HP is of potential utility for a system that transduces reactions otherwise difficult to follow into a measurable quantity. Our HP sensor is based on the decompositionn of HP by a catalyst such as the enzyme catalase or a metal. When HP is decomposed, an increase in oxygen partial pressure is found, that is indicated by an oxygen-sensitive dye. Two sensors are needed. The one measures the actual oxygen pressure, and the other the increase in oxygen pressure as a result of the decomposition of HP.

Chloride, bromide and iodide (in increasing order) rather specifically quench the fluorescence of quaternized heterocycles, a finding that has been exploited for the development of a fibre-optic halide sensor (14). Potassium (and probably other cations as well) can be determined in the milli-molar range using on ion-selective optode based on the fluorimetric determination of an electric potential created at a water-lipid interface such as a LB-bilayer (15). A schematic of the sensing layer of an ion-selective optode (IOS) is shown in Fig. 3: Valinomycin acts as an ion carrier that creates a potential at the lipid-water interface, arachidic acid is used for the fabrication of a double-layer, and rhodamine B octadecyl ester is a potentiometric dye that responds to changes in the potential at the interface.

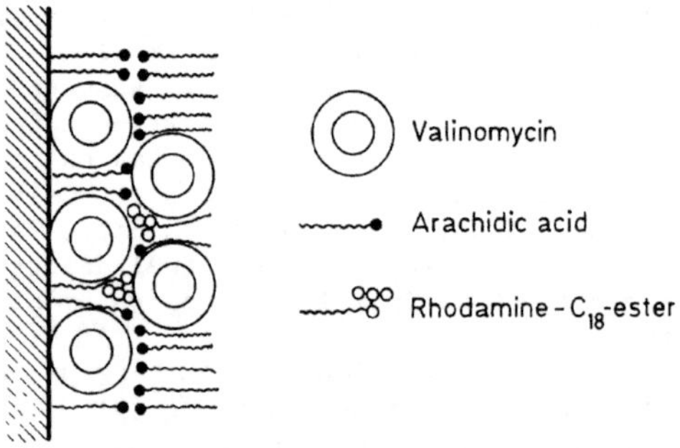

Fig. 3. Schematic of the LB layer used for fluorimetric determination of potassium

ENZYME ACTIVITY PROBES

Following this more general description of our activities in the field of optical sensors since 1979, I would like to present now in some more detail an interesting new approach for the application of fibre-optic sensors that is likely to be of interest for in-vivo sensing purposes.

The knowledge of enzyme concentration of a sample solution is of considerable practical interest in clinical analysis, and numerous methods have therefore been proposed for assays (16). Most of these are kinetic, based on the measurement of changes in the concentration of a natural or synthetic substrate. Assays have mainly been performed in solution or on "dry reagent" phases . None of them, however, is suitable for in-vivo application.

Arnold (17) has described the immobilization of the enzyme, alkaline phosphatase (together with light-scattering material) at the end of a fibre optic. When immersed into a substrate solution such as p-nitrophenyl phosphate, the second fibre sees the formation of yellow p-nitrophenolate.

Our interest in photometric and fluorimetric enzyme activity determination (18, 19) resulted in the idea to determine enzyme activities via fibre optics in very small sample volumes (20) with the aim to develop a method for an in-vivo assay in blood and related biomatter. In contrast to the method of Arnold (17), the substrate is immobilized in this case, and the formation of the hydrolysis product, which remains in immobilized form, is monitored.

A schematic of the experimental setup is shown in Fig. 4. Light from a xenon light source (L) of an Aminco SPF 500 fluorimeter was focused into the end of a 100 um single fibre). After passing the optical coupler OC (where fibres F_1 and F_2 are combined to one single fibre F), the exciting light of wavelength 465 nm was guided to a membrane with an immobilized enzyme substrate on it. The membrane was pulled over the ferrule of the fibre end (o.d. 20 mm) and fixed with an O-ring or epoxy.

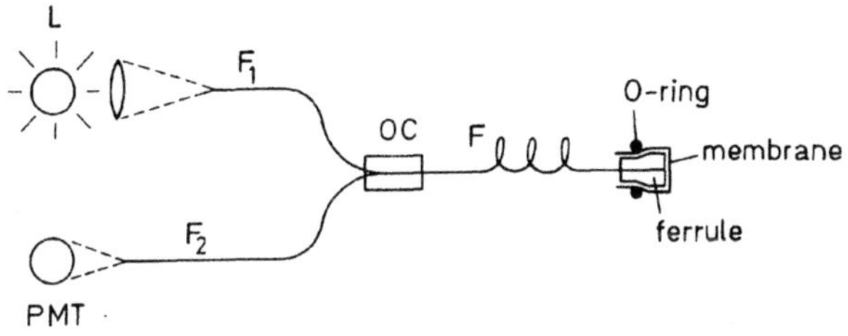

Fig. 4: Experimental set-up for remote determination of enzyme activities.

Green fluorescence of the membrane, which is almost zero in the absence of an esterase, and scattered blue light return through the same fibre. 50% of the total returning light are coupled into fibre F_2 in OC. After passing a secondary filter adjusted to 530 nm, fluorescence is detected with a photomultiplier and its intensity displayed in the SPF 500 instrument.

The membrane with immobilized substrates for cholinesterase and related carboxylesterases was prepared by electrostatic immobilization of 1-hydroxypyrene-3,6,8-trisulfonate (HPTS) and subsequent acetylation with acetic acid anhydride. The resulting HPTS acetate was shown to be a useful substrate for a direct and kinetic assay of esterases (18).

Enzymatic action results in the cleavage of the substrate to give immobilized HPTS which, because of its low pKa of 7.0, in phosphate buffer of pH 7.2 is dissociated by more than 50%. Both free HPTS and its phenolate fluoresce at 520 nm with quantum yields close to unity.

Fig. 5 shows how fluorescence increases when the tip of a fibre with the HPTS acetate membrane is immersed into solutions containing various concentrations of carboxylic ester hydrolase. A fresh piece of membrane, cut from a 2x2 cm membrane prepared as described before, was used in each experiment. The curves are typical for an enzymatic reaction in showing a rapid increase in signal per time in the inital phase and a slow-down of the signal change as the substrate gets more and more consumed. The initial slope varies linearly with enzyme concentration.

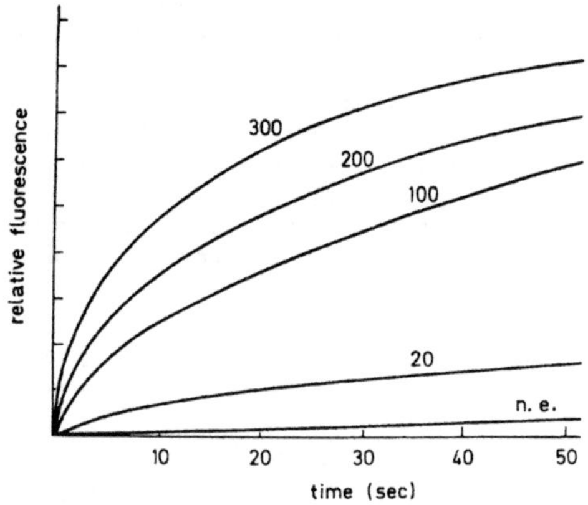

Fig. 5: Increase in fluorescence intensity with time when immobilized HPTS acetate is immersed into a pH 7.2 solution containing various amounts of carboxylesterase. The figures refer to microgram enzyme per ml solution, and n.e. is the increase in signal due to non-enzymatic hydrolysis.

The sensivity of the assay is limited by the rates of the enzymatic and the non-enzymatic reaction. It can be seen from Fig. 5 that the limit of detection for carboxylesterase is at about 0.02 mg protein per mL. The results show the feasibility of a fibre-optic determination of enzyme activities. The method is considered to have two decisive advantages over former optical methods: It allows a determination over distances of metres and in a rather small sample volume which, in principle, is limited only by the size of the fibre. It therefore provides a suitable means for invasive detection and determination of enzyme activities. Since assays can be performed in-vivo, real-time probing becomes possible.

The fluorescent dye which, conceivably, may be physiologically harmful, remains in the immobilized state so that it does not present a risk for in-vivo sensing. The acid part, in contrast, is released, but this is the "natural" part of the synthetic substrate, for instance a fatty acid anion, phosphate, sulfate, or a sugar molecule such as glucose.

Because the substrate is irreversibly consumed, the probe has a limited lifetime. Consequently, it does not lend itself to continous monitoring, which is in contrast to known invasive sensors for pH and oxygen (21). Therefore, this kind of device cannot be called a true sensor which, by definition, is able to indicate continuously and reversibly a chemical or physical parameter.

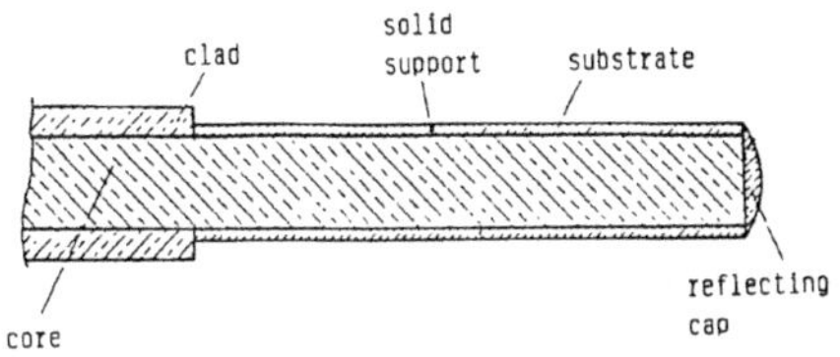

Fig. 6. Measurement of enzyme activities by the evanescent wave technique.

Various arrangements can be used for enzyme sensing. Figs. 6, 7 and 8 show alternatives: In Fig. 6 the clad of the light guide is removed and replaced by the enzyme-sensitive layer, whose fluorescence is measured by the evanescent wave technique. In Fig. 7, an enzyme-permeable membrane protects the substrate, which is immobilised to a high-molecular-weight polymer that cannot leave the membrane.

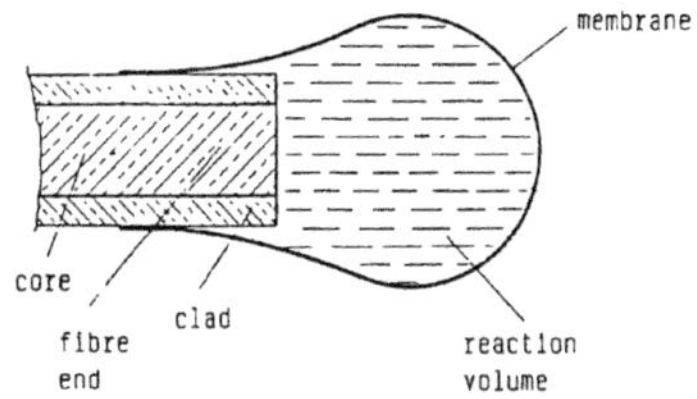

Fig. 7. Measurement of enzyme activities by the balloon method, with the substrate immobilised to a large polymer inside the balloon.

In Fig. 8, finally, the substrate is immobilised to the wall of a cyclinder. As long as the substrate is immobilised, i.e., as long as there is no enzyme present, it cannot be seen by the fibre with its angle of acceptance 2 alpha. If, however, the fluorophore is released from the wall by enzymatic action, it diffuses into the reaction volume and then will be seen by the fibre, resulting in an increase in fluorescence which is proportional to the actual enzyme activity.

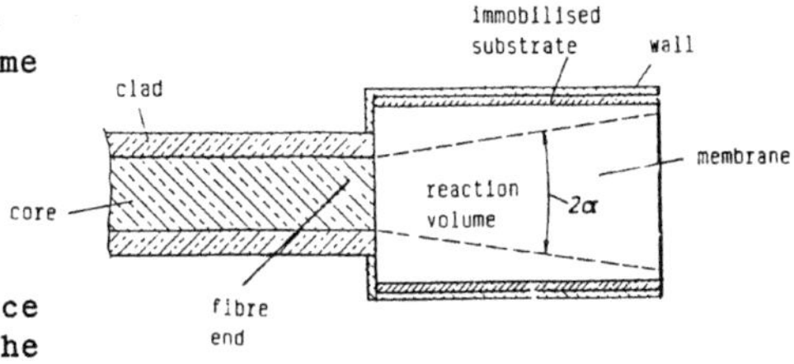

Fig. 8. Determination of enzyme activities exploiting the limited angle of acceptance of a fibre core. As long as the substrate is immobilised, it cannot by seen by the fibre.

REFERENCES

(1) T. Hirschfeld, T. Deaton, F. Milanovich and S. Klainer, The Feasibility of Using Fiber Optics for Monitoring Groundwater Contaminants, EPA Report AD-89-F-2A 074 (1983).
(2) W.A. Chudyk, M.M. Carraba and J.E. Kenny, Anal. Chem. 1985,57, 1237.
(3) B. J. Tromberg, J. F. Eastham and M. J. Sepaniak, Appl. Spectrosc. 1984, 38, 38.
(4) J. M. Schmitt, J. D. Meindl and F. G. Mihm, IEEE Trans. Biomed. Eng. 1986, 33, 98.
(5) H. Offenbacher, O. S. Wolfbeis and E. Fürlinger, Sensors & Actuators 1986, 9, 73.
(6) D. W. Lübbers and N. Opitz, Sensors and Actuators 1983, 4, 641; and refs. cited.
(7) O. S. Wolfbeis, H. Offenbacher, H. W. Kroneis and H. Marsoner, Mikrochim. Acta 1984, I, 153.
(8) J. I. Peterson, R. U. Fitzgerald and D. K. Buckhold, Anal. Chem. 1984, 56, 427.
(9) O. S. Wolfbeis, H. Posch and H. Kroneis, Anal. Chem. 1985, 57, 2556.
(10) US Pat. Appl. 470.920 (22. July 1983).
(11) Z. Zhujun and W. R. Seitz, Anal. Chim. Acta 1984, 160, 47.
(12) US Pat. 4.557.900; Eur. Pat. 105.070.
(13) O. S. Wolfbeis and H. E. Posch, Anal. Chim. Acta 1986, 185, 321.
(14) E. Urbano, H. Offenbacher and O. S. Wolfbeis, Anal. Chem. 1984, 56, 427.
(15) O. S. Wolfbeis and B. P. H. Schaffar, Anal. Chim. Acta (1987); in press.
(16) U. H. Bergmeyer, (ed.), Methods of Enzymatic Analysis, Verlag Chemie, Weinheim, Deerfield Beach, 1984.
(17) M. A. Arnold, Anal. Chem. 1985, 57, 565.
(18) O. S. Wolfbeis and E. Koller, Anal. Biochem. 1983,129, 365.
(19) E. Koller annd O. S. Wolfbeis, Anal. Biochem. 1984, 143, 146.
(20) O. S. Wolfbeis, Anal. Chem. 1986, 58, 2874.
(21) J. L. Gehrich, D. W. Lübbers, N. Opitz, D. R. Hansmann, W. W. Miller, J.K. Tusa and M. Yafuso, IEEE Trans. Biomed. Eng. 1986, 33, 117.

DEVELOPING BIOSENSORS

Elizabeth A H Hall
The Biotechnology Centre
University of Cambridge
Downing Street
Cambridge
CB2 3EF UK

ABSTRACT. Biosensors are currently being developed by the adaption of existing bioassay - most notably protometric methods, and through the application of newer physico-chemical techniques, among which surface plasmon resonance has demonstrated some promising results in preliminary experiments. Immobilisation of the biorecognition molecule is required whatever the technique employed, and some care is necesary in selecting the immobilisation method to be compatable with the type of sensor.

A biosensor is a multicomponent system which requires the union of biological and electronic units through transduction of a physico-chemical parameter. Solution to this union may be made by two different approaches. On the one hand some traditional methods are suitable for conversion from multi-step solution assays to solid-state one step reagentless sensors, while on the other hand physico-chemical parameters new to bioassays can be explored for their potential exploitation in biosensors (figure 1). The expectations of these two approaches can be entirely different, and the results should be judged accordingly.

Biosensors currently receive a great deal of publicity, and many claims have been made regarding their ability to 'revolutionise' analysis. While this will hopefully be proved to be true, 'revolutionise' should not be confused with 'replace'; the role of biosensors is not to replace traditional laboratory analysis. Indeed the development of biosensors primarily stems from a market where immediate assay is required, or no substantial laboratory facilities are available, and as such doctors' surgeries, some treatment of patients (eg glucose monitoring for diabetics) or any 'field' application can be considered as prime targets.

If the assay, which is to be developed as a biosensor, has been adapted from a traditional method, then the conditions and parameters have already been optimised and the problem is more a technical one, of immobilisation of the biorecognition molecule, transduction of the chosen parameter,

G. G. Guilbault and M. Mascini (eds.),
Analytical Uses of Immobilized Biological Compounds for Detection, Medical and Industrial Uses, 227–236.

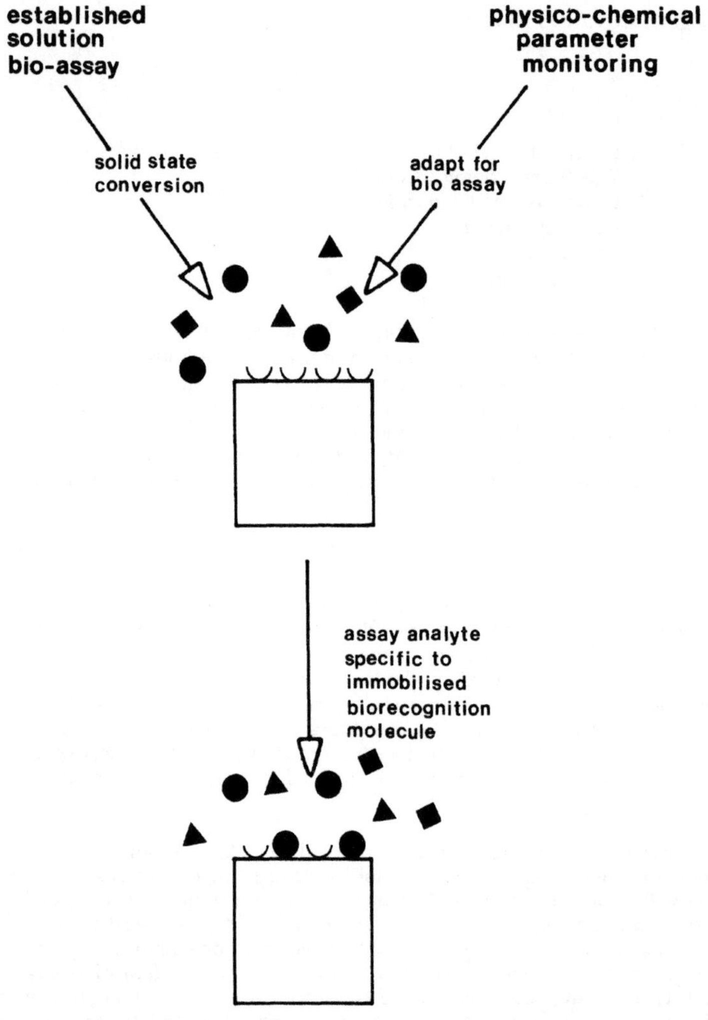

Figure 1 Development of a biosensor from existing bio-assay or by application of new techniques. Both approaches require the immobilisation of the biorecognition molecule on a surface capable of transduction of the biochemical event as an electrical signal.

and final optimisation of the system. While these are far from trivial problems their solution should be realisable on a much shorter time scale than that required for the development of entirely new assay techniques, where even the assay parameter is not established, and a significant effort must be devoted to data accumulation and feasibility studies, before actual sensor development can be seriously considered.

It is probably true to say that a majority of the traditional bioassays are based on a photometric techniques, and although many of these analyses are unsuitable for conversion to biosensors, any book of bioanalytical methods contains a wealth of potential optical biosensors.

The Dumas test for albumin, for example, relies on the chromogenic reaction that occurs when a complex is formed between albumin and bromocresol green, in a solution buffered near the 'end point' for the dye. Conversion of this solution assay to a solid state sensor has been achieved by forming the glutathionine conjugate of the dye and attaching it to an optically clear cellophane membrane (1) which is placed in the light path in the assay solution, conceivably forming the cell wall.

Examination of the UV-visible absorbance spectrum of the dye-conjugate in the presence and absence of albumin shows that the wavelength of interest is around 640nm, corresponding to the λ_{max} of emission for a red LED (light emitting diode), which may be used as a light source. Coupled with a suitable photodiode detector therefore, these are the basic components of an optical albumin biosensor, where the probe radiation is carried out to and from the cell by fibre optics, and the low power requirements of LED's and photodiode detectors are suitable for battery operation (2) (figure 2).

Exploitation of the pH indication features of these and many other dyes has also resulted in the development of other fibre optic probes. Peterson et al (3), for example, have developed a pH probe based on the indicator dye, phenol red bound to polyacrylamide microspheres. This is a light scattering technique, where irradiation is with white light and the pH is monitored by the ratio of the intensity of the transmitted light at the λ_{max} of the dye in its basic and acidic forms. This pH probe, which has been tested in-vivo is claimed to have an accuracy of 0.01pH units in the range pH 7.0-7.4.

Similarly, the pH indicator dye p-nitrophenol has been employed to construct a 'fibre optic ammonia sensor'(4), although in this example the step towards full immobilisation of the dye still has to be made since it is retained behind a teflon membrane in solution.

Rather more of a challenge perhaps, is to couple these indicators with an enzyme linked assay which involves a pH change. Penicillinase and urease for example react on their substrates to produce a decrease and an increase in pH respectively. For maximum response the pH indicator dye should be chosen to have a pk_a value close to the optimum pH value for the enzyme. Penicillinase co-immobilised with bromocresol green on

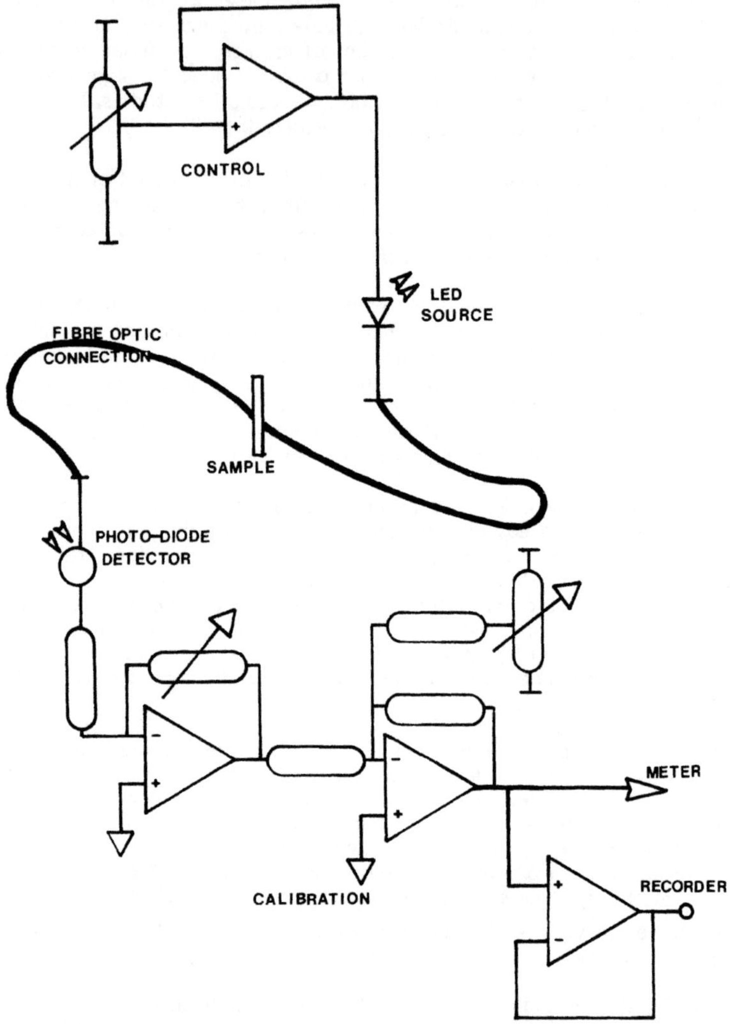

Figure 2 Simple circuit electronics required for the control and data handling of a LED based colorimetric solid state biosensor.

a cellophane membrane therefore, gives a solid state sensor that responds to penicillin G in the 0-10mM range. Similarly at the other end of the pH scale, urease co-immobilised with bromothymol blue gives a sensor for urea (5). So far in the examples cited, fibre optics have been employed entirely as a means of transporting the light to and from the sample cell, but they have not been participant in the assay itself. Examination of the construction of a fibre optic and the mechanism of light propagation reveals a transduction surface ideally suited to biosensor modification.

The fibre optic consists of a high refractive index core surrounded by a coating of different refractive index (figure 3(a)), such that the light passes down the fibre by a series of total internal reflections (figure 3(b)). At each reflection the interference between the incident and reflected radiation generates an electromagnetic wave-form. This standing wave (figure 3(c)) decays exponentially into the surrounding medium, with a penetration depth given by:

$$d_p = \frac{\lambda}{2\pi\,(n_1^2 \sin^2 \theta - n_2^2)^{1/2}}$$

The wavelength characteristics of this, the evanescent wave, are the same as the incident beam, so that any molecule that can absorb light at that wavelength and that is within the penetration depth of the evanescent wave will cause a reduction in the intensity of the reflected light in the fibre.

One can envisage therefore a family of true 'fibre optic sensors' where for example, the indicator dyes/and enzymes already mentioned are immobilised directly to the fibre, and the reaction with analyte monitored via the evanescent wave (figure 3(d)). Since the depth of penetration is less than a wavelength, this method has the advantage that it is specific to the reaction occuring on the surface of the fibre and is independent of the optical properties of the bulk solution.

A number of systems have been investigated using this technique (6-9), where an indirect assay is made of the analyte via a label. Giuliani for example (7), employed a glass capillary tube as a waveguide, coated with a thin solid film of an oxazine perchlorate dye to create an ammonia sensor in the range 60-1000ppm, while a polystyrene fibre with bromocresol green and penicillinase immobilised on the surface (6) acted as a penicillin sensor.

Exploitation of this phenomenon has also been reported for a capillary-fill immuno-sensor from Unilever, using a fluorochrome labelled antigen (8,9). In this device two glass slides sandwiching a capillary space form the body of the sensor, and a competitive immunoassay is performed between labelled and sample antigen and the antibody immobilised at one interface of the capillary space.

It would represent a major advance to be able to perform direct immunoassay, without the need for labels. However, in order to achieve this a physico-

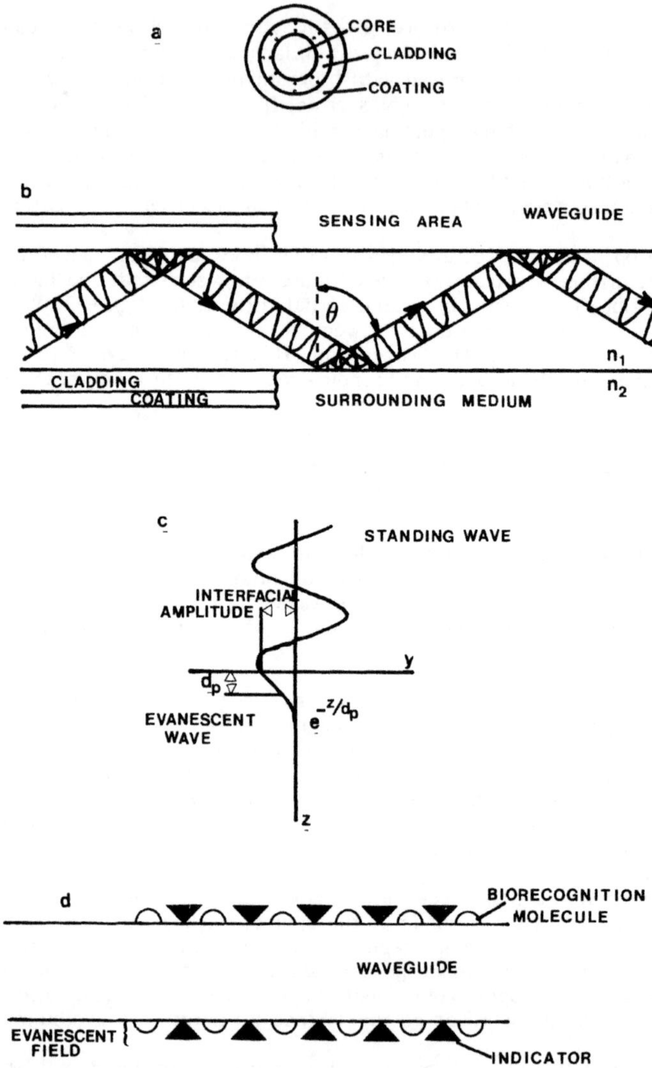

Figure 3 (a) cross section through a fibre optic (b) propagation of light along a fibre (or waveguide) and development of interference at each total internal reflection. Fibre stripped of cladding to allow interaction of the light passing down the fibre with the surrounding medium through the evanescent wave; (c) development of the standing wave and decay of the evanescent wave at the surface of the waveguide; (d) construction of a colorimetric biosensor, activated through the evanescent wave.

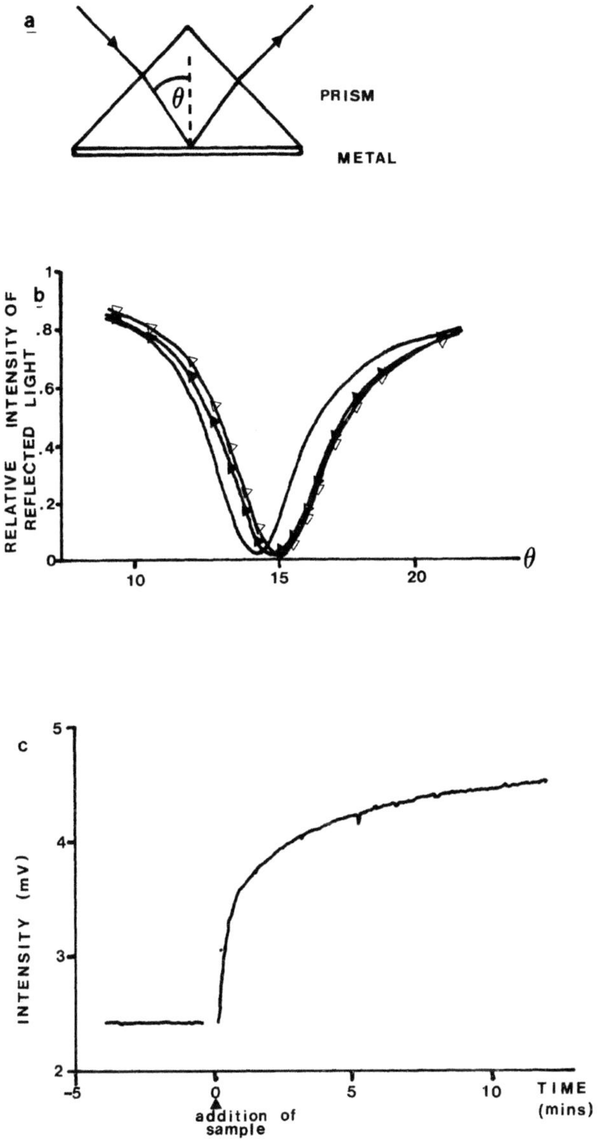

Figure 4 (a) excitation of plasmon resonance in a thin metal film; (b) observation of surface plasmon resonance as a minimum in the reflected light intensity for a metallised diffraction grating on addition of water (——), anti serum albumin (▲—▲); and serum albumin (△—△); (c) monitoring the addition of ovalbumin with time at a fixed wavelength and incident angle.

chemical parameter must be identified which can be monitored to follow the immuno-reaction. To this end domains of surface physics and chemistry that hitherto have not been considered for bioassay, are being explored for potential exploitation in biosensors. This, on the whole, is a comparatively slow process since large quantities of data must be accumulated and new techniques learned and manipulated for use under the 'non-model' conditions required for biomolecules.

One of these newer techniques is surface plasmon resonance. Surface plasmons exist in the boundary of a solid whose electrons behave like those in a quasi-free electron gas, and can be characterised at a given excitation wavelength by the resonance angle. Non radiative surface plasmons may not be excited directly by light, since the light wave vector is always smaller than that of the plasmon. This may be overcome (figure 4(a)) if the plasma is bounded by two media of different dielectic constants - ie by coupling light through prisms (compare evanescent wave), or if the light travels along a surface with a sinusoidal profile, where a change of momentum is induced of $2\pi\hbar/a$ (where a is the period of the grating).

If the metal film is sufficiently thin then the induction of the surface plasmon is sensitive to the dielectic constant of the medium immediately adjacent to the surface of the film. Application of this phenomenon was investigated for the anaesthetic agent, halothane, where the reversible dissolution of halothane in silicon oil, spread on a thin silver film, was monitored via the induced change in the refractive index of the oil, and effect on the plasmon resonance (10).

More recently the technique has been applied to immunoassay, and in a preliminary study employing metallised diffraction gratings, encouraging results were obtained for serum albumin/antiserum albumin models, where the antibody was immobilised on the metal film by adsorption (11) (figure 4(b)).

Surface plasmon resonance is observed as a minimum in the reflected light for a given wavelength. Such initial explorations must accumulate data across the whole experiment rather than monitoring an isolated parameter, so that plots of intensity of reflected light against incident angle for the above system show shifts in the intensity minimum of the order of 1°. In practice the change in light intensity would be monitored at one angle which lay on the rapidly changing portion of the intensity/incident angle curve (figure 4(c)).

This experiment has demonstrated the feasibility of using surface plasmon resonance as a bioanalytical tool. It has also highlighted the need for a more vigorous approach to the method of immobilisation of the biomolecule on the metal surface and the role that mathematical modelling may play in optimising the experiment and identifying the parameter most suitable for exploitation in the development of a sensor device.

Employing a full solution to the theory of surface plasmon resonance the response curves for various plasmon materials with different surface coverings may be predicted. Early results (12) using such a theoretical model have shown how metal thickness and light frequency can be optimised for maximum response.

The problem of immobilisation of the biorecognition molecule on the metal surface with the correct orientation is not so easy to deal with via theoretical models. Interaction of a protein molecule with a surface is known to involve the same forces as those between individual peptide chains - ie hydrophobic, hydrophilic ionic and polar interactions. Obviously the protein can interact in a variety of different ways depending on the respective domains on protein and surface and the orientation of approach to the surface. The probable randomness of this latter feature is largely overcome by LB (Langmuir Brodgett) film technology. But even with ordered LB films the possibility of adsorption and readsorption cannot be ignored. Indeed in an unrelated study involving adsorption of antibody on a metal surface, and its reaction with solution antigen (13), there is evidence to suggest that under some conditions the antibody-antigen complex 'falls off' the metal surface. In view of these considerations some effort should be devoted towards the direct covalent bonding of the antibody to an ordered reactive surface.

The correct method of immobilisation of the biomolecule is essential to all forms of biosensors, and as such preoccupies their development. In mediated amperometric biosensors, for example, mediator and redox enzyme must be immobilised such that electron transfer is possible along the chain: electrode-mediator-enzyme. In this instance orientation of the enzyme (or the mediator) in an ordered monolayer at the electrode surface is probably not required, but the enzyme must be positioned such that it is available to the substrate, and may participate in the 'electron-chain' to the electrode.

An electrochemical method of immobilisation would be particularly attractive for these electrochemical amperometric sensors, since one could anticipate that the species would be deposited exclusively at the sensor surface - ie on the electrode.

Such an immobilisation procedure is described by the incorporation of the redox enzyme - glucose oxidase into the polymer matrix of polypyrrole. Electrochemical polymerisation of pyrrole from aqueous media in the presence of glucose oxidase gives a polymer film whose enzyme activity is proportional to the concentration of enzyme in the polymerisation solution (14).

The reaction between glucose oxidase and its substrate follows the scheme:

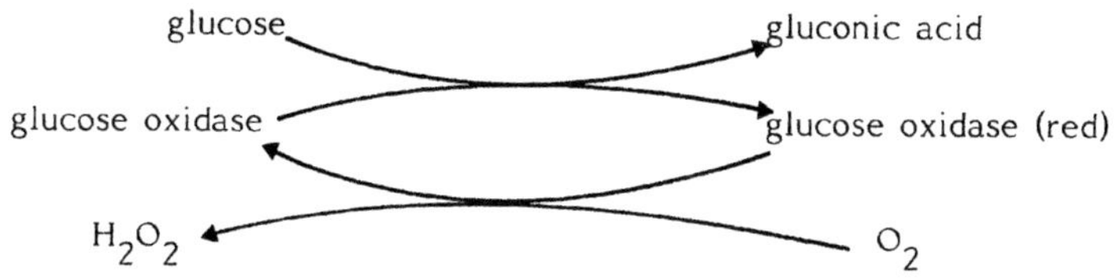

so that glucose can be monitored on the above electrode via the oxidation current due to hydrogen peroxide. The problems inherent to employment of this oxidation dependent measurement have been established (15) and ferrocene has been proposed as an alternative synthetic mediator for the reaction. N-substituted pyrrole monomers have therefore been prepared with the ferrocene redox functionality:

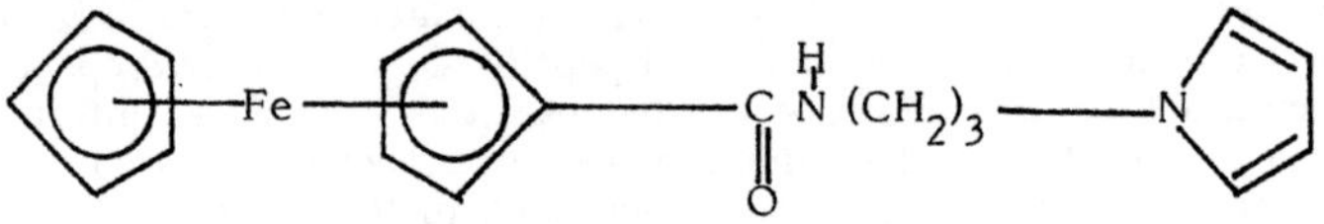

and co-polymerised with pyrrole in the presence of glucose oxidase (16). The resulting electrode shows all the signs of being a reagentless amperometric biosensor for glucose, and demonstrates an immobilisation technique apparently ideally suited to amperometric electrodes.

References

1. M J Goldfinch & C R Lowe. Analytical Biochemistry. 1980, **109**, 216.
2. E A H Hall & P Maynard. Unpublished instrument development.
3. J I Peterson, S R Goldstein, R V Fitzgerald & D K Buckhold. Anal.Chem. 1980, **52**, 864.
4. M A Arnold & T J Ostler. anal.Chem. 1986. **58**, 1137.
5. M J Goldfinch & C R Lowe. Anal. Biochem. 1984, **138**, 430.
6. E A H Hall. Unpublished result.
7. J F Giuliani, H Wohltjen & N L Javis. Opt.Lett. 1983.
8. R A Badley, R A L Drake, I A Shanks, A M Smith & P R Stephenson. 'Optical biosensors for immunoassays'. The Royal Society discussion meeting. May 86, London.
9. A M Smith. World Biotech Report (Online). 1986, 1, D9.
10. C Nylander, B Liedberg & T Lind. 'Sensors and Actuators'. 1982/83, **3**, 79.
11. D C Cullen, R G W Brown & C R Lowe, in press.
12. M J Liley, E A H Hall & C R Lowe. Unpublished results.
13. E A H Hall & J Cooper. Unpublished results.
14. N C Foulds & C R Lowe. J.Chem.Soc. Faraday Trans 1, 1986, **82**, 1259.
15. A Turner. NATO ARW. Florence 1987.
16. N C Foulds, Abstr.Spinks Symposium on Biosensors RSC/RSA London May 86; PhD thesis Cambridge University 1987.

IMMOBILIZED BIOLOGICAL COMPOUNDS IN BIO- AND CHEMILUMINESCENCE ASSAYS

P. R. Coulet and L. J. Blum
Laboratoire de Génie Enzymatique - Atelier de Biotechnologie
CNRS - Université Claude Bernard Lyon 1 - 43, Bd du 11 Novembre 1918
69622 Villeurbanne Cedex
France

ABSTRACT. Bio- and Chemiluminescence reactions are now gaining more and more attention due to their very high sensitivity especially for trace analysis. Both firefly and bacterial bioluminescence for the microdetermination of ATP and NAD(P)H respectively, including substrates or enzymes involved in related reactions, have been extensively studied with free enzymes. Enzyme-mediated chemiluminescence for H_2O_2 detection appeared also to be very valuable in a series of hydrogen peroxide-producing enzymatic reactions. Concerning these three systems, only a few studies on immobilized enzymes or other compounds involved in the reactions, have been reported so far. The potential of such sensing elements however, deserves real attention when the design of photobiosensor of the optode type is considered. Current progress in the field will be summarized in this paper.

1. INTRODUCTION

The very high sensitivity and specificity of light-producing enzymes from both the firefly and marine bacteria have long been recognized and used for analytical purposes, employing free enzymes. The development in the last decade of immobilized enzymes allowing their reusability and generally an enhancement of their stability led different groups in the world to consider the immobilization of bioluminescent enzymes as a potential real improvement for microdetection enzymatic methods.

Some chemiluminescent reactions are also very valuable for the determination of H_2O_2 which is a product of several enzymatic reactions and recently, a few papers appeared in the literature dealing with the detection of new analytes.

The aim of this paper is to briefly review what has been done in the past decade with immobilized biological compounds, and what are the problems still to be solved. Three main points will be considered : - the enzymes used - the immobilization on suitable matrices - the adaptation of available instrumentation.

G. G. Guilbault and M. Mascini (eds.),
Analytical Uses of Immobilized Biological Compounds for Detection, Medical and Industrial Uses, 237–248.

2. BIOLUMINESCENT ANALYSIS

2. 1. Enzymatic reactions of bioluminescence

Firefly luciferase in the presence of luciferin (LH_2), ATP, Mg^{2+}, and molecular oxygen catalyzes the production of light according to the global following reaction :

$$ATP + LH_2 + O_2 \longrightarrow AMP + oxyluciferin + PPi + CO_2 + h\nu\ (\lambda_{max} = 560\ nm) \qquad (1)$$

Under appropriate reaction conditions the intensity of the light produced is proportional to the ATP concentration.

In the luminescent marine bacteria system used for analytical purposes, the light is produced by two consecutive enzymatic reactions. In the first reaction, catalyzed by the NAD(P)H:FMN oxidoreductase, $FMNH_2$ is produced [reaction (2)] and then utilized in the second reaction catalyzed by a luciferase to produce light [reaction (3)] in the presence of molecular oxygen and a long-chain aldehyde (R-CHO).

$$NAD(P)H + H^+ + FMN \longrightarrow NAD(P)^+ + FMNH_2 \qquad (2)$$

$$FMNH_2 + R\text{-}CHO + O_2 \longrightarrow FMN + R\text{-}COOH + H_2O + h\nu\ (\lambda_{max} \cong 490\ nm) \qquad (3)$$

When NAD(P)H is the limiting substrate of this bi-enzymatic system, the light intensity is proportional to the NAD(P)H concentration.

The two most useful light-emitting enzyme systems were isolated from *Beneckea harveyi* (*B. harveyi*) and *Photobacterium fischeri* (*P. fischeri*).

In addition of the direct measurement of either NAD(P)H or ATP by the bioluminescent reactions described above, the analysis of some NAD(P)H or ATP dependent enzymes and their substrates can be performed via the bacterial or the firefly luminescent systems.

2. 2. Properties and analytical use of immobilized bioluminescent enzymes

Insoluble bacterial luciferase was first prepared and used in the early 1970's by Erlanger *et al.* (1) for investigating the mechanism of bacterial luminescence. Since then, various supports have been used in the last decade for bioluminescence enzyme immobilization (Table I).

The bacterial luciferase-oxidoreductase system from *Beneckea harveyi* was covalently bound to arylamine porous glass beads via diazotization by Jablonski and DeLuca (2). The beads were then glued onto glass rods. Although the amount of recovered active enzyme was very low, properties of the immobilized enzymes were found very similar to the soluble enzymes for pH and substrate optima. Linearity was obtained for NADH and NADPH in the range 1 pmol - 50 nmol and 10 pmol - 200 nmol, respectively. Such rods have been used to monitor enzymic reactions producing NAD(P)H (4). Dithiothreitol (DTT) necessary for maintaining enzyme activity had to be removed prior to immobilization in order to prevent a dramatic reduction in retained activity. Light production was initiated by immersing the immobilized enzyme rod into a reaction mixture containing optimal concentrations of components except the enzyme to be assayed and the light output was integrated for 60 sec. Picomole levels of malate dehydrogenase, lactate dehydrogenase, alcohol dehydrogenase and glucose-6-phosphate dehydrogenase plus hexokinase have thus been assayed.

Glass rods have also been used for immobilizing the firefly luciferase from *Photinus pyralis* (3). The immobilized enzyme had a lower pH optima than the soluble enzyme and emitted light with a major peak at 615 nm instead of 562 nm. The peak light intensity was linear with respect to ATP concentration in the range of 1 x 10^{-5} to 1 x 10^{-8} M. The rate of ATP production catalyzed by creatine phosphokinase have also been measured using the firefly luciferase rods.

TABLE I. Supports used for the immobilization of bioluminescent enzymes.

SUPPORT	ENZYME SOURCE	REFERENCE
Arylamine porous glass beads	bacterial & firefly	2 - 6
BSA + glutaraldehyde	bacterial & firefly	7, 8
Cellophane	firefly	8 - 10
Cellulose/$NaIO_4$ or Cl triazine	firefly	11
CNBr-sepharose	bacterial & firefly	6, 8, 9, 12 - 22
Collagen	bacterial & firefly	23 - 25
Nylon coils or membranes	bacterial & firefly	26 - 29
Polyacrylamide gel	firefly	8
Polyacrylic acid	bacterial	1

Glucose-6-phosphate dehydrogenase co-immobilized on arylamine glass beads with the light-emitting enzyme system from *B. harveyi* allowed to quantitate 1 pmol of glucose-6-phosphate (5). The bioluminescent system alone, can detect as little as 0.2 pmol of NADH per assay sample. By co-immobilizing a fourth enzyme, hexokinase, it was possible to detect 20 pM Glucose. Compared with free enzymes, a more efficient conversion of NADH or glucose-6-phosphate occured apparently after immobilization.

In a further study, the bacterial enzymic system was covalently bound to sepharose 4B by Ford and DeLuca (14). Compared to the previously reported data obtained with glass beads as a support, the sepharose preparation had a much higher activity and NADH could be determined in the range 1 pmol - 2 nmol. Furthermore, it was possible to extend the range of assays by using other enzymes co-immobilized with the bioluminescent system. Low amounts of glucose-6-phosphate could be determined in the range 10 pmol - 20 nmol with immobilized glucose-6-phosphate dehydrogenase. With the immobilization of 3α- or 3β-hydroxysteroid dehydrogenase, it was possible to determine 0.8 pmol of either andro- or testosterone.

Properties of both firefly and bacterial bioluminescent systems immobilized onto sepharose were studied by Wienhausen *et al.* (15). Both systems could be reproducibly immobilized onto sepharose 4B and CL 6B. Both are subject to analytical interferences from proteins. For NADH, light intensity was proportional from 0.2 to 1 x 10^3 pmol. The intra-precision was 3.5 % and 5.8 % at 2 x 10^{-5} M and 2 x 10^{-6} M respectively. Dehydrogenases may also interfere in the assays. For ATP, the assay was linear from 0.2 to 2 x 10^3 pmol and the intra-precision was equal to 4.8 % and 3.2 % at 2 x 10^{-4} and 1 x 10^{-5} M respectively. Immobilized enzymes rapidly frozen in the presence of glycerol and DTT could be stored for at least two months without loss of activity. The purified bacterial two-enzyme system has also been immobilized on sepharose 4B with other enzymes for the bioluminescent assay of various metabolites including D-glucose, L-lactate, 6-P-gluconate, L-malate, L-alanine, L-glutamate, NAD and NADP (16). Although the detection limit was different for each of them, the measurement range was 10-100 pmol with linear light output with respect to analyte concentration.

Roda *et al.* (17) developed a rapid, sensitive and specific method for the measurement of 7α-hydroxy bile acids based on bacterial 7α-hydroxysteroid dehydrogenase co-immobilized on sepharose 4B with the bacterial luminescent system. The detection limit was 0.5 pmol per tube with a linear range comprised between 0.5 and 50 pmol. The intra and inter-assay precision was 6-8 % and 8-10 % respectively.

A packed-bed reactor of commercial bacterial bioluminescence enzymes immobilized on sepharose was used by Kurkijärvi *et al.* (18) as a high-sensistivity detector for auto-analyzers. The column which consisted of a glass tube filled with the gel, was placed in front of the photomultiplier tube. About 400 NADH measurements could be done with a single immobilized enzyme reactor.

Bacterial luciferase was immobilized onto sepharose 6B and its reactivity and subunit functions were studied by Watanabe *et al.* (19). The study confirmed that the two different subunits (alpha and beta) are required for luciferase activity even after immobilization.

Kricka *et al.* (20) have proposed an automated flow-system for the assay of NADH, glucose-6-phosphate, primary bile acids and ATP using the bioluminescent enzymes immobilized on sepharose and packed into small flow-cells. The detection limit of the system was at the picomole level. A good reproducibility was obtained with coefficient of variations comprised between 2 and 5 % and 30 samples could be assayed per hour. Each packed-bed flow-cell was used to perform several hundreds of assays during a period of 3-4 weeks. The main problem affecting the analytical performances of the system was microbial growth despite the presence of gentamicin in the packed bed after use for a long period at room temperature. This resulted in a very high background light level. Additional drawbacks concerning packing, disruption of the gel and channeling have also to be mentioned

After isolation of all bile acids, Rossi *et al.* (22) performed bioluminescence measurement of the unconjugated 7α-hydroxy bile acids using sepharose beads containing co-immobilized 7α-hydroxysteroid dehydrogenase, diaphorase and luciferase. The method appeared superior to previously reported methods for determining the bioequivalence of chenodiol preparations and may be extended to other bile acids provided the suitable dehydrogenase is available.

Brovko, Ugarova and coworkers have taken an interest in the immobilization of the luciferase from the firefly *Luciola mingrelica*. After immobilization on CNBr-Sepharose 4B, ATP could be determined in the range 50 x 10^{-9} M - 1 x 10^{-6} M (12). Activity measurements of pyruvate kinase and ATPase have also been performed using the immobilized luciferase. The inactivation and reactivation mechanisms of the luciferase, when immobilized, and the role of sulfhydryl groups in these processes were investigated (13). The activity and the stability of this luciferase immobilized on various polysaccharide carriers have been studied (9). The better activity was obtained with Ultrodex-bound luciferase whereas the sepharose-bound enzyme exhibited a better stability. Cellophane films were also used for firefly luciferase immobilization by Ugarova et al. (10). ATP detection in the range 0.1 pM - 1 mM could be performed as well as activity determination of pyruvate kinase and ATPase. The same cellophane bioactive film could be used 40 times for ATP quantification but lost 50% of its activity after only 5 h at 25°C. An other study of the firefly luciferase from *Luciola mingrelica* immobilized on different supports was performed by Ugarova *et al.* (8). The supports used were : albumin and polyacrymamide gel, AH-, CH-, and CNBr-Sepharose 4B as well as Ultrogel, Ultrodex and Cellophane films. Only immobilization on cyanogen bromide- activated polysaccharide carriers resulted in highly active immobilized luciferase. Kinetic properties of immobilized luciferase hardly differed from those of the soluble enzyme. A 1000-fold stabilization effect was noted for the luciferase immobilized on CNBr-Sepharose 4B at pH 7.5 and 25 °C. A stabilization mechanism for the immobilized luciferase was also discussed.

Kinetic properties and diffusional hindrances for the bioluminescent enzyme immobilized on cellulose films activated by cyanuric chloride or sodium periodate were studied (11). The inner and outer diffusion had no appreciable effect on the kinetic parameters. Thermo-inactivation of the immobilized enzyme occured in two steps, rapid and slow. DTT and cysteine stabilized luciferase at the rapid step whereas salts stabilized at both steps. Firefly luciferase immobilized on cellulose films could be used repeatedly to measure ATP microquantities down to 0.1 nM.

Using CNBr-Sepharose, Ivanova *et al.* (21) developed a bioluminescent method for creatine kinase (CK) assay with immobilized firefly extract containing the bioluminescent co-immobilized system : adenylate + luciferase. ADP for the CK reaction was produced from the initial mixture of AMP and ATP. The ATP formed in the reaction with CK was quantified using firefly luciferase. The lowest detection limit for CK activity was 0.5 ± 0.2 U/l in the sample. A linear range of the determined CK activities was 0.5-1000 U/l.

Rodriguez and Guilbault (7) have immobilized in a BSA gel the commercially available bacterial luciferase/oxidoreductase system from *P. fischeri*. The system was coupled to an NADPH-producing commercial assay for microscale analysis of creatine kinase activity. Although the range of linearity covers the expected values of normal CK activity, improvements were needed to lower the limit of detection of the system and enhance reproducibility.

More recently, Worsfold and Nabi (6) have developed a FIA system using the firefly luciferase from *Photinus pyralis* immobilized on sepharose 4B and controlled-pore glass (CPG). The sepharose system had a limit of detection three orders of magnitude better and a sensitivity two orders of magnitude greater than the CPG system. The determination of ATP concentration was possible over the range 1×10^{-12} - 1×10^{-5} M. Creatine phosphokinase and creatine phosphate assays could also be performed from 10 to 400 U/l and from 1×10^{-5} to 1×10^{-1} M, respectively.

A continuous flow bioluminescent method for NADH analysis has been developed by Girotti *et al.* (26) using the luciferase/oxidoreductase system from *P. fischeri* covalently immobilized to a nylon coil. NADH determinations were performed from 1 to 2500 pmol/tube. Inter and intra assay precision of the method was 5-10 %. More than 20 samples per hour could be analyzed without carryover. The nylon immobilized enzymes were stable for more than two months and over 500 samples could be analyzed. This continuous-flow bioluminescence method was extended to primary bile acid measurements in serum (27). 7α-hydroxy steroid dehydrogenase and the bacterial bioluminescent system were covalently co-immobilized on a nylon coil (1 m x 1.0 mm i.d.). The assay was highly specific for 7α-hydroxy bile acids. The continuous-flow determination of ethanol, glycerol, aldehyde and 3α-, 7α- and 12α-hydroxy bile acids with nylon tube-immobilized bioluminescent enzymes have also been performed by the same group, using co-immobilized NAD(P)H generating enzymes (29). The specific assays were at picomole levels and the precision of the method as well as its correlation with other methods was satisfactory.

Recently another continuous-flow system with firefly luciferase immobilized on a nylon coil positioned inside a LKB-Wallac luminometer in front of the photomultiplier was developed by Carrea et al. (28) for the assay of ATP. According to the authors, activity recoveries ranged from 0.21 to 0.65 % depending on enzyme concentrations. The immobilized enzyme with a half-life of 15 days at 25°C exhibited a strongly improved stability compared to the free enzyme. The assay was linear between 0.3 and 100 pmol of ATP and 20 samples/h could be processed by manual injection with no carry-over when washing was performed between samples. The relative standard deviation for both intra and interassay was 5-8 %. The described system has been utilized to measure the ATP and ADP content of platelets from healthy subjects and from patients with blood deseases.

Blum and Coulet (23) used collagen strips having undergone an acyl-azide activation process for the immobilization of the bacterial bioluminescent system from *P. fischeri*. These strips were easier to handle than particulate supports for analytical purposes. NADH was determined in the range 1×10^{-9} - 2×10^{-5} M with a precision of 5 %. The stability of the system was greatly improved by immobilization. After two weeks of storage, 70 % of the initial activity still remained whereas the soluble preparation was not usable after this period.

The immobilization of firefly luciferase from *Photinus pyralis* using the same procedure was also reported by Blum *et al.* (24). Immobilization performed in the presence of both ATP and luciferin allowed to increase the activity retained on the strip. ATP measurements could be performed in the range 1×10^{-11} - 3×10^{-6} M. The strips could be stored in a dehydrated form at 4°C and then retained 20 % of their initial activity after 8 months. Human blood ATP could be measured with such strips and results were found in good agreement with those obtained with free luciferase.

The kinetic properties of collagen-bound firefly luciferase have been also investigated by Blum and Coulet (25). Atypical kinetics were observed. After immobilization, the time necessary to reach the steady-state increased from 300 ms to 3 min and the two substrates, luciferin and ATP behave differently with respect to the enzyme. The nature of the saturating substrate first in contact with the bound enzyme is not indifferent suggesting that immobilization can reveal behaviors or mechanisms which are not visualized with the free enzyme.

3. CHEMILUMINESCENT ANALYSIS

3.1. Hydrogen peroxide-requiring chemiluminescent reactions

The best known chemiluminescent synthetic molecule is luminol (5 - amino - 2, 3 - dihydro - 1, 4 - phtalazinedione), which can produces light in the presence of H_2O_2 and a catalyst [reaction (4)]. Peroxidase and ferricyanide are the most common catalysts used.

$$2\,H_2O_2 + \text{luminol} \longrightarrow \text{3-aminophtalate} + N_2 + 3\,H_2O + h\nu\ (\lambda_{max} = 430\text{ nm}) \tag{4}$$

Diaryl oxalates such as TCPO (bis [2, 4, 6 - trichlorophenyl] oxalate) also undergo a chemiluminescent oxidation reaction with hydrogen peroxide. The diaryl oxalate reaction, sensitized with a fluorophore, can be then used for H_2O_2 determination.

Enzyme-catalyzed systems producing hydrogen peroxide can be coupled to chemiluminescence detection. Some exemples are given below.

$$\beta\text{-D-Glucose} + O_2 + H_2O \xrightarrow{\textit{Glucose oxidase}} \text{Gluconic acid} + H_2O_2 \tag{5}$$

$$\text{Cholesterol} + O_2 + H_2O \xrightarrow{\textit{Cholesterol oxidase}} \Delta\text{-4-Cholestenone} + H_2O_2 \tag{6}$$

$$\text{L-Lactate} + O_2 + H_2O \xrightarrow{\textit{Lactate oxidase}} \text{Pyruvate} + H_2O_2 \tag{7}$$

3. 2. Immobilized reagents in chemiluminescent analysis

Variable matrices have also been used for solid phase chemiluminescent assays supporting either the catalysts or a substrate (Table II).

TABLE II. Supports used for the immobilization of compounds in chemiluminescent analysis.

SUPPORT	REFERENCE
Cellulose	30
Collagen	31
Controlled-pore glass	32 - 35
Photocrosslinkable resins	30
Polyacrylamide gel	36, 37
Polyamide	38
Sepharose	39
Silane	35

A blood glucose measurement was developed by Bostick and Hercules (39) using an immobilized glucose oxidase column. Chemiluminescence proportional to β-D-glucose concentration was produced by using mixed luminol-ferricyanide reagent. A linear range was obtained between 1 x 10^{-8} and 1 x 10^{-5} M for hydrogen peroxide and 1 x 10^{-8} and 1 x 10^{-4} M for glucose. For urine glucose analysis, prior adsorption of uric acid had to be performed.

The chemiluminescent reaction of TCPO with H_2O_2 in the presence of perylene was tested for the measurement of hydrogen peroxide (32). The chemiluminescence intensity was found linear in the range 7 x 10^{-8} - 1 x 10^{-3} M. Sensitive analysis was possible between pH 4 and 10. By using immobilized glucose oxidase, and the chemiluminescent reaction of TCPO, glucose could be determined in urine without interference from uric acid.

Freeman and Seitz developed a new concept in the luminescent measurements. Chemiluminescent H_2O_2 detection based on the luminol reaction was achieved using peroxidase immobilized in a polyacrylamide gel on the end of a fiber optic probe. The fiber optic probe transmitted the chemiluminescent signal to a detector. The probe shows response which is second order in peroxide. The detection limit was close to 1 x 10^{-6} M and at 1 x 10^{-5} M H_2O_2, steady-state was reached in about 4 sec. The rate of mass transfer from solution to the surface of the enzyme phase was the limiting factor for chemiluminescence intensity.

Horseradish peroxidase was immobilized with glutaraldehyde on a collagen matrix by Ikariyama *et al.* (31) to studying the reaction mechanism of the luminol-hydrogen peroxide system which undergoes luminescent reaction in two steps : an initial burst decaying rapidly, followed by the delay luminescence. The authors found that the delayed luminescence took place even in the absence of peroxidase if the instantaneous luminescence was initiated by the enzyme.

An immobilized enzyme sensor using the chemiluminescence of a luciferin analog was described by Kobayashi *et al.* (30) for the determination of different substrates. Xanthine oxidase, peroxidase, glucose oxidase, uricase and cholesterol oxidase were immobilized using photocrosslinkable prepolymer or ion exchange cellulose beads. H_2O_2, xanthine and hypoxanthine could be measured within 100 sec but for glucose, cholesterol and uric acid, 10 min were necessary for a sensitive assay.

A novel biosensor combining matrix-bound peroxidase and photodiode as transducer was designed for the determination of hydrogen peroxide by Aizawa *et al.* (37). The photocurrent of the sensor was correlated with the hydrogen peroxide concentration ranging from 1mM to 10mM. The feasability of a biophotodiode for glucose, using a glucose oxidase - peroxidase bienzyme membrane, was also demonstrated.

A miniaturized flow injection system for the determination of D-glucose, L-lactic acid, creatinine and free cholesterol has been proposed by Petersson *et al.* (33). All the substrates tested were enzymatically degraded by oxidases immobilized on controlled porosity glass. H_2O_2 generated was determined with the luminol - hexacyanoferrate system. The detection limits were 0.03 mg/dl, 0.03 mg/dl, 0.3 mM and 0.5 mg/dl for glucose, lactate, creatinine and cholesterol, respectively. Practically no loss of activity was observed over a three-month period.

Sucrose was also determined in a flow injection system involving four enzymatic conversion steps and using a microporous membrane flow cell in order to insure two different optimum pH : one for the enzymatic production of hydrogen peroxide (pH 5), the other for a maximum light production at pH 10.5 (34). First, the hydrolysis of sucrose by invertase produced α-D-glucose which was converted in β-D-glucose by mutarotase. The combined use of both dissolved and immobilized invertase and mutarotase was investigated. The oxidation of α-D-glucose by soluble glucose oxidase allowed hydrogen peroxide to be generated and then determined with the luminol-peroxidase system free in solution. The working range, obtained with a precision of 2-3 %, was 5μM - 1mM with an analysis time equal to 2 min.

Hool and Nieman proposed a new approach of utilizing luminol by immobilizing it to a solid support (35). Glutaraldehyde was used to bind luminol to an aminoalkylsilane through amino groups in an ethanol/dimethylsulfoxide solvent mixture. A flow-cell packed with the bioactive matrix was used in a flow injection system. Over 500 peroxide assays can be done with one gram of silica containing immobilized luminol. The working range and detection limit were similar to those obtained with soluble reagents.

In a recent study, Blum *et al.* (38) have immobilized horseradish peroxidase on a new polyamide membrane (Biodyne ImmunoAffinity Membrane) supplied in a preactivated form by Pall. H_2O_2 could be detected in the range 1×10^{-8} - 1×10^{-4} M by measuring the maximum light intensity emitted. Cholesterol could be determined in the range $2 \times 5\ 10^{-8}$ M - $2 \times 5\ 10^{-5}$ M with coupled cholesterol oxidase free in solution. A compromise between the conditions of enzymes activity expression, a sufficient light emission and a low background luminescence was accepted in the choice of pH.

4. CONCLUSION

Immobilized bio- and chemiluminescent reagents are gaining in popularity because they offer the sensitivity of luminescence techniques together with the economy and convenience of matrix-bound compounds. They also create the opportunity for designing novel analytical devices. In this connection, optical fiber biosensors and FIA systems appear very promising.

REFERENCES

1. **Erlanger, B. F., Isambert, M. F., and Michelson, A.M.,**
   ' Insoluble bacterial luciferases : a new approach to some problems in bioluminescence ',
   Biochem. Biophys. Res. Commun., **40**, 70-76 (1970).
2. **Jablonski, E., and DeLuca, M.,**
   ' Immobilization of bacterial luciferase and FMN reductase on glass rods ',
   Proc. Natl. Acad. Sci., **73**, 3848-3851 (1976).
3. **Lee, Y., Jablonski I., and DeLuca M.,**
   ' Immobilization of firefly luciferase on glass rods. Properties of immobilized enzyme ',
   Anal. Biochem., **80**, 496-501 (1977).
4. **Haggerty, C., Jablonski, E., Stav, L., and DeLuca, M.,**
   ' Continuous monitoring of reactions that produce NADH and NADPH using immobilized luciferase and oxidoreductases from *Beneckea harveyi* ',
   Anal. Biochem., **88**, 162-173 (1978).
5. **Jablonski, E., and DeLuca, M.,**
   ' Properties and uses of immobilized light-emitting enzyme systems from *Beneckea harveyi* ',
   Clin. Chem., **25**, 1622-1627 (1979).

6. **Worsfold, P. J., and Nabi, A.,**
' Bioluminescent assays with immobilized firefly luciferase based on flow injection analysis ',
Anal. Chim. Acta, **179**, 307-313 (1986).
7. **Rodriguez, O., and Guilbault, G. G.,**
' Immobilized bacterial luciferase for microscale analysis of creatine kinase activity ',
Enzyme Microb. Technol., **3**, 69-72 (1981).
8. **Ugarova, N. N., Brovko, L. Y., and Kost N. V.,**
' Immobilization of luciferase from the firefly *Luciola mingrelica*. Catalytic properties and stability of the immobilized enzyme ',
Enzyme Microb. Technol., **4**, 224-228 (1982).
9. **Brovko, L. Yu., Kost, N. V., and Ugarova, N. N.,**
' Immobilized luciferase from the fireflies *Luciola mingrelica*. Change in the pH dependence of the catalytic activity and stability of the enzyme after immobilization on various polysaccharide carriers ',
Biochemistry SSR, **45**, 1199-1204 (1980).
10. **Ugarova, N. N., Brovko, L. Yu., and Berezin, I. V.,**
' Immobilized firefly luciferase and its use in analysis ',
Anal. Lett., **13**, 881-892 (1980).
11. **Ugarova, N. N., Brovko, L. Y., and Beliaieva, E. I.,**
' Immobilization of luciferase from the firefly *Luciola mingrelica*: catalytic properties and thermostability of the enzyme immobilized on cellulose films ',
Enzyme Microb. Technol., **5**, 60-64 (1983).
12. **Brovko, L. Yu., Ugarova, N. N., Vasileva, T. E., Domroski, V. A., and Berezin, I. V.,**
' Use of immobilized firefly luciferase for quantitative determination of ATP and enzymes that synthesize and destroy ATP ',
Biochemistry SSR, **43**, 633-639 (1978).
13. **Brovko, L. Yu, and Ugarova N. N.,**
' Kinetics and mechanism of the inactivation and reactivation of immobilized luciferase of fireflies *Luciola mingrelica* and the role of sulfhydryl groups in these processes ',
Biochemistry SSR, **45**, 607-613 (1980).
14. **Ford, J., and DeLuca, M.,**
' A new assay for picomole levels of androsterone and testosterone using co-immobilized luciferase, oxidoreductase and steroid dehydrogenase ',
Anal. Biochem., **110**, 43-48 (1981).
15. **Wienhausen, G. K., Kricka, L. J., Hinkley, J. E., and DeLuca, M.,**
' Properties of bacterial luciferase/NADH:FMN oxidoreductase and firefly luciferase immobilized onto sepharose ',
Applied Biochem. Biotechnol., **7**, 463-473 (1982).
16. **Wienhausen, G., and DeLuca, M.,**
' Bioluminescent assays of picomole levels of various metabolites using immobilized enzymes ',
Anal. Biochem., **127**, 380-388 (1982).
17. **Roda, A., Kricka, L. J., DeLuca, M., and Hofmann, A. F.,**
'Bioluminescence measurement of primary bile acids using immobilized 7 α - hydroxy-steroid dehydrogenase : application to serum bile acids ',
J. Lip. Res., **23**, 1354-1361 (1982).

18. **Kurkijärvi, K., Raunio, R., and Korpela, T.,**
' Packed-bed reactor of immobilized bacterial bioluminescence enzymes : a potential high-sensitivity detector for automated analyzers ',
Anal. Biochem., **125**, 415-419 (1982).
19. **Watanabe, H., Hastings, J.W., and Tu, S-C.,**
' Activity and subunit functions of immobilized bacterial luciferase ',
Arch. Biochem. Biophys., **215**, 405-413 (1982).
20. **Kricka, L. J., Wienhausen, G. K., Hinkley, J. E., and DeLuca, M.,**
' Automated bioluminescent assays for NADH, glucose-6-phosphate, primary bile acids and ATP ',
Anal. Biochem., **129**, 392-397 (1983).
21. **Ivanova, L. V., Brovko, L. Yu., Shekhovtsova, T. N., Ugarova, N. N., and Dolmanova, I. F.,**
' Bioluminescent method of determining the creatine phosphokinase activity using immobilized firefly luciferase ',
J. Anal. Chem., **41**, 593-599 (1986).
22. **Rossi, S. S., Clayton, L. M., and Hofmann, A. F.,**
' Determination of chenodiol bioequivalence using an immobilized multi-enzyme bioluminescence technique ',
J. Pharm. Sci., 75, 288-290 (1986).
23. **Blum, L. J., and Coulet, P. R.,**
' Bioluminescent determination of reduced nicotinamide adenine dinucleotide with immobilized bacterial luciferase and flavin mononucleotide oxidoreductase on collagen film ',
Anal. Chim. Acta, **161**, 355-358 (1984).
24. **Blum, L. J., Coulet, P. R., and Gautheron, D. C.,**
' Collagen strip with immobilized luciferase for ATP bioluminescent determination ',
Biotechnol. Bioeng., **27**, 232-237 (1985).
25. **Blum, L. J., and Coulet, P. R.,**
' Atypical Kinetics of immobilized Firefly luciferase ',
Biotechnol. Bioeng., **28**, 1154-1158 (1986).
26. **Girotti, S., Roda, A., Ghini, S., Grigolo, B., Carrea, G., and Bovara, R.,**
' Continuous flow analyses of NADH using bacterial bioluminescent enzymes immobilized on nylon ',
Anal. Lett., **17**, 1-12 (1984).
27. **Roda, A., Girotti, S., Ghini, S., Grigolo, B., Carrea, G., and Bovara, R.,**
' Continuous-flow determination of primary bile acids, by bioluminescence, with use of nylon-immobilized bacterial enzymes ',
Clin. Chem., **30**, 206-210 (1984).
28. **Carrea, G., Bovara, R., Mazzola, G., Girotti, S., Roda, A., and Ghini, S.,**
' Bioluminescent continuous-flow assay of adenosine 5'-triphosphate using firefly luciferase immobilized on nylon tubes ',
Anal. Chem., **58**, 331-333 (1986).
29. **Roda, A., Girotti, S., and Carrea, G.,**
' Flow systems utilizing nylon-immobilized enzymes ',
in *Methods in enzymology*, Vol. 133, DeLuca, M. A., and McElroy, W. D., Eds., Academic Press, New York, 1986, pp 238-248.

30. **Kobayashi, T., Saga, K., Shimizu, S., and Goto, T.,**
' The application of chemiluminescence of a Cypridina luciferin analog to immobilized enzyme sensors ', *Agric. Biol. Chem.*, **45**, 1403-1408 (1981).
31. **Ikariyama, Y., Suzuki, S., and Aizawa, M.,**
' Delayed luminescence of luminol initiated by a membrane-bound peroxidase ', *Applied Biochem. Biotechnol.*, **6**, 223-235 (1981).
32. **Williams III, D. C., Huff, G. F., and Seitz, W.R.,**
' Evaluation of peroxyoxalate chemiluminescence for determination of enzyme generated peroxide ', *Anal. Chem.*, **48**, 1003-1006 (1976).
33. **Petersson, B. A., Hansen, E. H., and Ruzicka, J.,**
' Enzymatic assay by flow injection analysis with detection by chemiluminescence : determination of glucose, creatinine, free cholesterol and lactic acid using an integrated FIA microconduit ', *Anal. Lett.*, **19**, 649-665 (1986).
34. **Koerner, C. A., and Nieman, T. A.,**
' Chemiluminescence flow injection analysis determination of sucrose using enzymatic conversion and a microporous membrane flow cell ', *Anal. Chem.*, **58**, 116-119 (1986).
35. **Hool, K., and Nieman, T. A.,**
' Chemiluminescence analysis in flowing streams with luminol immobilized on silica and controlled-pore glass ', *Anal. Chem.*, **59**, 869-872 (1987).
36. **Freeman, T. M., and Seitz, W.R.,**
' Chemiluminescence fiber optic probe for hydrogen peroxide based on the luminol reaction ', *Anal. Chem.*, **50**, 1242-1246 (1978).
37. **Aizawa, M., Ikariyama, Y., and Kuno, H.,**
' Photovoltaic determination of hydrogen peroxide with a biophotodiode ', *Anal. Lett.*, **17**, 555-564 (1984).
38. **Blum, L. J., Plaza, J. M., and Coulet, P. R.,**
' Chemiluminescent analyte microdetection based on the luminol H_2O_2 reaction using peroxidase immobilized on new synthetic membranes ', *Anal. Lett.*, **20**, 317-326 (1987).
39. **Bostick, D. T., and Hercules D. M.,**
' Quantitative determination of blood glucose using enzyme induced chemiluminescence of luminol ', *Anal. Chem.*, **47**, 447-452 (1975).

IMMOBILIZED BIOSPECIFIC PROTEINS IN ANALYTICAL CLINICAL CHEMISTRY

A. Roda*, S. Girotti, G. Carrea**,
P. Simoni, M.A. Angellotti, R. Rizzoli.
* Institute of Analytical Chemistry,
University of Messina, Institute of Chemical
Science, University of Bologna, ** Institute
of Chemistry of Hormones, CNR Milan
Italy

ABSTRACT.

Immobilized proteins are widely used in the clinical laboratory, e.g., immobilized enzymes in assay for analytes such as glucose, uric acid and immobilized antigen and antibodies in immunoassay.
In the enzymatic analysis the use of immobilized enzymes make these reagents recoverable and thus reusable and in most cases increases their stability and catalytic activity. In conjuction with bioluminescent enzymes (firefly and bacterial luciferases) we set up high sensitive flow methods based on the use of nylon tube coil as solid support.
All the NAD(P)/NAD(P)H dependent dehydrogenases (bacterial luciferase), ATP dependent enzymes (firefly luciferase) and oxidases producing $H_2 O_2$ (peroxidase) can be immobilized and a large variety of analytes have been sensitively measured.
In the immunoassay the immobilization of an antibody facilitate the separation of the protein from other components in solution i.e. For example an antigen-antibody complex. The combination of immobilized antibody with luminescent detection of the antigen-antibody reaction gave very specific and sensitive immunoluminescent metodologies.

INTRODUCTION

The use of immobilized biospecific proteins such as enzymes or antibodies is a very popular and attractive tool in applied analytical chemistry.
The complexity of a biological matrix and the relative low

G. G. Guilbault and M. Mascini (eds.),
Analytical Uses of Immobilized Biological Compounds for Detection, Medical and Industrial Uses, 249–265.

concentration of analytes present in serum or urine speciments requires the use of high specific, selective and sensitive methods.

In the clinical chemistry field there is also a continuous need of automation in order to reduce the analytical variability and to control accurately all the parameters of a particular test. Temperature, pH, time of the specific reaction, an appropriate elaboration of the signal, and an adequate data processing are important factors to be controlled in order to reduce coefficent of variation of the assay which in many methods is often above 10-20%.

Two approaches have been so far utilized according to the expected concentration of the analyte in the sample:

1) a specific enzyme for the analyte with "end point" measurement of the direct product or coupled with other enzymes able to catalitically amplifly the production of a compound that can be sensitively measured. These systems include in the majority of the methods a spectophotometric fluorescent or bio and chemiluminescent measurement of the product which is directly related to the concentration of the analyte (1-5).

2) if a subconcentration measurement is needed i.e. at pmoles or femtomoles level, the widest methods used are those which utilize antibodies. The high specificity and affinity of the antibody for its antigen (analyte) in order of 10^9 - 10^{10} L/M in conjuction with an appropriate detection of the antigen-antibody reaction permit a potential measurement of all proteins and also molecules with a molecular weight up to 150-200 such as steroids drugs, etc.

The high sensitivity of the immunological methods such as RIA, EIA, IRMA depends mainly on two factors: high affinity constant of the antibody and high specific activity i.e. high signal/mass ratio of the "tracer" used. In addition the imprecision of the methods which is high at low dose in the competitive one greatly affect the performance of the assay.

The heterogeneous immunological methods require a separation of the antigen-antibody complex from the free antigen, which can be easly achieved if the antibody in the system is immobilized on a solid support.

Many "tracers" have been developed and applied and starting from the radioisotope ^{125}I, now the most popular include enzymes, fluorescent and chemiluminescent targets. Both in the enzymatic and immunological methods the use of immobilized reagents offer an enormous advantage allowing to reduce the cost and increase the performance of the assay.

In the present paper we report our experience on the use of immobilized enzymes and antibodies in the development of sensitive methods applied on biological matrix with enphasis on the use of luminescence as a detection system.

IMMOBILIZED LUMINESCENT SYSTEMS

Firefly Bioluminescence

Firefly luminescence is the most extensively studied bioluminescent system (6). The light-producing reaction requires the enzyme luciferase, luciferin, Mg^{2+}. ATP, and molecular oxygen, wich give an oxyluciferyl adenylate-enzyme complex in the excited state. After emission (560 nm) the ground-state complex dissociates to form enzyme, AMP, pyrophosphate (PPi), oxyluciferin , and carbon dioxide, the last being derived from the carboxyl group of luciferin.

The firefly luciferase allows the determination of ATP and ATP dependent systems by catalizing the reaction:

$$ATP + Luciferin + O_2 \xrightarrow{\text{Luciferase}} AMP + oxyluciferin + PPi + CO_2 + LIGHT$$

where PPi: Pyrophosphate.

Since ATP is such an ubiquitous and important metabolite, the firefly bioluminescence reaction is widely used for the assay of ATP concentrations in various types of material (Table 1).
In addition the concentration of other metabolites or the activity of enzymes participating in coupled reactions leading to the formation or consumption of ATP can be assayed with the aid of the firefly luciferase dependent reaction (1-6).

The bioluminescent enzyme system from marine bacteria (7) consist of an NAD(P)H:FMN oxidoreductase and a luciferase which emit light at 490 nm in the presence of FMN, NAD(P)H, a long chain aliphatic aldehyde and molecular oxygen according to reactions (1) an (2).

$$1.\ NAD(P)H + FMN + H^+ \xrightarrow{\text{NAD(P)H:FMN oxidoreductase}} NAD(P)^+ + FMNH_2$$

$$2.\ FMNH_2 + RCHO + O_2 \xrightarrow{\text{luciferase}} FMN + RCOOH + H_2O + LIGHT$$

Luciferase acitvity is highly specific for $FMNH_2$, but the enzyme also shows weak activity towards other flavins and flavin analogs. Only aliphatic aldehydes with a chain length of eight or more carbon atoms are effective in the luminescent reaction.

TABLE I - Analytical applications of ATP bioluminescent assay

A) Oxidative phosphorilation and electron transport
Mitochondria
Bacterial chromatophores
B) Bacteriuria screening
Urine
C) Energy metabolism
Epidermal cells
Spermatozoa
Bacterial cells
Pancreatic islets
D) Hematology
Erythrocytes
Platelets
E) Cell lysis
Various mammalian cells
F) Antibiotic determination
G) Antibiotic susceptibility testing
H) Vitamin determination
I) Biomass determination
Marine water
Wastewater
Beer
Milk
Drinking water
Yeast
Soil
Marine and lake sediment

Bacterial Bioluminescence

Various biologically important substances and enzyme activities can be analyzed by coupling the luciferase and the oxidoreductase to a third reaction which produces NADH or NADPH.

$$(3)\ XH_2 + NAD(P)^+ \xrightarrow{E} X + NAD(P)H + H^+$$

Either XH_2, the reduced substrate or the dehydrogenase (E) catalyzing the oxidation of XH_2 , can be measured.

Chemiluminescence

One of the oxidizing systems that reacts with luminol to generate chemiluminescence involves hydroge peroxide as the oxidizing agent and peroxidase (POD). The amount of H_2O_2 which derives from the catalysis of a specific oxidase may be assayed by adding excess POD and luminol and measuring steady state chemiluminescence intensity according the scheme:

$$\text{Luminol} + 2H_2O + OH^- \xrightarrow{POD} \text{Aminophthalate} + 3H_2O + N_2 + \text{LIGHT}$$

The use of enhancer such as luciferin or p-iodophenol in the system allows to further increase the light emission and to transform the signal from a flash to a steady state output.

IMMOBILIZED FLOW SYSTEMS

These luminescent reactions are of special interest since they can be utilized as detector systems for a lot of other reactions catalyzed by dehydrogenases, oxidases and other enzymes, which lead to the production of NAD(P)H, ATP and H_2O_2 by means of stable intermediates.

Enzyme immobilization

The extreme sensitivity and specificity of the light producing enzymes from marine bacteria and firefly gave rise to many analytical applications of the soluble systems (1). Furthermore the use of immobilized enzymes for analytical purposes adds other advantages.

In general the immobilized enzymes are more stable than the soluble forms. The sensitivity of the assay is greatly increased due to the fact that the immobilized luciferase is present in a micro-environment with locally high concentrations of ATP or $FMNH_2$. For example, coimmobilized NADH:FMN oxidoreductase and bacterial luciferase produce about 100 times more light per pmole of NADH than the comparable amounts of soluble enzymes (8).

A variety of procedures are available for coupling proteins to insoluble solid supports and these may be polyacrylic hydrazide, porous glass, amino alkylated glass rods, cellulose films, collagen, Sepharose 4B and Nylon. The chemical methods of immobilization give better yields of active immobilized luciferases, and Sepharose, collagen and Nylon (9-11) have proved to be the most effective of the different solid supports which have been investigated. Another important aspect is that immobilized enzymes can be incorporated into flow cells where they can be used for multiple assays.

In our laboratories we developed flow systems for the assay of several metabolites (Table 2) in biological samples (serum and saliva) using bioluminescent enzymes and specific dehydrogenases immobilized on nylon tubes activated with triethyloxonium tetrafluoroborate.

TABLE II - Bioluminescent flow systems

Analyte	linear range	solid support	reference
ATP	1-10 nmol	BSA Gel	17
NADH	1-10 nmol	BSA Gel	"
NADH	1 pmol-10 nmol	Sepharose	18
ATP	3-3000 pmol	"	15
NADH	6-600 pmol	"	"
Cholylglycine	8-325 pmol	"	"
Chenodeoxycholyl-glycine	8-325 pmol	"	"
Cholyltaurine	4-750 pmol	"	"
Glucose-6-phosphate	1.5-1500 pmol	"	"
NADH	1-2500 pmol	Nylon	11
7α-bile acids	10-1500 pmol	"	12
3α-bile acids	1-1500 pmol	"	19
12α-bile acids	10-1500 pmol	"	19
Ethanol	50-1000 pmol	"	13
Glycerol	50-500 pmol	"	19
Acetic Aldeyde	50-1000 pmol	"	19
L-Alanine	10-500 pmol	"	19
ATP	0.3-100 pmol	"	14
ADP	0.3-100 pmol	"	14
NADH	0.10-1000 pmol	"	16

The manifold developed for bioluminescent continuous-flow determination of ATP is shown in Fig.1. The flow system involves two streams: the first is the working bioluminescent solution and the second a continuous flow of air into which a known volume (5-100 μl) of sample is intermittently added.
A similar manifold has been developed using bacterial luciferase for the assay of NAD(P)H or bile acid with coimmobilized 3 α-hydroxysteroid dehydrogenase.

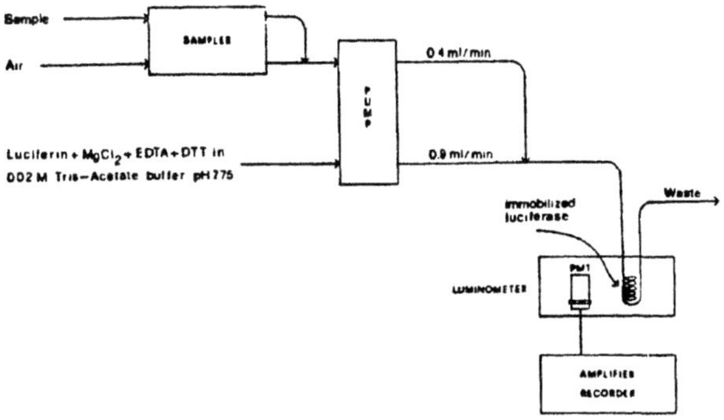

Fig. 1: Manifold for the assay of ATP using nylon immobilized firefly luciferase.

The use of coimmobilized system i.e. with the analyte specific enzyme toghether with the luminescent enzymes restrict the pH value compatible for the three enzymes thus reducing the sensitivity and practicability of the test .
In order to over come this problem we develop a system which utilize two separate nylon coil: the first containing the specific enzyme and the second one the luminescent enzymes.

For the analysis by means of separately immobilized enzymes (Fig. 2), the flow system involves four streams: the first supplies the bioluminescent enzymes with the working bioluminescent solution, the second and third supply the immobilized dehydrogenases, placed outside the luminometer, with the substrates. The fourth is a continuous flow of air into which a known volume of sample is intermittently added.

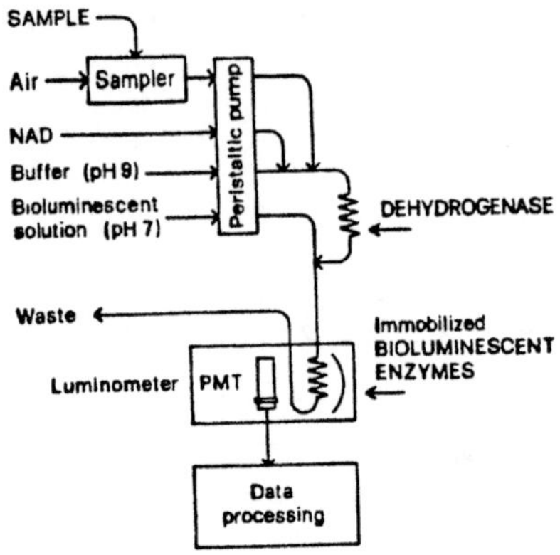

Fig. 2: Continuous flow system using two separate coils: a specific analyte dehydrogenase and Bioluminescent enzymes (bacterial luciferase, FMN oxidoreductase

The bioluminescent reactor - a 0.5-1m, coil of Nylon tube containing the luminescent enzymes - is wound around a plexiglass support and positioned inside the luminometer in front of the photomultiplier window. The luminometer we use is the Model 1250 (LKB, Wallac), which required only slight modificatons of the original light-recording system. The sample, is aspirated uniformely without fragmentation. Operational steady-state is reached in about 5 min after a preliminary washing with phosphate buffer containing dithiothreitol.

Serum bile acids analyzed by the bioluminescent method gave results in good agreement with those obtained by radioimmunoassay, enzyme-immunoassay, and high performance liquid chromatography (12). Two serum samples with low and high concentration of bile acids were assayed to determine intra and interassay variations. The relative standard deviation values were lower than 8% or 10% at low or high concentration, respectively. Up to 25 samples per hour were analyzed with no carry-over. Also the alcohol concentration determined in saliva and serum was in good agreement with the literature data (13).

Similarly, the firefly luciferase, immobilized on Nylon coils, allowed the determination of ATP and ADP at pmol levels in human platelets in a simple and reliable way (14).

In addition the use of immobilized peroxidase allows to analyze many analytes by using oxidase enzyme (glucoseoxidase, amminoacidoxidase, oxalate oxidase etc.). By coupling luminol reaction to the reaction catalized at acidic pH (3.8) by oxalate oxidase (OXA) it is possible to determine oxalate:

$$\text{Oxalate} + O_2 \xrightarrow{\text{OXA}} 2\ CO_2 + H_2O$$

The analysis of oxalate in urine is important for the diagnosis of hyperoxaluria and kidney stone disorder.

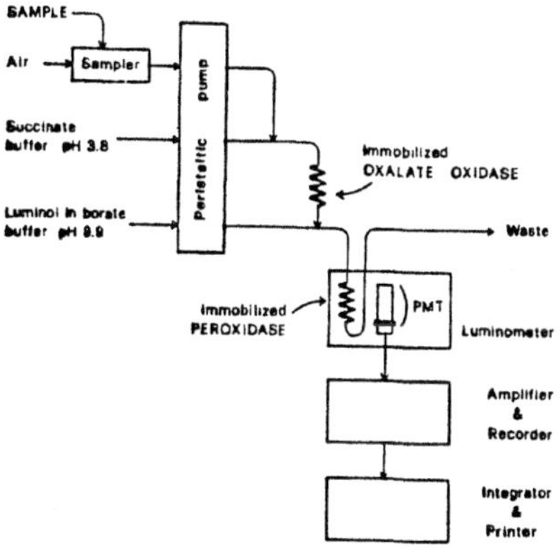

Fig. 3: Manifold for the assay of H_2O_2 or oxidase using immobilized peroxidase.

The use of these 3 systems permits to analyze a large variety of analytes and enzymes useful in clinical chemistry. The main advantage of these systems is the extremely high sensistivity and selectivity with the possibility to develop "real time" luminescent specific biosensor for in vivo analysis via optic fiber trasmission of the emitted light. Moreoven these flow system are easily suitable for a complete automation with a computer control af all the analytical variables (flow, pH,temperature etc.)

A typical time courses of light peaks using flow systems is reported in Fig. 4.

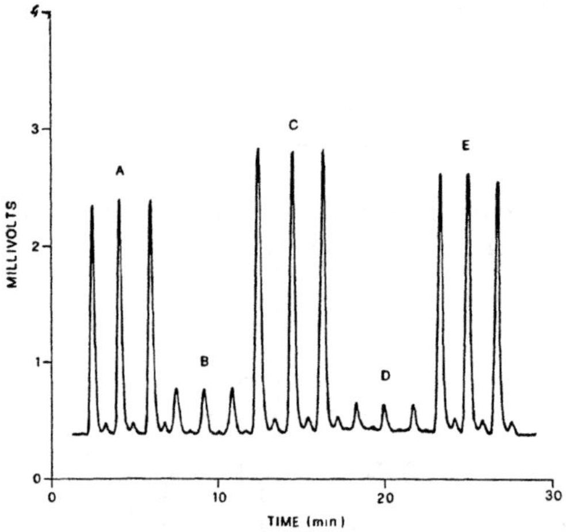

Figure 4: Time courses of light peaks for the bioluminescent assay of L-Alanine. A: L-alanine standard (570 pmol); B: sample (115 pmol); C: A + B; D: sample (66 pmol); E: A + D.

IMMOBILIZED ANTIBODIES

In recent years analytical methods using enzymes and fluorophores have been replaced radioactive tracers for heterogeneous high sensitivity immunoassay(20). One of the most common system used today is the enzyme immunoassay (EIA) in which horse radish-peroxidase or orther enzymes are used as a tracer(21-25).

The main disadvantage associated with colorimetric determination of enzyme activity in immunoassay is that the procedure must be accurately timed and generally require further 20-30 min. after the immunological reaction for an appropriate development of the coloured product.

The use of a chemiluminescent detection of POD have been widely used and enormous advantages have been noted, which includes an increased sensitivity associated with a short incubation time (26-27). The main disadvantage using luminescence is that the signal is time and pH dependent and its measurement require the use of an integrator device. More recently has been described a system which include not only the luminol/(H_2O_2) but also an additional component which acts as enhancer of the light intensity. These molecules include luciferin, 6hydroxy benzothiazole, p-iodophenol and their complex mechanism of action has not be so far a lucidated. Because the high light levels the reaction can be optimized to give a continuous output of light, rather than a flash.

Furthemore such enzymatic tracer require the use of a clear system in order to avoid interferences and quenching. This goal is reached using immobilized system in which the antibody is immobilized on a solid macro support.

When the analyte is a small molecule such as a steroid or a drug the immunological method is necessarly of competitive type which includes a limited and fixed amount of antibody in the assay tube.

Therefore the immobilization is a key factor in determining the analytical performance of the assay. The physical absorption on polystirene or polypropylene support is the most popular and used method, and in order to achieve a good immobilization, pH, ionic strength, temperature, protein concentration must be accurately controlled . Using this procedure we developed an hight sensitive assay for plasma progesterone based on the use of monoclonal antibodies to progesterone 6-HS (Kaff=10^{10}: 1/M) immobilized on polystyrene microtiter strips. As enzymatic tracer we synthesized a progesterone-3 CMO-POD i.e. the method presents heterology of derivative and position

The developed method require 30 min. of immunological reaction at 25°C in which 100µl of enzymatic tracer is mixed with 100µl of sample or standard followed by a washing step and 10 min. of colorimetric reaction using o-phenylendiamine /H_2O_2 as substrate and stopping the enzymatic activity with H_2SO_4 4N.

The sensitivity is high (0.01 ng/ml) and the assay can be performed directly on plasma samples (50 µl). The dose response curve range from 10 to 1000 pg/tube and the coefficent of variation is alway less than 5% both at low and high doses (Fig.5). The sensitivity can be increased using a luminometric detection of the enzymatic activity of the tracer using luminol / H_2O_2. In this system the time of immonological reaction can be reduce to 5-10 min. (non equilibrium conditions) and the detection of the enzymatic activity to few seconds(Fig.5). The method fulfill all the requisite of reproducibility, accuracy, precision and is suitable of a complete automation.

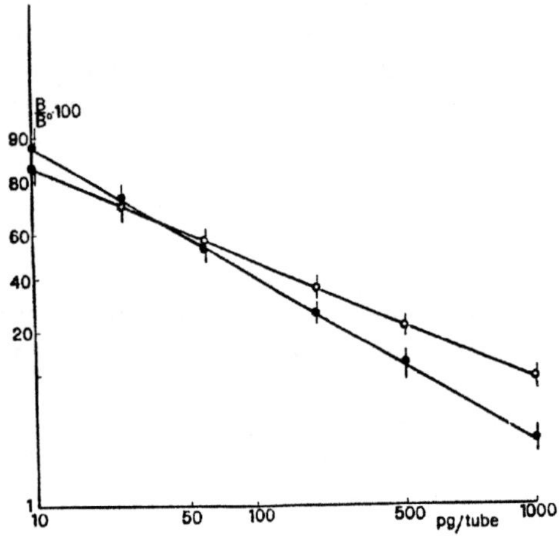

Figure 5: Dose response curve for the assay of progesterone (logit/log plot). Each point is the mean values ± SD. (-O-) colorimetric detection (30 min immunological reaction + 10 min /o-phenilendiamine/H_2O_2). (-●-) luminometric detection (10 min immunological reaction + 30 sec luminol/H_2O_2/ enhancer).

The immobilized antibody presents an elevate stability and the loss of immunoreactivity after one year, if stored dry at 4°C, is less than 10% as reported in Fig 6.The kinetic of the antigen-antibody reaction is slow when the antibody is immobilized on the macro phase as reported in Fig.7. The only problem associated with the use of immobilized antibody in the assay of small molecules such as steroides,is the aspecific binding which is drastically reduced by a second coating with bovine serum albumin. Moreover when a protein molecule must be analyzed a not ideal binding occurs particulary at high dose.

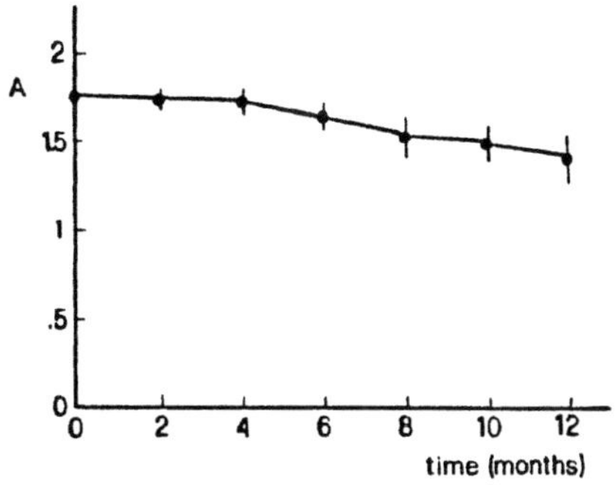

Figure 6: Stability of the polystyrene immobilized antibody when stored dry at 4°C. Each point represents the means values $\pm$ SD of the maximum absorbance value (A) i.e. the maximum binding capacity.

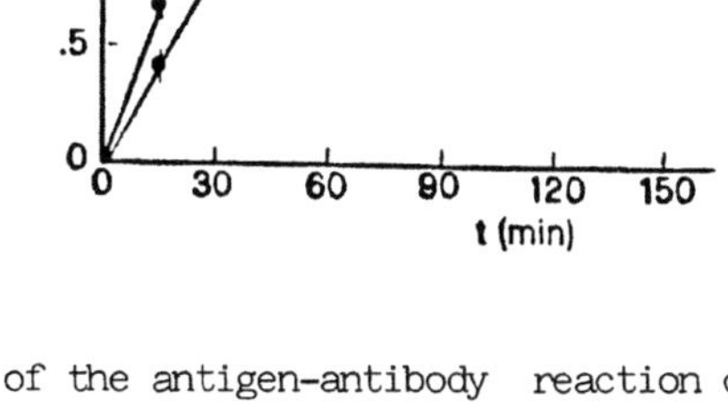

Figure 7: Kinetic of the antigen-antibody reaction of the antibody in solution and immobilized on polystyrene micro strip. The ordinate represents the maximum absorbance value (A) i.e. the maximum binding capacity.

A prozone phenomenon resulting in a hooked curve when absorbance is plotted versus antibody concentration has been observed. It is presumed that adsorption to polystyrene may destroy or makes unaivable same of the antigen site (steric hindrance). According to the structure of the antibody, i.e. of an Immunoglobulin IgG, is possible to drive immunologically the immobilization procedure in order to orient the binding sites (Fab portion).

The first coating on the polistyrene support is performed with an Anti FcIgG and the second one with the specic antibody. In this case the majority of the specific antibody is bound to the surface with its Fc portion and the Fab one is oriented toward the solution thus facilitate the binding with the antigen. In these conditions is possible to increase the performance of the assay, to eliminate the hook effect and to further reduce the time of the immunological reaction and to develop "real time immunoassay" or flow systems.

CONCLUSION

The bioluminescent enzymes immobilized on nylon tubes make possible to specifically assay NAD(P)H, ATP, ADP, bile acids, ethanol, glycerol and aldehyde at pmol levels. The precision of the method as well as its correlation with other methods such as radioimmunoassay, enzymeimmunoassay and high performance liquid chromatography is satisfactory (Table 2).

The adopted continuous-flow system is simple, requires only minor modifications of a commercial detector, and allows to analyze about 20-30 samples per hour. Unlike Sepharose column, Nylon reactors present no problems with packing or disruption of gel matrix or bacterial growth which markedly enhances background light level. This together with its handiness, makes the Nylon tubes very suitable enzyme support for continuous-flow analysis, inspite of the relatively low activity recovery of immobilized enzymes, which, however, is sufficient to achieve high sensitivity. Up to 700-900 samples can be analyzed with use of only a few milligrams of enzymes, and therefore the continuous-flow bioluminescent method appears highly competitive with other methods such as radioimmunoassay, enzymeimmunoassay, high performance liquid chromatography and fluorometry where radioactive materials, separation steps, sample manipulation or expensive equipment are needed.

Potentially, a variety of other NAD(P)H, ATP generating metabolites, H_2O_2 producing substrate could be analyzed using the bioluminescent reactor (immobilized bacterial or firefly luciferase or peroxidase) coupled with a proper immobilized enzyme.

As far as the use of immobilized antibody is concerned the high affinity of the recent produced monoclonal antibody allow to set up very rapid and sensitive methods. If a chemiluminescent detection of the enzymatic tracer is used the performance of these methodology can be further improved thank to the extremely high signal/mass ratio of this tracer associated with a very low background of the light.

The combination of two or more enzymes or antibodies are universal methodologies and thanks to their reusability, offer economic analytical methods not only in clinical chemistry but also in other fields of applied analytical chemistry such as food analysis, ecology and pharmacology.

REFERENCES

1) A. RODA, S. GIROTTI, B. GRIGOLO, S. GHINI . Giorn It. Chim. Clin, **8**,309,1983.

2) M.DELUCA, Method Enzymol., **57**,1978.

3) L.J.KRICKA, T.J.N. CARTER, 'Clinical and Biochemical Applications of Luminescence', Marcel Dekker, Inc, New York,1982.

4) L.J.KRICKA, P.E. STANLEY, G.H.G. THORPE, T.P. WHITEHEAD 'Analytical Applications of Bioluminescence and Chemiluminescence', Academic Press, London,1984.

5) T.P. WHITEHEAD, L.J. KRICKA, T.J. N. CARTER, G.H.G. THORPE, Clin. Chem., **25**,1531,1979.

6) F.R.LEACH, J.Appl. Biochem., **3**,473,1981.

7) J.W.HASTINGS, T.O. BALDWIN, M.Z. NICOLI, Method Enzymol., **57**,135,1978.

8) G. WIENHAUSEN, M. DELUCA, Anal.Biochem., **127**,380,1982.

9) L.J. KRICKA, K. GREEN, Trends Anal. Chem. **2**,244,1983.

10) L.J. BLUM, P.R. COULET, D.C. GAUTHERON, Biotechn. Bioengin. **27**,232,1985.

11) S. GIROTTI, A. RODA, S. GHINI, B. GRIGOLO, G. CARREA R. BOVARA, Anal. Letters, **17**,1,1984.

12) A. RODA, S. GIROTTI, S. GHINI, B. GRIGOLO, G. CARREA, R. BOVARA, Clin. Chem., **30**,206,1984.

13) S. GIROTTI, A. RODA, S. GHINI, A. PIACENTINI, G. CARREA, R. BOVARA, Anal. Chim. Acta, **183**, 187,1986 .

14) G. CARREA, R. BOVARA, G. MAZZOLA, S. GIROTTI, A. RODA,S. GHINI, Anal. Chem., **58**, 331,1986 .

15) L.J. KRICKA, G.K. WIENHAUSEN, J. HINKLEY, M. DELUCA, Anal. Biochem. **129**, 392, 1933

16) D. VELLOM, J. HINKLEY, A. LOUCKS, H. EGGHART, M. DELUCA, in 'Analytical Applications of Bioluminescence and Chemiluminescence'. L. J. Kricka Editor, p. 133, Accademic Press Inc., London 1984.

17) O. RODRIGUEZ, G.C. GUIBAULT, Enzyme Microbiol. Technol. **3**, 69, 1981.

18) K. KURKIJARVI, R. RAUNIO, T. KORPELA, Anal. Biochem.,**125**, 415, 1982.

19) A. RODA, S. GIROTTI, G. CARREA ,Method Enzymol., **133**, 238, 1986.

20) L.J. KRICKA,'Ligand-binder assays', New York Marcerl Dekker, 1985.

21) L.E.M. MILES, C.P. BIEBER, L.F. ENG, D.A. LIPSCHITZ, IEA, **149**, 1974.

22) E. ENGVALL, PEARLMANN, Immunochemistry **8**, 871, 1971.

23) G.B. WISDOM, Clin. Chem. Biochem., **18**, 197-208, 1980.

24) A.H.W.M. SCHUURS and B.K. VAN WEEMEN, Clin. Chim. Acta, **81**, 1-40, 1977

25) B.K. VAN WEEMEN and A.H.W.M., Febs Lett., **15**, 232-236, 1971.

26) A. RODA, S. GIROTTI, S. LODI, S. PRETI, Talanta, **31**, 895, 1984

27) A. RODA, S. GIROTTI, A.L. PIACENTINI, S. PRETI, S. LODI, Anal.Biochem., **156**, 267, 1986.

IMMUNOSENSORS

Isao Karube
Tokyo Institute of Technology, Nagatsuta
Midori-ku, Yokohama 227 JAPAN
and
Masao Gotoh, NOK Corporation,
Fujisawa 251 JAPAN

ABSTRACT The development of practical immunosensors is an important topic for biosensor research. Recently the authors have demonstrated novel immunosensors, such as the immuno-FET sensor, the piezoimmunosensor, the pulse immunoassay and the bio-image sensor. The immuno-FET sensor was constructed using an immobilized anti-HSA poly-vinylbutyral membrane and an ISFET. A linear relationship was obtained between the initial rate of response and the logarithm of HSA concentration from 0.01 to 1 mg/ml. Piezo-immunocrystals, surface coated with immobilized anti-*Candida* antibody, were used to detect *C. albicans*. The resonant frequency shift due to the reaction with *C. albicans* correlated with the concentration of *C. albicans* in the range 10^6 - 5 x 10^8 cells/cm^{-3}. A novel immunoassay using an electric pulse was proposed and applied to the detection of *C. albicans*, an antigen with a size of several micrometers. The estimation of *C. albicans* concentration was possible in the range of 10^7 - 6 x 10^7 cells cells/ml. Using the bio-image sensor, quantitative detection of guinea pig haptocarcinoma cells in a mixed cell suspension was carried out. Hepatocaricinoma cells could be detected quantitatively in the range 10-100 % (target cell content), a linear relationship being observed between the target cell content and image area in this same range.

1. INTRODUCTION

Many sensors consisting of bio-active substances (enzymes, micro-organisms, organelles, antigens or antibodies) and electrochemical devices have been developed for clinical analysis [1]. Substances such as glucose may be specifically determined by these devices, but few sensors have been developed for determining macromolecules. Usually these molecules are measured by immunoassay.

G. G. Guilbault and M. Mascini (eds.),
Analytical Uses of Immobilized Biological Compounds for Detection, Medical and Industrial Uses, 267–279.

Immunoassays are based on the specific binding reaction that occurs between antigens and their antibodies. Antigens, including proteins, peptides, drugs and microorganisms can be determined by immunoassay. Binding results in the formation of antigen-antibody complexes. The amount of agglutinated adduct can be measured by visual inspection, turbidimetry, or weighing after centrifugation. When the amount is too small to be detected by these methods, radioisotopes [2], enzymes [3], and fluorescent dyes are used as labels for these reactions. Widely used are enzyme immunoassays based on anitgens, haptens or antibodies labeled with enzymes, forming a group of binding assays in which the molecular recognition properties of antibodies are used. Recently such methods have been applied in histochemistry and cytochemistry, and they are regarded as alternatives to radioimmunoassay.

Immunosensors based on a similar principle have been demonstrated [4-6]. These sensors are simple and easy to use, but they require long incubation times, and washing steps to remove the unbound antigens. Therefore new immunosensors based on novel principles are required. In this chapter, we describe four types of novel immunosensors.

2. IMMUNO-FET SENSOR

The determination of HSA (Human Serum Albumin) is important for diagonostic testing in clinical fields. HSA is known as the main protein in blood and an indicator of protein metabolism disease. On the other hand, the ISFET (Ion Sensitive Field Effect Transistor) has developed as a micro device based on silicon fabrication technology. The ISFET is attractive because of its small size, ease of mass production and rapid response. The ISFET device was first reported by Bergveld in 1970 [7]. Later, Matsuo et al. improved the ISFET by using silicon nitride as the gate insulator, reporting that it could be used as a micro-ion sensitive device, for example, for proton, sodium ion, potassium ion, and surface charge [8,9]. In 1980, Janata et al. reported the enzyme FET, using an ISFET as the electrochemical device [10]. Since then, several studies on the application of ISFETs to enzymatic analysis have been reported [11-13]. We have also previously reported a micro-biosensor that utilizated an ISFET [14-17]. In this section, a novel bioelectrochemical system for HSA, consisting of the ISFET and the immobilized anti-HSA polyvinylbutyral membrane, is described.

The sensor system consists of two ISFETs: one, an ISFET having a cross-linked polyvinylbutyral-anti-HSA membrane (HSA-sensitive ISFET, AHSAFET) and the other an ISFET having only a cross-linked polyvinylbutyral membrane (referentce ISFET, REFFET). A polyvinylbutyral membrane was formed over the gate insulator of the two ISFETs using dropping method as follows: 0.1 g of polyvinylbutyral and 1 ml of 1,8 diamonino-4-aminomethyloctane were dissolved in 10 ml of dichloromethane. After stirring for approximately 30 minutes, this polymer solution was dropped over the gate insulator of the two ISFETs. The ISFETs were then immersed in 5% glutaraldehyde solution at room temperature for approximately one day to advance the cross linking reaction.

Anti-HSA was immobilized on the polyvinylbutyral membrane covering the gate insulator of the AHSAFET, by immersing the tip into a 5 mg/ml anti-HSA solution of 4°C for approximately one day. Response of the FET to HSA was performed in a differential voltage mode,vis-a-vis mesuring the difference in gate voltage output between the HSA sensitive gate and the reference gate. The circuit diagram is shown in Fig. 1. An Ag/AgCl reference electrode was placed directly in solution with the AHSAFET and the REFFET. The gate voltage was measured between the Ag/AgCl reference electrode and the source of the AHSAFET and the REFFET. Variation of surface charge effects a change in the surface potential on the silicon nitride insulator of the AHSAFET and the REFFET, giving rise to a proportional change in the gate output voltage. In this case, the voltage between source and drain is held constant at 4.0 V, the current between source and drain also being constant at 500μ A. The differential gate output voltage between the AHSAFET and the REFFET was displayed on a recorder (TOA Electronics, Model EPR-100A). The total volume of solution in the cell was 3.9 ml, and was composed of 10 mM Tris-HCl buffer. HSA smaples were also made up in this solution. The cell temperature was thermostatically controlled at 37 ± 1°C. The AHSAFET, REFFET and Ag/AgCl reference

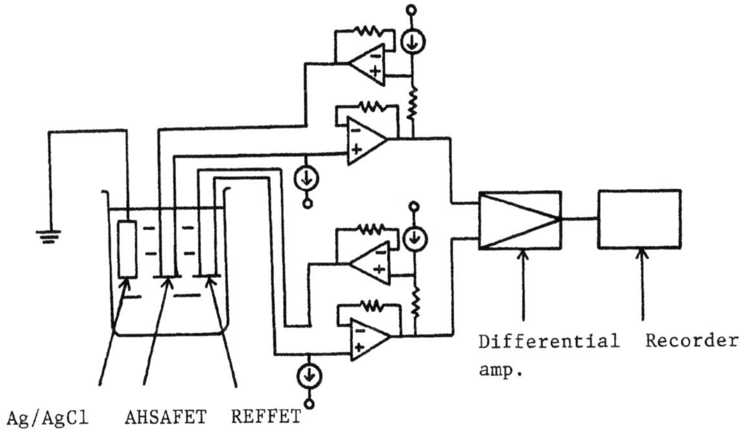

Figure 1. Circuit diagram of measuring system

electrode were allowed to sit in this solution for approximately 10-20 minutes to allow the differential gate output voltage to reach a steady state. A 100 μl aliquot of HSA sample solution was then injected into the cell and the differential gate output voltage change was recorded.

The differential gate output voltage change caused by binding the HSA to the anti-HSA was measured. Figure 2 shows the response curve of the immuno-FET sensor system. The experimental condition was pH 7.0 at 37°C. Since the isoelectric point of the HSA molecule is pH 4.8, it has a negative charge at pH 7.0. The negative shift in figure 2 was considered to be result of a complex formation of HSA and anti-HSA. The HSA negative charge bound to the anti-HSA alters the surface potential of the gate, producing a decrease in the gate output voltage.

The initial rate of the differential gate output voltage the change after injection was plotted against the logarithmic value of HSA concentration. Figure 3 shows the calibration curve. A linear relationship was obtained in the range 0.01-1 mg/ml HSA. No detectable response was observed below 0.01 mg/ml HSA compared to the buffer solution. Above 1 mg/ml, the response did not increase appreciably.

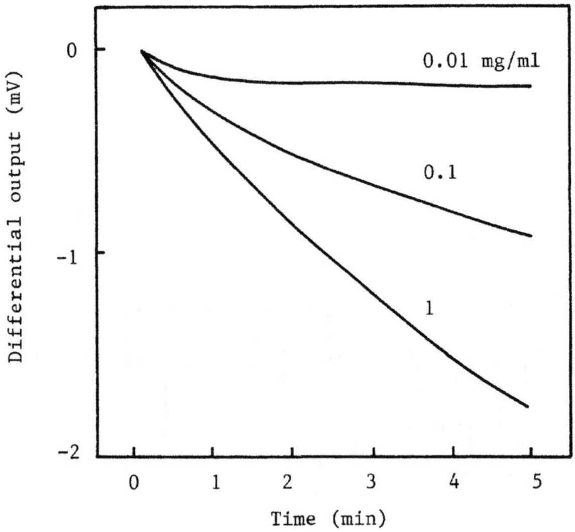

Figure 2. Responses to HSA

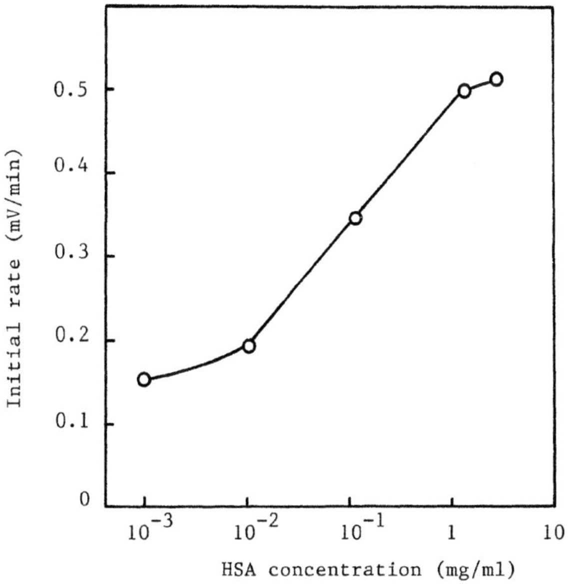

Figure 3. Calibration curve of immuno-FET sensor

The effect of pH on the output of the immuno-FET sensor system was investigated. The isoelectric point of the HSA molecule is pH 4.8. Below this point it has positive charge. Above this point, a negative charge. As the result of experimental conditions, the response increased in proportion to the pH value (pH 7.0, 9.0) because of the negative charge increasing. On the other hand, in acidic conditions (pH 3.0, 4.0), a response was not observed. Generally, the antibody is stable in alkaline condition and likely to denature in acidic conditions. The difference of responses is attributed to these phenomena.

The reusability of the immuno-FET sensor system was investigated. The HSA release from anti-HSA was done by immersion in 4 M $MgCl_2$. Between measurements, the AHSAFET and REFFET were stored at 4^oC. In five assay, the relative activity decreased by approximately 80%. This loss of activity is probably a result of insufficient release of HSA from anti-HSA, or anti-HSA denaturation during storage and operation. The immuno-FET sensor is suitable as a disposable sensor.

3. PIEZO-IMMUNOSENSOR

Piezo-electric sensors are based on the measurement of small mass changes on the surface of a piezo-electric crystal resulting from the specific adsorption of molecules onto the specially modified surface. The relationship between surface mass change, dm (grams), and resonant frequency, F (Hz), for a piezo-electric crystal is given by the Sauerbrey equation:

$$dF/F = -dm/A \rho t$$

where A is the crystal area covered by adsorbed material (cm^2), ρ is the density of the quartz (g/cm^3) and t is the thickness of the uncoated crystal (cm). Piezoelectric crystals have been used to detect water in gases, organic compounds in air [18] and traces of metals in solution [19]. Clinical applications include the surface acoustic wave device which has been used to detect Human IgG [20].

Piezo-immunocrystals, surface coated with immobilized anti-*Candida* antibody, were used to detect *C. albicans*. The feasibility and the first application of piezo-electric crystals to determine micro-organism concentration is described below.

The system consisted of an AT cut piezoelectric crystal, oscillator and frequency counter. The surface of the Pd plated electrodes was treated by anodic oxidation, then with γ -aminopro-oyltriethoxysilane (2% in acetone) for 1 h at 25°C. They were air-dried and the resonant frequency of the crystal was measured (F_1). The electrodes were placed in aqueous glutaraldehyde (5%, pH 7) for 3 h. Anti*Candida* antibody was immobilized on the electrodes via the surface aldehyde groups by dipping into the antibody preparation for 30 min. The unreacted aldehyde was then blocked with 0.1 M glycine. Prior to use, sensors were rinsed with 0.5 M sodium chloride to remove any effect resulting from non-specific adsorption.

The treated piezoelectric crystals were dipped in the microbial suspension for 30 min to allow reaction to occur between the immobilized antibody and the microbe. The crystals were then rinsed with 0.5 M sodium chloride and water and dried in air. The resonant frequency, F_2 was measured. The frequency difference, $\Delta F = F_1 - F_2$, was calculated.

The resonant frequency shift due to the reaction with *C. albicans* correlated with the concentration of *C. albicans* in the range 10^6 - 5×10^8 cells/cm^3. Furthermore, *Candida* adsorbed on the surface of the electrodes were also monitored by microscopy, confirming that the magnitude of dF was related to the actual numbers of adsorbed species. The crystal sensor showed no increase in response with microbe concentration for other yeast species, such as *Saccharomyces cerevisiae*, eliminating the possibility of any non-specific interference. The anti-serum used here (*Candida* check No. 4) had high specificity to *C. albicans*. The low value of dF caused by *S. cerevisiae* is a result of non-specific adsorption.

The piezo-immunosensor has great advantages in both its convenience of measurement and its wider effective concentration range. Improvement in the techniques used for antibody immobilization could hopefully improve the piezo-immunosensor sensitivity.

4. PULSE IMMUNOASSAY

Agglutinated adducts can be monitored by visual inspection, turbidimetry, or weighing after centrifugation. However, the reaction proceeds very slowly, sometimes requiring 1-2 days to complete. Therefore, the acceleration of the aggultination is required for rapid immunoassay.

Increased rates of agglutination can be obtained by increasing the contact between antigen and antibody. Mechanical stirrring may not be effective for a minute amount of reaction solution. Increase in temperature is also not effective for biologically active substances including antibodies.

It has been observed that conducting or non-conducting particles, including aluminum powder, carbon powder, potato starch, polystyrene particles, red blood cells, and yeast cells suspended in various fluids, formed linear linkages under the influence of electric fields [21]. When the size of antigens is micrometer-order, agglutination occurs in two steps: 1, binding of an antibody to an antigen; and 2, binding of another antigen to the antibody bound antigen. The electric pulse probably accelerates the second step and increases the total reaction rate. Similar effects were

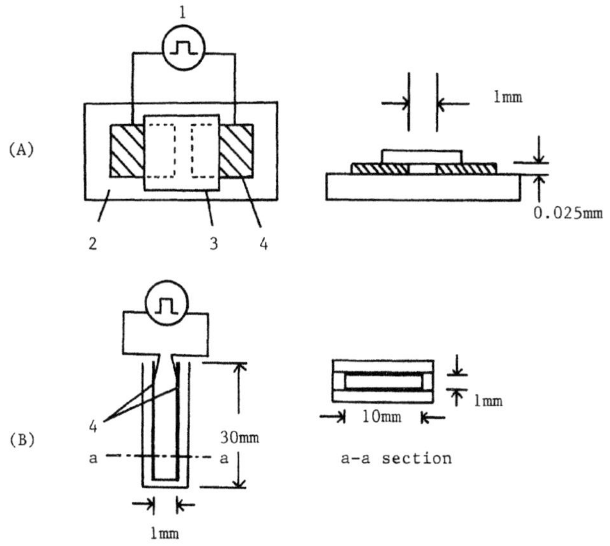

Figure 4. Cells for pulse immunoassay:
(A) slit type; (B) cuvette type.
1, pulser; 2, slide glass;
3, cover glass; 4, electrode.

expected in the case of small antigens (e.g. protein, peptide) and antibodies immobilized on a large particle (e.g. latex bead).

A novel immunoassay using an electric pulse was proposed and applied to the detection of Candida albicans, an antigen with a size of several micrometers. Since yeast cells were known to form linear linkages under electric fields, the immunoreaction of C. albicans and antibody was expected to be accelerated.

Figure 4 shows the cells used for the immunoreaction. The slit cell (A) was composed of two electrodes (electrode distance, 1 mm, electrode thickness, 25μm) and a glass slide. The cuvette cell (B) was constructed with glass plates. The electrodes (3 x 1cm^2) were attached on the inside suface of the cuvette, the distance between them being 1 mm.

A drop of C. albicans cell suspension was put on the slit reactor, covered with a cover glass, and observed under a microscope. Then electric pulses were applied with a pulse generator. Reversible agglutination and dispersion of microbial cells were observed in the presence and absence of an electric field. C. albicans cells were homogeneously distributed. When electric pulses were applied to the suspension, linear linkages were formed within several seconds. The extent of linear linkage formation depended on electric pulse conditions as described below. Linear agglutinations, however, dispersed immediately after the removal of the electric pulse. The effects of electric pulse height on the formation of linear agglutination at 8 KHz was determined. No agglutination was observed below 20 V. Above 20 V, the extent of agglutination increased with increasing pulse height. The effect of the frequency was also determined. The formation of linear agglutation increased with increasing frequency. Similar agglutination of latex beads was observed under electric pulse.

In order to estimate the extent of binding, the agglutination rate (AR), defined as follows, was determined by analyzing five photographs for each sample:

$$AR = \frac{\sum_{h=5}^{x} Nn}{\sum_{h=1}^{x} Nn} \times 100\ (\%)$$

where Nn is the total number of beads forming n-bead agglutinations or n-agglutinated cells.

Binding was performed in the presence and absence of an electric pulse. Without the electric pulse, the AR increased gradually and reached about 10% in 20 min as shown in Fig. 5. In contrast, the AR increased sharply and reached 50% in 5 min with the electric pulse. On the other hand, without the antibody, AR also increased under the electric field, though it was about 10% in 20 min. These results clearly demonstrate that the electric pulse

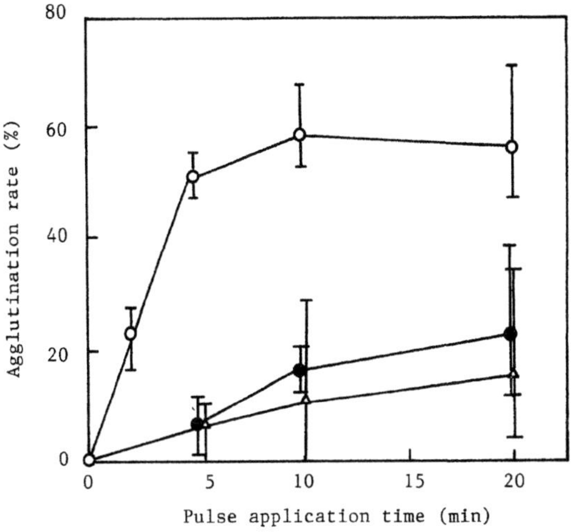

Figure 5. Time courses of agglutination in pulse immunoassay C.albicans:
(○) Antibody with electric pulse;
(△) Antibody without electric pulse;
(●) No antibody with electric pulse.

accelerated the antibody-antigen interaction. From the time courses obtained above, a reaction time of 5 min was recorded.

The effect of Human Serum Albumin on agglutiantion under an electric field was also determined. In the presence of the field, a certain amount of protein is probably denatured, due to the heat generated. Since denatured protein often causes irreversible agglutination, the effect of HSA on the cell agglutination under an electric field was investigated.

When a C. albicans cell suspension containing 2.9 mg of protein/ml of HSA was exposed to electric pulses, a small amount of agglutination appeared. The extent of agglutination was similar to that obtained without antibody as shown in Fig. 5. Therefore the presence of HSA did not affect the non-specific agglutination.

The reaction rate depends substantially on the relative concentration of antibody and antigen. Therefore the effect of antibody concetration on the AR was examined. Antigen (C. albicans) concentration was fixed at 6.0 x 10^7 cells/ml. Without the electric pulse, no agglutination was observed below 2.4 mg of protein/ml, but the AR increased slightly above this concentration. On the other hand, under the electric field, the AR increased sharply above 2.4 mg of protein/ml and reached 50% at 2.9 mg of protein/ml.

protein/ml. From these results, antibody concentration was fixed to 2.9 mg of protein/ml in the following experiments.

The present method was applied to various concentrations of *C. albicans*. The relationship between the AR and cell concentration was investigated. Without the electric pulse, the AR was at the most 10% in the range 3.8×10^7 -3.6×10^8 cells/ml. In contrast, the AR sharply increased with increasing cell concentration below 6.0×10^7 cells/ml. However the AR decreased above this concentration, possibly because the cell concentration was too high as compared with the antibody concentration. From these results, it is concluded that an estimation of the *C. albicans* concentration is possible by the present method in the range of 10^7 -6×10^7 cells/ml.

Although these results demonstrated the value of the electric pulses in immunoassay, the visual observations and manual calculations of the agglutination rates were very tedious and time-consuming. Therefore an on-line image analyzing system was introduced for the assay system (see below).

5. BIO-IMAGE SENSOR

Analysis at the cellular or tissue level is developing as one of the most important topics for future clinical diagnosis. Cell diagnosis is mainly performed by the visual inspection of trained experts or using an automated cell analyzer. These methods give us valuable information about cells, but both methods are still time consuming. The cell sorter [22], which gives densitometric informtion, can analyze cells very rapidly, but cannot be applied to tissue slices or agglutinated samples. Using the two-dimensional imaging sensor system and cell-recognizable monoclonal antibodies, we have developed a novel cell analyzing system which gives us morphometric and densitometric information on cells.

The specific detection of tumor cells in this sensor was based on the complement-mediated cytotoxic reaction. Cells that carry anti-bodies bound to cell surface antigens may be lysed by complement. Lysis was followed using a phase-lag of transmitted light. Normal cells, therefore, look bright under the phase-contrast microscope, whereas damaged cells that have lost membrane integrity appear darker. In this study, applying the decrease in brightness of lysed cells, specific and non-specific are distinguished.

The image sensor system was set as follows. A phase-contrast transmitted light microscope was fitted with a dark-field condenser to obtain contrasting cell images. Video images of cell suspensions contained in the counting chamber of the hemocytometer were obtained using a CCD video camera mounted onto the microscope with a standard C-mount adapter. Video images of cells were displayed on a 12 inch black and white monitor and the image was focused by adjusting the focal length between the microscope objective lens and the hemacytometer. The video display was fed into an image memory board connected to the expanded bus of a 16 bit personal computer (Model PC-9801E, NEC, Tokyo, Japan). This image memory board has a 6 bit A/D converter, 8 bit D/A converter and a 64 KB RAM.

The image-digitizing program had the main function of scanning a defined frame area (256 x 256 picture elements (pixels); scan time 1/60 s) of the image, binarization of the frozen image, and recording the cell image counts (number of white level pixels) for different 10 frame scans obtained by manually moving the field of view. To obtain one data set (scanning, freezing, binarization and counting of the cell image area) takes approximately 2 s. Using this system, quantitative detection of guinea pig hepatocarcinoma cells in a mixed cell suspension was carried out. The cell concentrations of standard suspensions containing hepatocarcinoma cells (target cells) and non-specific guinea pig leukemia cells were corrected so as to retain the same image counts throughout. The two suspensions were then mixed (the target cell content being the volume ratio of the two suspensions). Cell suspension (100 μl), 100 μl of the hepatocarcinoma specific monoclonal antibody solution and 50 μl of rabbit serum (complement source) were mixed. After 30 min incubation at 37°C, the sample was transferred into a hemacytometer and analyzed using the imaging sensor system. Hepatocarcinoma cells could be detected quantitatively in the range 10-100% (target cell content), a linear relationship being observed between the target cell content and image area in this same range.

This system was also applied to the detection of mouse T cell leukemia cells. Similar results were obtained. Target cells were detected in only 15 min.

6. CONCLUSIONS

Immunoassays provide a useful method of trace level analysis in analytical and clinical chemistry. Although enzyme sensors are commercially applied in clinical analysis, and microbial sensors also applied in environmental control and fermentation processes, practical immuno-sensors have not been developed.

Four types of novel immunosensors based on novel principles are described here. These immunosensors have been studied at a fundamental level to data. However it will be possible to construct practical immunosensors according to such principles in the near future.

7. REFERENCES

1. I.Karube and S.Suzuki, 'Immobilized enzymes for clinical analysis', in: Enzymes and Immobilized Cells in Biotechnology, (Ed. A.I.Laskin), The Benjamin/Cumming Publishing Co.Inc., California, p. 209 (1985).
2. R.S. Yallow, Methods in Radioimmunoassay of Peptide Hormones, North Holland, Amsterdam (1976).
3. G.B.Wisdom, 'Enzyme-immunoassay' Clin.Chem., 22. 234 (1976).

4. M.Aizawa, M.Morioka and S.Suzuki, 'Enzyme Immunosensor, 2. Electrochemical determination of IgG with an antibody bound membrane' J.Membrane Sci., 4, 221 (1978).
5. M.Aizawa, A.Morioka, S.Suzuki and Y. Nagamura, 'Enzyme immunosensor, 3.Amperometric determination of human chorionic gonadotropin by membrane-bound antibody' Anal.Biochem. 94 22 (1979).
6. M.Aizawa, A.Morioka and S.Suzuki,'An enzyme immunosensor for the electrochemical determination of the tumor antigenfetoprotein' Anal.Chim.Acta, 115, 61 (1980).
7. P. Bergveld, 'Development of an ion-sensitive solid-state device for neurophysiological measurements' IEEE Trans. on BME., BME-17, 70 (1970).
8. T.Matsuo and K.D.Wise,'An integrated field-effect electrode for biopotential recording' IEEE Trans. on BME, BME-21, 485 (1974).
9. T.Matsuo and M.Esashi,'Methods of ISFET fabrication' Sensors and Actuators, 1, 77 (1981).
10. S.D.Caras and J.Janata,'Field effect transistor sensitive to penicillin' Anal.Chem. 52, 1935 (1980).
11. S.D.Caras, J.Janata, D.Saupe and K.Schmitt,'pH-based enzyme potentiometric sensors. Part 1. Theory' Anal.Chem. ,57 1917 (1985).
12. S.D. Caras, D.Petelenz and J.Janata, 'pH-based enzyme potentiometric sensors. Part 2. Glucose-sensitive field effect transistor' Anal.Chem, 57, 1920 (1985).
13. S.D.Caras and J.Janata,'pH-based enzyme potentiometric sensors. Part 3. Penicillin-sensitive field effect transistor' Anal.Chem., 57, 1924 (1985).
14. I.Karube, E.Tamiya, J.M.Dicks and M.Gotoh,'A microsensor for urea based on an ion-selective field effect transistor' Anal. Chim.Acta, 185, 195 (1986).
15. M.Goth, E.Tamiya and I.Karube,'Polyvinylbutyral resin membrane immobilization to an ISFET microbisensor' J.Mol.-Catal., 37, 133 (1986).
16. M.Gotoh, E.Tamiya, I.Karube and Y.Kagawa, 'A microbiosensor for adenosine-5'-triphosphate pH-sensitive field effect transistor'Anal.Chim.Acta, 187, 287 (1986).
17. M.Gotoh, M.Momoi, Y Kagawa, E.Tamiya and I.Karube, 'Acetylcholine sensor based on ion sensitive field effect transistor and acetylcholine receptor' Anal.Lett., in press.
18. G.G. Guilbault, 'Determination of formaldehyde with an enzyme coated piezoelectric crystal detector' Anal.Chem., 55, 1682 (1983).
19. T.Nomura and T.Nagamune, 'Internal electrolytic determination of silver in solution with piezoelectric quartz crystal' Anal.Chim.Acta, 155, 231 (1983).
20. J.E.Roederer and G.J. Bastiaans, "Microgravimetric immunoassay with piezoelectric crystals' Anal.Chem., 55, 2333 (1983).

21. U.Zimmerman, J. Vienken and G.Pilwat, Z.Naturforsch., C: Biosci., 36C, 173 (1981)
22. R.M.Melamed, P.F.Mullaney and M.L.Mendelsohn (Eds), Flow Cytometry and Sorting, John Wiley and Sons, New York (1979).

ELECTROCHEMICAL IMMUNOSENSORS

William R. Heineman and H. Brian Halsall
Department of Chemistry
University of Cincinnati
Cincinnati, Ohio 45221-0172, U.S.A.

ABSTRACT. The excellent detection limits of modern electroanalytical techniques makes electrochemistry an attractive methodology to combine with the selectivity of an immunoassay. One type of electrochemical immunoassay involves covalently linking a chelating agent such as diethylenetriaminepentaacetic acid to a protein such as human serum albumin to serve as a chelon for a metal label. The metal ion label is released by acidification after competitive equilibrium with antibody that is immobilized on Protein A from Staphylococcus aureus. The released metal ion is then determined by anodic stripping voltammetry. Heterogeneous enzyme immunoassays in which antibody is adsorbed on the walls of polystyrene cuvettes have been developed for orosomucoid, digoxin and rabbit immunoglobulin G. Alkaline phosphatase, which catalyzes the conversion of electroinactive phenylphosphate to phenol, is the enzyme used as the label. The rate of phenol production is measured by either liquid chromatography/electrochemistry or flow injection analysis/electrochemistry with a thin-layer carbon paste electrode that oxidizes phenol. Homogeneous enzyme immunoassays based on the detection of NADH have been demonstrated on phenytoin and digoxin. Detection levels in the low pg/mL range on 20 μL samples (ca. 10^{-18} moles of analyte) have been achieved.

1 INTRODUCTION

A problem of paramount importance to analytical chemistry is selectivity, particularly at low detection limits where potential interferents might be present at substantially higher concentrations than the analyte(s) of interest. An extraordinarily selective reagent is the antibody, Ab. As a part of the immune defense system in animals, antibodies with high specificity can be synthesized by an organism in reasonable quantity within weeks of injecting a foreign species called an antigen, Ag (1).

Immunoassay is based on the use of an antibody as a selective chemical reagent for an antigen (molecular weight > ca. 1,500) or a hapten (molecular weight < ca. 1,500) analyte. The predominant immunoassay has been radioimmunoassay (RIA), which is based on the detection of a radioactive label. RIA has had a dramatic impact on the clinical laboratory as a result of the ease with

G. G. Guilbault and M. Mascini (eds.),
Analytical Uses of Immobilized Biological Compounds for Detection, Medical and Industrial Uses, 281–290.

which assays can be conducted by automated equipment at low concentration due to the very low detection limits of radioactivity. Minimal sample preparation is required due to antibody selectivity. A significant advantage of RIA is the absence of interfering radioactivity in clinical samples. However, the desire to avoid the inconveniences of handling radioactivity has stimulated the development of non-isotopic immunoassays that are based on other labeling strategies. Thus far, the most successful of these are the enzyme and the fluorescent labels.

The excellent detection limits that can be achieved on small sample volumes with modern electrochemical techniques such as liquid chromatography with electrochemical detection (LCEC), flow injection analysis with electrochemical detection (FIAEC) and differential pulse anodic stripping voltammetry have stimulated the development of electrochemical immunoassay.

Immunoassays based on electrochemical detection can be implemented in numerous ways depending on label type, assay format and electrochemical technique (2). This article is organized according to the two main types of labels that have been used -- (a) a label that is electroactive or that renders a molecule electroactive, e.g. the addition of a nitro group to an aromatic ring, and (b) an enzyme label that catalyzes the production of an electroactive product. Immunoassays are commonly categorized as heterogeneous, in which antibody-bound antigen is separated from free antigen at some point in the procedure, or homogeneous, which has no such separation step.

2 NON-ENZYMATIC ELECTROACTIVE LABELS

One class of electrochemical immunoassays involves labeling an antigen with a group that renders the antigen electroactive. The labeled antigen (Ag*) would be reducible or oxidizable in a potential range over which Ag is electroinactive thereby enabling Ag* to be electrochemically distinguishable from Ag.

2.1 Homogeneous Assays

The homogeneous assay is based on a competitive equilibrium between Ag and Ag* for a limited amount of antibody as described by the competition reaction shown in Figure 1. The assay relies on a decrease in the current signal for reduction/oxidation of Ag* when bound to Ab.

This format has been demonstrated with estriol, an estrogen whose plasma and urine levels are related to pregnancy (3). Estriol is made electroactive by labeling with nitro groups in the 2 and 4 positions to give estriol-$(NO_2)_2$. A differential pulse polaragram of estriol-$(NO_2)_2$ gives distinctive reduction peaks in a potential window in which estriol is electroinactive. The addition of Ab causes peak height to decrease as estriol-$(NO_2)_2$ is bound to Ab. Consequently the magnitude of the peak height can be used to monitor the binding, since it is determined by the distribution of estriol-$(NO_2)_2$ between the free and antibody-bound forms. This labeling strategy has also been used with other systems such as morphine labeled with ferrocene which was detected by FIAEC at a glassy carbon electrode (4), ovalbumin labeled with electroactive diazotized p-aminobenzoic acid that was detected polarographically by reduction at a DME (5,6),p-(p-aminophenylazo)- phenylarsonic acid which was monitored by cathode-ray polarography (7); and estriol labeled with mercuric acetate detected by differential pulse polarography (8).

Homogeneous immunoassays of this type are apparently based in part on the difference in diffusion coefficient between the free labeled antigen and the labeled antigen bound to the much larger antibody (7). Consequently, this is a general approach that should be applicable when the antigen is sufficiently smaller than the Ab to give a large difference in diffusion coefficient.

Immunoassays for human serum albumin (HSA) have been based on HSA labeled with Pb^{2+} (10), Co^{2+} (11), and Zn^{2+} (12). The metal ion bound to the HSA can be detected with differential pulse polaragraphy by reduction at a mercury electrode. The peak current for the labeled HSA decreases upon binding with Ab, which is the basis for the immunoassay.

An electrochemical luminescence-based immunoassay has been presented with HSA labeled with the aromatic hydrocarbon amino-pyrene (13). The pyrene-labeled HSA luminesces during reduction at a platinum electrode, and on binding to Ab the intensity of luminescence is decreased. A detection limit of 10^{-6} M for HSA was obtained. Labels with greater luminescence efficiency should provide lower detection limits.

2.2 Heterogeneous Assays

A releaseable electroactive metal ion may be used as the label for a heterogeneous assay (14). Release of the metal ion label after competitive equili-

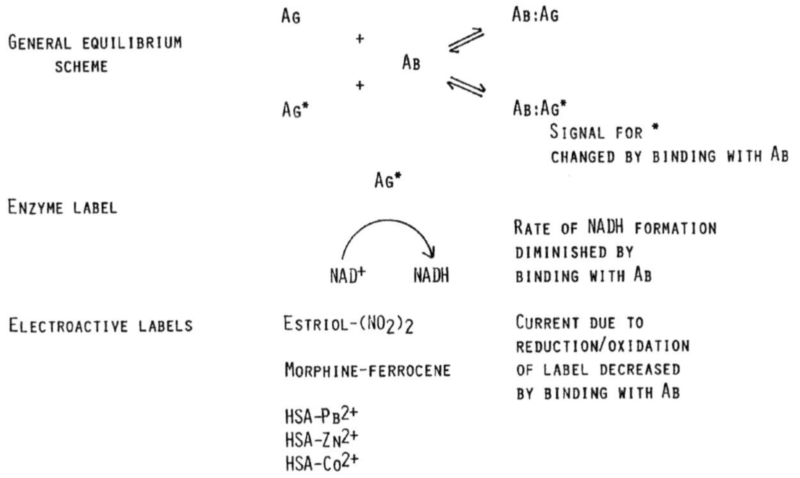

Figure 1. General reaction scheme for homogeneous immunoassay with examples of enzyme and electroactive labels.

bration with antibody enables the very sensitive technique of anodic stripping voltammetry (ASV) to be used as the electroanalytical method. In^{3+} has been used as a label for this purpose since it is easily determined by ASV and forms strong complexes with bifunctional chelates such as diethylenetriaminepentaacetic acid (DTPA), which can be used to attach the label to an antigen. In^{3+} is as yet unfound in human tissue, and hence in clinically relevant analyses interference from the sample itself is avoided. This strategy has been demonstrated with HSA. Indium-labeled HSA consists of HSA to which is covalently attached DTPA, which strongly complexes the In^{3+} label ($K_f = 10^{29}$). After equilibration with HSA and Ab (which in this case is adsorbed to insolubilized Protein A of Staphylococcus aureus), Ab:HSA-In^{3+} is separated by centrifugation of the immunoadsorbent complex. The In^{3+} label is then released by acidification and determined by differential pulse ASV. A calibration curve shows peak height of the stripping wave for In^{3+} decreasing as increasing concentrations of HSA displace more HSA-In^{3+} from Ab.

3 ENZYME LABEL CATALYZING PRODUCTION OF ELECTROACTIVE PRODUCT

Another strategy for electrochemical immunoassay is based on labeling the antigen with an enzyme that catalyzes the production of an electrochemically detectable product. Enzyme immunoassays rely on the antibody for specificity and the enzyme label for sensitivity through chemical amplification.

3.1 Heterogeneous Enzyme Immunoassays

Several heterogeneous enzyme immunoassays with electrochemical detection have been demonstrated. These are based on the Enzyme Linked Immunosorbent Assay (ELISA) technique in which antibody is typically immobilized on the walls of small volume (ca. 500 μL) plastic cuvettes. The general procedure is outlined in Figure 2 for the determination of antigen, (Ag), which will represent the analyte in this and other examples.

Reagent cuvettes are first prepared by attaching specific antibody to the inside walls by passive adsorption or covalent bonding. The cuvettes are then rinsed with a non-ionic surfactant such as Tween 20 to cover any exposed surface between antibody molecules. Reagent cuvettes of this type can be prepared in advance and stored. Enzyme-labeled antigen (Ag*) and samples or standards containing Ag are then added to the tubes for competitive equilibration with the antibody (see competitive assay scheme in Figure 2). Equilibration may take several minutes or hours, depending on the analyte and the configuration of the reaction used. Adequate results may be obtainable before equilibrium is reached. In the competition between Ag and Ag* for a limited number of antibody binding sites, a greater concentration of Ag in the sample results in less Ag* being bound to adsorbed Ab. Unbound Ag and Ag* are then rinsed from the tubes and substrate (S) is added. At a fixed time, sample is withdrawn and analyzed for the electroactive product (P). In a typical standard plot of current from the electrochemical measurement of P vs. concentration of Ag standard, the electrochemical signal diminishes with increasing Ag concentration since less Ag* is able to interact with Ab and hence is rinsed out of the cuvette.

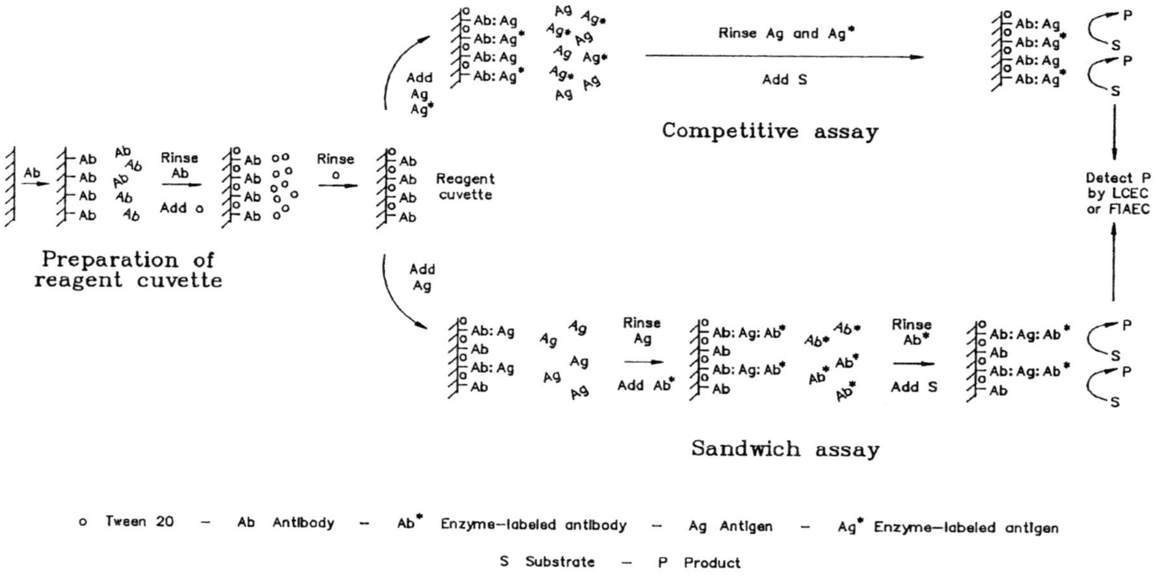

Figure 2. General protocol for heterogeneous enzyme immunoassay. Preparation of reagent cuvettes by coating with Ab, competitive assay format, and sandwich assay format.

An enzyme immunoassay of this type has been demonstrated for digoxin (15), a stimulant commonly used for the treatment of chronic heart disease. The therapeutic range is both low and very narrow (0.8 to 2.0 ng/mL in blood). Lower concentrations are not efficacious, whereas higher concentrations are toxic. Alkaline phosphatase, which catalyzes the hydrolysis of the phenyl phosphate ester to phenol and phosphate, is used as the enzyme label. The enzyme-generated phenol is easily detectable by either FIAEC or LCEC whereas phenyl phosphate is electroinactive and hence noninterfering (15,16). A schematic diagram of the instrumentation for these two methods is shown in Figure 3. In both cases, injected sample (typically 20µL) flows through a thin-layer electrochemical cell where phenol is oxidized at a carbon-paste working electrode at +870mV vs. Ag/AgCl. Although FIAEC has a higher sample throughput, a lower detection limit can be achieved by LCEC in which phenol is separated from the capacitance peak accompanying injection by retention on a short C-18 reversed phase column. A detection limit of 50 pg/mL has been achieved for digoxin, which is well below the therapeutic range

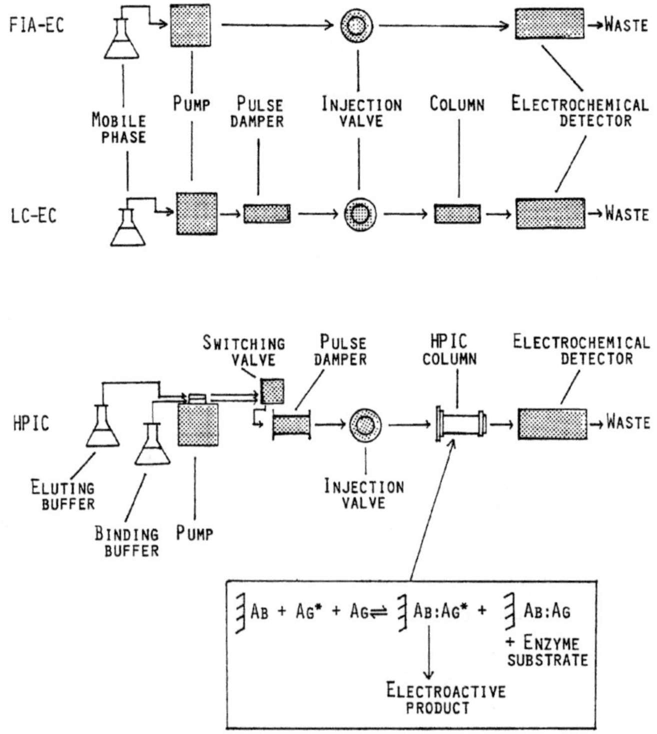

Figure 3. Schematic diagram of instrumentation for FIA-EC, LC-EC, and high-performance immunoaffinity chromatography (HPIC).

for the drug and illustrates the low detection level capability of the technique. Results of analyses on patient samples are in good agreement with results by an accepted radioimmunoassay procedure. Assays have also been developed for α_1-acid glycoprotein (17) and IgG, detection limit of 5 ng/mL (18).

The sandwich immunoassay is another widely-used assay format. This assay derives its name from the fact that the analyte is "sandwiched" between two different antibodies, one of which is enzyme-labeled. It can only be applied to molecules that are sufficiently large to bind to two antibodies simultaneously and have an appropriate topographical distribution of epitopes (antigenic determinants). As shown in Figure 2 (schematic for sandwich assay), Ab specific for analyte Ag is first adsorbed on the cuvette walls. A sample containing analyte Ag is then added to the cuvette, and Ag in the sample binds with Ab on the cuvette walls. The amount of Ag bound is measured by adding a second antibody also specific for Ag that has been labeled with enzyme (Ab*). After incubation, unbound Ab* is rinsed from the cell and substrate added. The rate of formation of P is proportional to the amount of Ab* bound to the walls which is proportional to the concentration of Ag in the sample. A sandwich assay has been demonstrated for IgG in which alkaline phosphatase is the enzyme label and the rate of phenol production is measured by LCEC (18). A detection limit of 10 pg/mL was achieved.

The heterogeneous enzyme immunoassay with electrochemical detection has several advantageous features. The detection limit, as demonstrated for digoxin and IgG, is typically in the low pg/mL range, and was a function of the antigen-antibody binding constant, rather than the ability to detect phenol by LCEC. Consequently, even lower limits should be attainable. Since the sample is rinsed out of the reagent tubes before adding substrate, problems with interferences by electroactive constituents in the sample and possible fouling of the electrode by the adsorption of protein films are eliminated. These concepts were demonstrated with the simple instrumentation for FIAEC and LCEC shown in Figure 3. Automation of many steps in the procedure and the treatment of data is needed for its ready application to the routine analysis of many samples.

Electrochemical immunoassays can also be based on high performance immunoaffinity chromatography (HPIC) with electrochemical detection of the column effluent (19). In this technique antibody is immobilized on a chromatography column. Antigen labeled with enzyme (Ag^*) is mixed with the sample Ag and then injected into the immunoaffinity column where competitive binding occurs with Ab as shown in Figure 3. Enzyme substrate is then injected. The amount of electroactive product formed is proportional to the amount of Ag^* bound to the column, which is inversely proportional to the amount of Ag in the sample. The electroactive product flows through a thin-layer detector where it is oxidized/reduced and quantitated from the area of the resulting current peak. A switching valve is then used to pass an acidic buffer through the column to displace bound Ag and Ag^* in preparation for the next sample. The Ab/Ag reaction is sufficiently fast that reactions are more than 90% complete in 8-10 s (20). The flow injection/chromatography mode enables carefully controlled conditioning of the immunosorbent (e.g., washing) which is essential to good experimental precision. The immunosorbent columns are stable for at least 3 months (500-600 assays). It has been demonstrated that the immunoaffinity column can be used to perform a sandwich assay for IgG with a detection limit of 1 femtomole (21) in less than 30 minutes. The second antibody is a glucose oxidase-goat anti-IgG conjugate, and hydrogen peroxide is detected following injection of glucose into the affinity column.

3.2 Homogeneous Enzyme Immunoassay

Immunoassays of the homogeneous type rely on a change in the intensity of the label signal that occurs when Ag* binds with Ab to form Ab-Ag*. The general equilibrium scheme for a homogeneous immunoassay is shown in Figure 1. The ability to distinguish between Ag* and Ab:Ag* enables the immunoassay to be done without a separation step.

When the label is an enzyme, then the immunoassay relies on a reduction in the rate of enzyme catalysis occurring when the antibody binds to the enzyme-labeled antigen. This enables free Ag* to be distinguished from Ab:Ag* on the basis of the rate of production of product. As shown in the equilibrium scheme in Figure 1 Ag* catalyzes, for example, the reduction of NAD^+ to NADH, whereas the catalytic activity of Ab:Ag* is substantially diminished as a consequence of steric hindrance and limited substrate access. Since a larger concentration of Ag results in more Ag* being in the unbound and, consequently, more active form, a calibration plot can be based on the rate of NADH formation versus Ag concentration. An assay of this type is called an Enzyme Multiplied Immunoassay Technique (EMIT^{TM}) and is commonly used with spectrophotometric detection of NADH (22).

Phenytoin, an antiepileptic drug, has been used as an initial test system for developing enzyme immunoassays with amperometric detection of NADH (23). The phenytoin is labeled with glucose-6-phosphate dehydrogenase, which catalyzes the reduction of NAD^+ to NADH. This immunoassay scheme is amenable to the electrochemical detection of NADH by LCEC. The concentration of NADH is monitored by oxidation at 0.75 V vs Ag/AgCl at a glassy carbon electrode in a flow-through thin layer cell. Antibody and proteins in the sample are removed by a reversed-phase C-18 liquid chromatography precolumn to prevent adsorption on the electrode and consequent fouling. Immunoassays are performed in a timed sequence of mixing fixed amounts of antibody, enzyme-labeled phenytoin, and enzyme substrate with each standard and unknown phenytoin sample. A standard curve can be constructed from the rate of NADH formation versus phenytoin concentration. Assays of serum samples from patients on phenytoin treatment gave results that were in good agreement with a standard spectrophotometric procedure at therapeutic levels (10-20 μg/mL).

Enzyme labels of this type can take advantage of enzyme amplification as a means of achieving a low detection limit. In this particular system NADH is generated so rapidly that samples must be diluted ca. 100-fold immediately prior to injection into the flow system for the optimum detection of NADH. This suggests that the electrochemical technique should be extendable to the determination of species at concentrations in the low ng/mL range, as has been demonstrated with digoxin (24).

A homogeneous enzyme immunoassay based on antibody inhibition of enzyme conversion from the apo- to the holo- form has been developed for DNP-aminocaproic acid, DNP-ACA (25). A competitive equilibrium is established between the analyte DNP-ACA and DNP-conjugate apoglucose oxidase, DNP-CAGO, which is the labeled hapten. Flavin adenine dinucleotide, FAD, added to the mixture binds to free DNP-CAGO to give DNP-CAGO:FAD, which is enzymatically active and catalyzes the production of H_2O_2 from O_2. Since FAD cannot bind to Ab:DNP-CAGO, the rate of H_2O_2 generation is a measure of the concentration of free DNP-CAGO, which in turn reflects the concentration of DNP-ACA in the sample. The rate of production of H_2O_2 is measured amperometrically by oxidation at 700mV. This assay has a range of ca. 2-40 μg/mL.

The main advantage of a homogeneous immunoassay, compared to the heterogeneous assay, is the absence of a step for separating the antibody-bound antigen from the free antigen. This usually translates into a simpler assay procedure. However, homogeneous assays are usually restricted to higher detection levels than heterogeneous assays and are more susceptible to interferences from other sample constituents which would otherwise be removed in a separation step.

4 CONCLUSIONS

The combination of antibody selectivity with the low detection limits of modern electrochemical techniques results in a powerful analytical tool. The relatively simple instrumentation requirements for LCEC and FIAEC are advantageous. Detection levels in the low pg/mL range on 20 μL samples (ca. 10^{-18} moles of analyte) have been achieved. In these cases, the limit of detection was restricted by the antibody antigen binding constant, rather than the electrochemical technique (15,18). Smaller amounts of analyte should be detectable by working with smaller volumes of sample.

ACKNOWLEDGEMENT

The authors acknowledge support provided by the National Science Foundation (CHE-8217045).

REFERENCES

(1) I. Roit, Essential Immunology, 4th Ed., Blackwell Scientific Publications, Boston 1980.

(2) W.R. Heineman, H.B. Halsall, Anal. Chem. 57 (1985) 1321A-1328A.

(3) K.R. Wehmeyer, H.B. Halsall, W.R. Heineman, Clin. Chem. 28 (1982) 1968-1972.

(4) S.G. Weber, W.C. Purdy, Anal. Lett. 12 (1979) 1-9.

(5) B. Breyer, F.J. Radcliff, Nature 167 (1951) 79.

(6) B. Breyer, F.J. Radcliff, Austral. J. Exp. Biol. **31** (1953) 167-172.

(7) H. Schneider, A.H. Sehon, Trans. N.Y. Acad. Sci. **24** (1961) 15-22.

(8) W.R. Heineman, C.W. Anderson, H.B. Halsall, Science **204** (1979) 865-866.

(9) J. Steinhardt, J.A. Reynolds, Multiple Equilibria in Proteins, Academic Press, New York 1969, p. 58.

(10) I.A. Alam, G.D. Christian, Anal. Lett. **15** (1982) 1449-1456.

(11) I.A. Alam, G.D. Christian, Z. Fresenius, Anal. Chem. **318** (1984) 33-36.

(12) I.A. Alam, G.D. Christian, Z. Fresenius, Anal. Chem. **320** (1985) 281-284.

(13) Y. Ikariyama, H. Kunoh, M. Aizawa, Biochem. Biophys. Res. Commun. **128** (1985) 987-992.

(14) M.J. Doyle, H.B. Halsall, W.R. Heineman, Anal. Chem. **54** (1982) 2318-2322.

(15) K.R. Wehmeyer, H.B. Halsall, W.R. Heineman, C.P. Volle, I. Chen, Anal. Chem. **58** (1986) 135-139.

(16) K.R. Wehmeyer, M.J. Doyle, D.S. Wright, H.M. Eggers, H.B. Halsall, W.R. Heineman, J. Liq. Chromatogr. **6** (1983) 2141-2156.

(17) M.J. Doyle, H.B. Halsall, W.R. Heineman, Anal. Chem. **56** (1984) 2355-2360.

(18) K.R. Wehmeyer, H.B. Halsall, W.R. Heineman, Clin. Chem. **31** (1985) 1546-1549.

(19) G.S. Sittampalam, G.S. Wilson, Trends Anal. Chem. **3** (1984) 96-99.

(20) J.R. Sportsman, J.D. Liddil, G.S. Wilson, Anal. Chem. **55** (1983) 771-775.

(21) W.U. de Alwis, G.S. Wilson, Anal. Chem., **57** (1985) 2754-2756.

(22) A.F. Rosenthal, M.G. Vargas, C.S. Klass, Clin. Chem., **22** (1976) 1899.

(23) H.M. Eggers, H.B. Halsall, W.R. Heineman, Clin. Chem. **28** (1982) 1848-1851.

(24) D.S. Wright, H.B. Halsall, W.R. Heineman, Anal. Chem. **58** (1986) 2995-2998.

(25) T.T. Ngo, J.H. Bovaird, H.M. Lenhoff, Appl. Biochem. Biotech. **11** (1985) 63-70.

FLOW INJECTION ANALYSIS - RECENT DEVELOPMENTS AND FUTURE TRENDS

Elo Harald Hansen
Chemistry Department A
The Technical University of Denmark, Building 207
DK-2800 Lyngby
Denmark

ABSTRACT. Combining three basic features - sample injection, controlled dispersion, and reproducible timing - into an integrated experimental entity, Flow Injection Analysis has led to uniquely useful analytical developments. The basic principles of Flow Injection Analysis are outlined, and a review of FIA gradient techniques is presented. Recent developments and future trends are briefly discussed, demonstrating how FIA has developed from a method primarily concerned with automation of methods for serial assay on a laboratory scale to a link between chemistry and instruments, thus constituting a novel and versatile tool for solution handling and information gathering, providing its use in entirely new areas of application such as continuous monitoring and process control and in biotechnological sensor development.

INTRODUCTION

The first paper on Flow Injection Analysis (FIA) appeared in 1975 [1]. Over the intervening period of time, the method has developed from an approach for merely rapidly conducting serial chemical assays into a novel concept in solution handling in analytical chemistry and a diagnostic tool to be exploited for general analytical studies. Having promoted new and exiciting developments in the field of continuous flow analysis [2, 3], this broad acceptance of FIA is amply reflected in the wealth of papers published in the international literature [4] which, counting contributions from more than 35 different countries, has increased exponentially and by the end of 1986 exceeds 1400 [5].

THE PRINCIPLE OF FLOW INJECTION ANALYSIS

The principle of Flow Injection Analysis is deceptively simple, being based on the injection of a definite volume of a liquid sample solution into a moving, nonsegmented continuous carrier stream of a suitable liquid (Fig. 1a). The injected sample forms a zone, which while dispersing and reacting with the carrier stream then is transported towards a detector which continuously records the absorbance, electrode potential, or any other physical parameter as it continuously changes as a result of the passage of the sample material through the flow cell. A typical recorder output has the form of a peak (Fig. 1b), the height (H) - or for certain applications the width

G. G. Guilbault and M. Mascini (eds.),
Analytical Uses of Immobilized Biological Compounds for Detection, Medical and Industrial Uses, 291–308.

(W), or even the area (A)- of which is related to the concentration of the analyte [2, 3, 6]. As the residence time (T) of the sample in the FIA system normally is less than 30 s, at least two samples can be analyzed per minute. The injected volume may be between 1 µl and 200 µl (typically 25 µl), which in turn usually requires no more than half a ml of reagent per analysis. This makes FIA an automated microchemical technique, capable of having a sampling rate of at least 100 determinations/h, with minimum reagent consumption, thus offering facilities for practically continuous monitoring in real time, being furthermore compatible with a wide range of detection means (Table I).

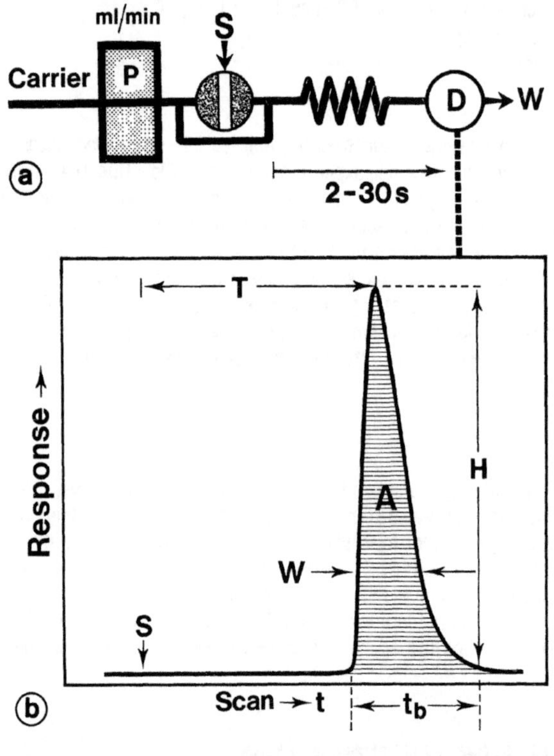

Figure 1. (a) Single-line FIA manifold utilizing a carrier stream of reagent (propelled forward by pump P) into which the sample solution (S) is injected; D - flow-through detector; W - waste. (b) The analog output has the form of a peak, the recording starting at S (time of injection); H - peak height; W - peak width at selected level; A - peak area. T is the residence time corresponding to the maximum peak height measurement.

Thus in contrast to any other technique utilizing continuous flow, FIA is based on a *combination* of three features: (1) **sample injection**; (2) **controlled dispersion** of the injected sample zone during its transport from the point of injection to the point of detection, which dispersion can be manipulated to exactly suit a given analytical procedure; and (3)

Table I. Detectors Used in Flow Injection Analysis

Atomic emission spectroscopy (incl. ICP)	Amperometry
Atomic absorption spectroscopy	Conductometry
Atomic fluorescence spectroscopy	Coulometry
Chemi- and bioluminescence	Polarography
Spectrophotometry (UV, VIS, IR)	Potentiometry
Reflectometry/optosensing	Chronopotentiometry
Refractometry	Ion-selective electrodes/ISFETS
Tubidimetry/nephelometry	Voltammetry
ICP mass spectrometry	Photoelectrochemistry
Piezoelectric quartz crystal	Radioactivity

reproducible timing of all events within each assay cycle. This approach yields a highly reproducible readout even when the mixing is incomplete, the chemistry does not reach equilibrium, and the signal is transient.

While dispersion is defined as the physical dilution process of the injected sample zone in the FIA system, the *dispersion coefficient* D is a numerical expression of the dispersion process in a particular element of fluid of the dispersed sample zone, being defined as the ratio of the concentrations before before and after the dispersion process has taken place in that element of fluid that yields the analytical readout, that is:

$$D = C^0/C, \text{ which for } C = C^{max} \text{ yields } D = C^0/C^{max} \qquad (0 < D < \infty)$$

where C^0 is the original concentration of the injected sample solution, and C is the concentration of any element of fluid along the gradient of the dispersed sample zone (Fig. 2), which element especially may be that one corresponding to the peak maximum of the recorded curve. The parameters governing the dispersion coefficient D has been the subject of detailed studies [2], the most powerful means of manipulating D being the physical dimensions of the FIA system (coil length, internal diameter), the residence time, the flow velocity and the injected volume, to which come the added possibilities inherent in using a single line versus a confluence manifold, and – as mentioned below – in selecting the element of measurement on any other part of the gradient of the dispersed sample zone than that corresponding to the peak maximum.

Depending on which type of assay is to be performed in a FIA system, different dispersion values are exploited. Thus, *limited* dispersion (D = 1 to 2) is used when the FIA system is to serve merely as a means of rigorous and precise transport of the sample material to the flow cell in undiluted form (e.g., measurement of pH or conductance) or as a means of reproducible sample introduction (as in atomic absorption spectrometry). *Medium* dispersion (D = 2 to 10) has the widest range of applications as it allows chemical reactions to take place after the mixing of reagent, contained in the carrier stream, with the sample components contained in the dispersed sample zone. *Large* dispersion (D > 10) is used when the sample material to be analysed is too concentrated to be accommodated within the detector range.

Recently, even FIA system operating at *reduced* dispersion (D < 1) have been designed, e.g., for trace analysis where enhancement of sensitivity is imperative. This is accomplished by injecting a large sample volume and by preconcentrating the analyte on a small column filled with an ion-exchanger. By subsequently injecting a small volume of eluent the analyte is desorbed into

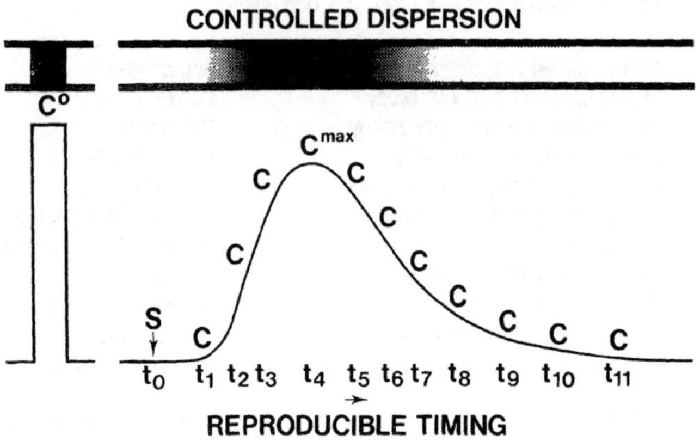

Figure 2. Dispersed sample zone, of original concentration C^0 injected at position S, and the corresponding recorder output. To each individual concentration C corresponds a specific dispersion value – each of which can be related to a fixed delay time t – which is minimum at C^{max} and increases to larger values along the continuous gradient.

a smaller zone which then is directed into the detector [4, 7]. In this manner up to 100 times preconcentration of analyte has been attained.

As mentioned above, any FIA readout is the result of two simultaneously occuring kinetic processes, i.e., that of the physical dilution process and that of chemical reactions. Provided that the reaction rates of the chemical processes taking place vary significantly, these might be exploited for the purpose of *kinetic discrimination* [5]. One such example is the determination of chlorate by the reduction of titanium(III) in the presence of leukomethylene blue (LMB), according to the following reaction scheme:

$$2\,ClO_3^- + 10\,Ti^{3+} + 12\,H^+ \longrightarrow 10\,Ti^{4+} + Cl_2 + 6\,H_2O$$
$$Cl_2 + LMB \longrightarrow MB$$
$$MB + Ti^{3+} \longrightarrow LMB + Ti^{4+}$$

The assay is performed by injecting a sample of chlorate into a acidic carrier stream of titanium(III), which is subsequently merged with a second stream of leucomethylene blue. While the first two of these reactions are very fast, the reduction of the blue coloured species MB by the third reaction is slow, i.e, the chlorate concentration might readily be quantified via the absorbance due to the colour of the MB species generated by reaction 2.

Although the simple FIA system depicted in Fig. 1 utilizes continuous pumping of the reagent stream, it should be emphasized that besides accommodating several reagent and/or carrier stream lines, a given FIA manifold might benefit from a number of additional features. Thus, intermittent pumping of sample, carrier stream and/or reagent streams allows a versatile approach to selection of optimum reagent/sample ratios for detector readout; allows expensive reagents to be injected as short zones, rather than being pumped continuously; allows introduction of standards into sample zones for automated standard addition in the reverse flow mode in the channel (sample used as carrier); and allows valveless, socalled hydrodynamic, injection to be performed [4, 8].

FIA GRADIENT TECHNIQUES

As is apparent from most FIA publications the analytical readouts are overwhelmingly derived from evaluation of peak maximum signals - possibly because these are the easiest to identify, but perhaps also for lack of realization of the enormous potential of the concept at hand. Yet, one may ask, why go through all the pain of creating a well-defined concentration gradient of the injected sample zone within the carrier stream - corresponding to and containing a virtually infinite number of elements of different concentration ratios of sample and reagent - and then only exploit a single one of them? In fact, by doing so we are constraining and thus homogenizing (at least conceptually) all the information contained in the gradient into a single piece of information - exactly as in the air-segmented continuous flow systems where physical homogenisation of sample and reagent is attained and the readout taken at steady-state, and similarly in those procedures where peak areas are measured. This is indeed a valid question, but then again it should be remembered that the full potential of FIA was only gradually being realized as research progressed. In 1977 the first step was taken towards the *gradient techniques* with the development of the FIA titration procedure, which was soon followed by the stopped-flow techniques, and recently a number of other FIA gradient techniques have been described which truly reveal the vast potential of FIA. More than anything else, the gradient techniques inherently distinguish FIA from all other continuous flow techniques - not in the least because the gradient approach yields the capability to perform procedures not feasible by conventional continuous flow analysis. Furthermore, the potential of the gradient approach suggests that FIA techniques are not restricted to providing convenient analytical methods, but are becoming generally applicable means of solution handling.

Taking advantage of the fact that any FIA curve represents a continuum of concentrations (between zero and C^{max} - cf. Fig. 2), and that there is no single element of fluid that has the same concentration as the neighbouring one, these gradient techniques will in the following briefly be described separately. What they do have in common, however, is that since the carrier stream is noncompressible, and its movement hence can be strictly controlled and reproduced from one measuring cycle to the next with high precision, they *all* rely on the principle that any element of fluid along the gradient corresponds to a fixed dispersion value. Thus the dispersion coefficient of the injected sample, $D_S = C_S^0/C_S$, and hence the dispersion coefficient of the reagent, $D_R = C_R^0/C_R$, within each of these elements can be related to a fixed delay time (t_1 to t_{11} on Fig. 2), elapsed from the moment of injection, providing us with the ability to reproducibly select and exploit each of these delay times for analytical purposes.

The principle of *gradient dilution* [8] is based on selecting for the analytical readout any other point than the peak maximum (C^{max}). Since each element along the dispersed sample zone will be characterized by a fixed D-value, corresponding to a fixed delay time t (Fig. 2), the nature of the physical process of dispersion on which FIA is based will ensure that the response, and the calibration curves obtained from data collected at fixed delay times, are linear, provided the analytical readouts are always taken at exactly the same time after the point of injection. Linear relationships will also be found if the physical dispersion is accompanied by a resulting chemical reaction provided the reaction is linear in nature. The advantage of this approach is that by obtaining the readouts after preselected delay times, manual predilution of the sample material can be avoided and a large concentration range can be accommodated within the dynamic range of the detector used. Gradient dilution can experimentally be performed in two ways: electronically [8] by programming a microprocessor to collect the readouts at selected time delays, whereby dilutions up to 2000 times might be obtained; or mechanically [9] by grabbing a selected section of the dispersed sample zone by means of a valve and by reinjecting it into a second FIA system for further dilution and processing, allowing dilution factors up to 100.000 with an RSD better than 1 % [10].

The *gradient calibration* technique [5, 11] is a logical extension of the gradient dilution procedure as it relies on the strict reproducibility of the same phenomenon: physical dispersion accompanied by chemical reactions. Its main goal is to avoid the usual repetitive calibration by means of serially diluted solutions, as the information sought is in fact already contained within some of the segments of fluid originating from a sample zone of the most concentrated standard sample material. Thus, by identifying on this curve the time delays associated with those segments of fluid which have concentrations corresponding to the concentrations that would provide the same readouts if the signals were collected from the individual peak heights (e.g., maxima), as obtained by separate injections of serial dilutions, one can obtain a multipoint calibration curve from a single injection. In other words, the highest (C^{max}) value can be supplemented by a series of sequential C values spaced along the gradient and identified through increasing delay times. Most conveniently the empirical formula describing the relationship C = f(delay time) is obtained from a normal calibration evaluation and stored by a computer.

The FIA *stopped-flow* approach is based on a combination of stopped-flow measurement and of gradient dilution [2, 11, 12]. There are two different purposes for operating a FIA system in the stopped-flow mode: (a) to increase the sensitivity of measurement by increasing the residence time T - and thus the yield of the measured species - or (b) by arresting a selected section of the dispersed sample zone in the observation field of the detector to measure a reaction rate which

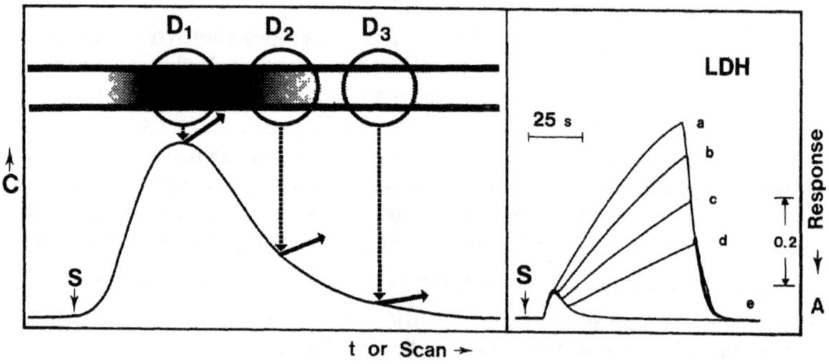

Figure 3. The FIA stopped-flow technique as demonstrated by the determination of the activity of lactate dehydrogenase (LDH). To the left is shown the dispersed sample zone, the slopes of the individual reaction rate curves measured at increasing delay times decreasing because of increasing dispersion. To the right is shown a series of actually recorded reaction rate curves as obtained by injecting identical concentrations of LDH but stopping at different delay times (a to d; e, continuous pumping).

then serves as the basis for the analytical readout. The stopped-flow technique is based on one of the most important findings of the FIA theory: to increase the residence time, keep the reaction coil short, and decrease the velocity of the carrier stream. In the extreme, when the carrier stream ceases to move, the dispersion of the sample zone will stop (except for a negligible contribution from molecular diffusion) and D will become independent of time. Thus, by an appropriate choice of length of the stop-go intervals, one can gain reaction time during the stop period when the carrier stream does not move. Provided the chemical conditions are that of a pseudo-zero-order reaction, the reaction rate will be proportional to the concentration of analyte (Fig. 3). Reaction rate measurements, where the rate of formation (or consumption) of a certain species is measured on a larger number of data points, not only improve the reproducibility of the assay but also secure its reliability. This benefit occurs because interfering phenomena such as blank values, existense of a lag phase, and nonlinear rate curves may be readily identified and eliminated. An added advantage of the FIA stopped-flow technique is that it allows a fine adjustment of the reagent-to-sample ratio by electronic means, i.e., by choosing an appropriate delay time. By so doing, the nonsegmented stream can be manipulated exactly in space and time so reproducibly that the same section of the dispersed sample zone can always be held still in the detector while the measurement takes place.

While the previously described gradient techniques are aimed at various improvements in assays of a diversity of samples, the *gradient scanning* technique, originally suggested by Janata and Ruzicka [13], provides a very convenient means of studying the chemical and physical phenomena which take place when solutions are mixed. Being a natural extension of the philosophy of FIA, namely to combine the *dynamic* method of FIA with a *dynamic* detector, the principle of the method is as simple as that of all FIA techniques: a physical parameter is measured continuously within a certain range (e.g., absorbance vs. wavelength (Fig. 4, left) or current vs. potential) along the dispersed sample zone. Provided that the rate of the scanning is much faster than the movement of the carrier stream, the dispersed zone may be advanced during the scan and still appear as to be stationary. If this is not the case, or if the chemical reaction occuring is too slow, then selected sections of the zone can be stopped consecutively within the detector. The resulting

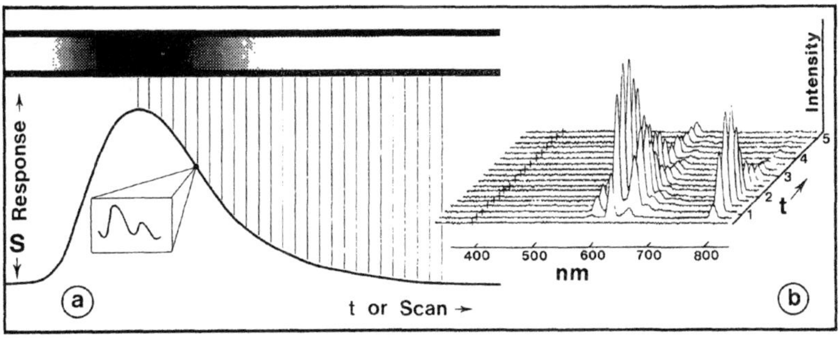

Figure 4. The principle of the FIA gradient scanning technique, where individual segments of the dispersed sample zone are sequentially monitored. To the right is shown a series of successive emission spectra recorded on the ascending and descending part of the dispersed zone containing Na, K and Ca injected into an atomic emission spectrometer and using a fast scanning monochromator.

record will have a multidimensional form, like the three-dimensional emission spectrum depicted in Fig. 4, right, which shows the response obtained when injecting a 40 µl sample solution, containing salts of the three elements Na, Ca and K, into an atomic emission spectrometer, using a fast scanning monochromator capable of scanning from 350 to 800 nm in 5 msec every 100 msec and starting the data collection 300 msec after sample injection [14].

The combination of rapid scanning instruments (both optical and electrochemical) with the FIA gradient approach is natural. The tedious manual handling necessary to prepare solutions for batchwise measurement by means of these advanced instruments is not flexible and does not exploit their full potential. Seen from a purely analytical viewpoint, the advantages of gradient scanning should become even more apparent when multicomponent analysis will be performed routinely on unknown samples, using spectrometry. Work has already been done by combining FIA with inductively-coupled plasma emission spectrometry [15-17]. An emission spectrum generated by an ICP contains an unavoidable combination of weak and strong lines, which cannot be accommodated within the optimum detector range by conventionally aspirating a single solution. If, however, sample is injected into a FIA channel connected to an ICP and the dispersed sample zone is repeatedly scanned while passing into the plasma torch, all lines can, regardless of their intensity and analyte concentrations present in the original sample solution, be evaluated on the basis of a single injection experiment. Combination of the gradient scanning technique and electronic calibration would be a logical extension of this approach.

The FIA *titrations* are the oldest gradient technique [2, 6, 8, 18, 19]. Although technical developments have advanced through various stages since the method was first described [18], the basic principles remain the same: If a sample zone, for example, of an acid, is injected into a carrier stream of a base, the dispersed zone will become gradually neutralized by the base penetrating through the interfaces with the carrier stream at the leading and trailing sections (Fig. 5, left). Therefore, both at the front and at the tail of the dispersed sample zone an element of

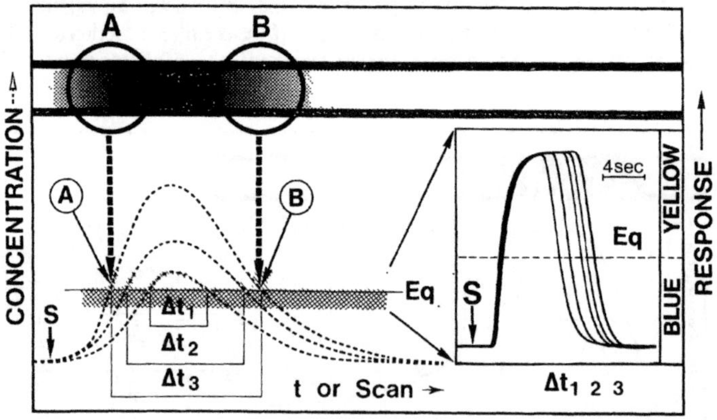

Figure 5. Flow injection titrations. The dispersed sample zone (top left) contains two equivalence points (A and B) the distance between them Δt being proportional to the acid concentration in the injected sample zone. By selecting the indicator the equivalence level (Eq.) is set and in this way decreasing values of Δt are obtained for decreasing concentrations of injected acid. Actual titration curves are shown on the right.

fluid exists within which the acid is exactly neutralized by the base. These two equivalence points (A and B), as identified via an acido-basic indicator contained in the carrier stream, form a pair, having the same D value, and their physical distance (measured as Δt for a constant flow rate Q) will increase with increasing concentration of the injected acid, and decrease with increasing concentration of the base, contained in the carrier stream. The concentration dependence of the time span Δt depends on the concentration profile formed by the dispersion of the zone on the way from the injector to the detector. It has been shown [2, 18, 19] that as long as the concentration profile conforms with the model known as 'one mixing stage' Δt will be proportional to the logarithm of the concentration of the sample. Originally [18], the mixing stage was actually a small mixing chamber (equipped with a magnetic stirrer), the volume of which (ca. 1 ml) dominated the flow design. Further studies have shown, however, that the injection process itself conforms with the 'one mixing stage' model. This dicovery allowed miniaturization and simplification of the FIA system, resulting in elimination of the mixing chamber and implementation of high-speed titrations [19].

Recently a number of gradient techniques based on *penetrating zones* have been developed. These techniques are based upon exploitation of the concentration gradients formed when two (or several) zones are injected simultaneously, synchronously or asynchronously, penetrating into each other and thereby yielding response curves which either completely or partially overlap each other. Since the dispersion coefficient D for each sample zone will have a given value for any given delay time t, the *ratio* of the D values for the sample zones for any value of t will be constant. This approach has been employed for developing a generalized method for quantifying the extent of interference of foreign species on a given chemical assay through introduction of the term selectivity coefficient [20]. It has spawned a novel standardization method to achieve multi-component standard addition over a wide, controllable range of standard/analyte concentration ratios [14], and has served as a basis for devising a new, economical way of introducing sample and reagent solutions into a FIA system via the split-loop injection technique [21]. Each of these techniques are explained in details in the original publications.

RECENT DEVELOPMENTS IN FLOW INJECTION ANALYSIS

When reviewing the development of FIA over the past ten years several patterns emerge, of which the variety of and versatility in methods and techniques used (Table II), the trends towards miniaturization, exploitation of the possibilities inherent in the concept of controlled dispersion, and application and introduction of new flow-through detectors are the most striking. While the first FIA system used more than 10 ml of reagent and 0.5 ml of sample for a single measurement, contemporary designs require ten to twenty times smaller volumes. This process of miniaturizing the FIA system has recently taken a great step forward with the introduction of *integrated microconduits* (cf. Fig. 6) which permit the integration of traditional manifold components, such as coils, connectors, injection valve, dialysis and gas diffusion units as well as columns containing ion-exchangers or immobilized enzymes, into a single unit approximately of the size of a credit card [22]. The purpose of miniaturization and integration has been twofold: (a) by reducing the overall volume of the FIA channel system, the economy of sample and reagent consumption is improved; and (b) by integrating also the *detector* into the FIA channel itself the detection can be executed exactly within that section of the channel where the dispersion and other conditions of an individual assay are optimal. This is in contrast to the conventional constructions of FIA manifolds, where the physical size and geometry of an available, external flow-through detector limits the potential of the method, giving an undesired additional dispersion and consequently a loss of signal and sampling frequency. Applications of integrated microconduits comprise for example systems incorporating packed reactors (with ion-exchangers [7] or immobilized enzymes [23]) and detection via chemiluminescence [23] or via optosensors exploiting reflectance as it is changed by chemical reactions taking place at or in close proximity to a surface surrounded by a flowing stream [21]. In this context the recently developed split-loop injection technique has proven most valuable [21].

Table II. Methods and Techniques Exploited in Flow Injection Analysis

* Automatic solution-handling based on precise timing, dilution and mixing of reagents makes the method generally applicable to optical and electrochemical based assays
* The concentration gradient formed allows a wide selection of sample/reagent ratios to be utilized
* Kinetic measurements feasible by stopped-flow
* Solvent extraction - continuous or in a closed loop
* Gas diffusion via hydrophobic membranes allows selective separation and kinetic discrimination
* Dialysis via hydrophillic membranes allows separation and dilution
* Column preconcentration increases sensitivity up to 100-fold
* Conversion on packed reactors allows indirect assays
* Titrations (acidobasic, redox, complexing) feasible within 15 seconds
* Impulse response FIA allows automatic testing of selected target materials such as sensors

Constituting a *link* between chemistry and instruments FIA allows solutions to be handled more reproducibly than by hand and in a protected environment. Since solutions are moved and mixed in closed tubular conduits, any influence of light, oxygen, or other elements of the laboratory environment are eliminated. Unstable reagents can be generated *in situ*, and volatile, radioactive or toxic materials that are handled in microvolumes are more easily contained. Such a link also facilitates multidetection, being an ideal vehicle for accommodation of several different detectors through which an injected sample material can be consecutively transported.

One of the great advantages of the FIA technology is the ease with which additional components can be added to the system to achieve a particular analytical objective. This is perhaps most notably exemplified by the incorporation of small packed reactors aimed at subjecting the injected sample to appropriate on-line *pretreatment* in order to facilitate or even empower the ultimate detection of analyte. Whatever the purpose, a packed reactor in a FIA system therefore is always designed to perform the same chemistry on each individual sample.

Of all the FIA methods using packed reactors, systems with *immobilized enzymes* are by far the most common. They offer the selectivity of enzymatic reactions and the economy gained by immobilizing the often costly catalysts. Thus Petersson et al. [23] recently described a miniaturized flow injection system for the determination of four different substrates, which each could be be degraded enzymatically by means of appropriate oxidases according to the following reaction schemes:

$$(1)\quad \beta\text{-D-glucose} + O_2 + H_2O \xrightarrow{\text{D-glucose oxidase}} \text{gluconic acid} + H_2O_2$$

$$(2)\quad \text{Creatinine} \xrightarrow{\text{creatininase}} \text{creatine}$$

$$\text{Creatine} + H_2O \xrightarrow{\text{creatinase}} \text{sarcosine} + \text{urea}$$

$$\text{Sarcosine} + O_2 + H_2O \xrightarrow{\text{sarcosine oxidase}} \text{glycine} + \text{formaldehyde} + H_2O_2$$

$$(3)\quad \text{Cholesterol} + O_2 + H_2O \xrightarrow{\text{cholesterol oxidase}} \Delta^4\text{-cholestenone} + H_2O_2$$

$$(4)\quad \text{L-lactate} + O_2 + H_2O \xrightarrow{\text{L-lactate oxidase}} CH_3COCOOH + H_2O_2$$

In each case the ultimate reaction product is hydrogen peroxide, which subsequently is determined by chemiluminescence via reaction with luminol and hexacyanoferrate(III). Thus the inherent lack of selectivity of the chemiluminescence reaction is compensated for by exploiting the substrate specificity of the enzymes. The manifold used for these procedures is depicted in Fig. 6a, while the actual FIA microconduit system is shown in Fig. 6b. The sample (S) is injected by means of the valve and then directed to the incorporated reactor containing the enzyme immobilized on controlled pore glass beads (the actual volume of the reactor is 80 µl - the authors having given a detailed description of the immobilization procedure employed), where hydrogen peroxide is generated. Then the sample zone is guided to the merging point where it is mixed with the luminol (R1) and the hexocyanoferrate reagent (R2) before entering the flow cell which embodies two photo diodes recording the generated light. Note that since the degree of conversion might be adjusted by the *time* the analyte is exposed to the enzyme reactor, the amount of converted analyte, and hence fulfilment of pseudo-first order reaction conditions, therefore can be regulated by adjusting the flow rate of the FIA system.

More elaborate systems incorporating a series of reactors have been described [24, 25]. In the author's lab work is presently under way on the determinations of creatinine in serum samples using the enzyme creatinine iminohydrolase which degrades the substrate according to the reaction scheme:

$$\text{creatinine} + H_2O \xrightarrow{\text{CIH}} \text{N-methylhydantoin} + NH_4^+$$

The ammonium formed is subsequently mixed with base, and the liberated ammonia determined via optosensing in a miniaturized combined gas diffusion/detection cell where the ammonia generated is diffused into an acceptor stream of an acido-basic indicator, the colour of which is monitored photometrically by means of an optical fibre bifurcated at the remote end to accept the incident light from an external light source and to acommodate a spectrophotometer [21]. Since the amount of ammonium generated by the enzymatic reaction is of the same order of magnitude as the endogenous concentration it is necessary to remove this prior to analysis which is facilitated by means of the enzyme glutamate dehydrogenase (GLDH):

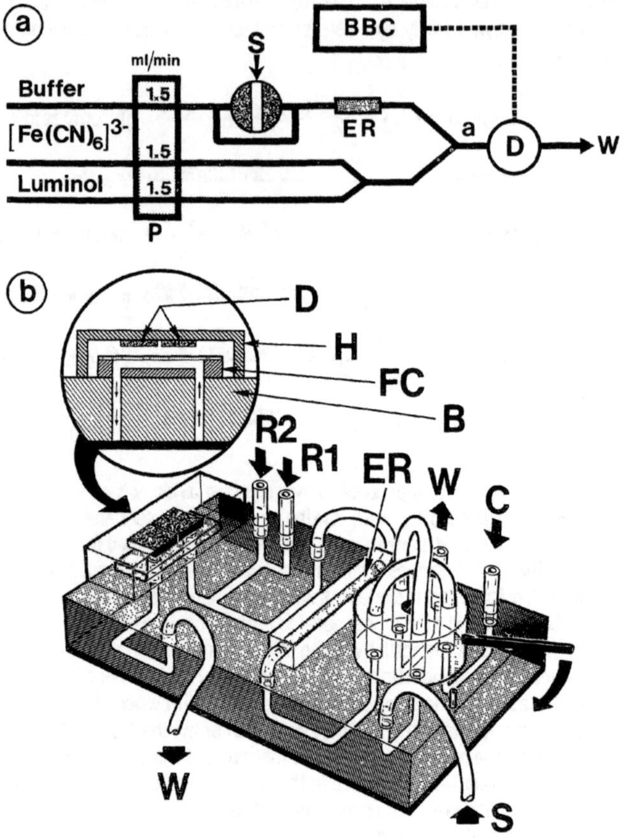

Figure 6. (a) Manifold for the determination of glucose, free cholesterol, creatinine and lactic acid with detection by chemiluminescence. The sample (S) is injected into a carrier stream of buffer propelled forward by pump P and carried to reactor (ER) containing immobilized enzyme (oxidase), in which the substrate of the sample is degraded leading to the formation of hydrogen peroxide. Being confluenced with luminol and hexacyanoferrate(III), the sample zone is finally guided through a short channel a (2 cm, corresponding to 16 μl) into the light detector D, the output from which is led to a computer (BBC). W, waste. (b) Integrated FIA microconduit accommodating the maniold components shown for the system in (a). C represents the carrier stream of buffer, R1 is luminol and R2 is hexacyanoferrate(III). The flow cell (FC) comprises two photo diodes (D) contained in house (H) mounted on the microconduit base plate (B).

$$NH_4^+ + \alpha\text{-ketoglutarate} + NADH \xrightarrow{GLDH} \text{glutamate} + NAD^+$$

Thus the FIA manifold comprizes two consecutive reactors with immobilized enzymes, the first one containing GLDH which effectively removes all ammonium present whereupon the sample zone is led to the second reactor containing creatinine iminohydrolase before being merged with the alkaline stream immediately prior to detection.

An intriguing exploitation of packed reactors in FIA is represented by the FIA *conversion* techniques [4, 26], which utilizes the fact that in the highly reproducible time-concentration domaine of the dispersed sample zone kinetic chemical exchange between any two phases (e.g., a liquid sample and a solid surface) can be executed reproducibly. In this manner a nondetectable species can be converted through a heterogeneous chemical reaction into a detectable component. Early examples of this approach are the assay of nitrate via conversion of nitrate to nitrite by a reductor column [2], and the analysis of anions via conversion of these to hydroxid ions by means of an anion exchange column, the hydroxid ions subsequently being determined spectrophotometrically through the generation of colour of an added acid-base indicator [2]. A disadvantage of these conversion techniques is the lack of selectivity, and therefore it is not surprising that considerable ingenuity has been invested in designing systems which could remedy this. Although an intriguing task, several selective conversion methods have recently been devised and published, e.g., for the determination of cyanide, sulphide [26] and calcium [27].

Since the FIA conversion technique allows the execution of a full measurement cycle to be concluded typically within less than one minute, this approach opens not only new avenues for detection of species normally non-detectable by conventional detection procedures such as spectrophotometry, AAS or ICP, but renders them of very practical importance.

This versatility of FIA is further demonstrated in the variety of fields of applications. Thus in fundamental physical chemical research FIA has been employed for measurements of reaction rate constants, extraction constants, diffusion coefficients, and solubility products. In the area of applied chemistry FIA is emerging as a new tool for continuous monitoring aimed at implementing the technique for on-line control of industrial processes. For *process control* applications (Fig. 7) FIA offers the benefit of real time continuous monitoring by aspirating regularly sample solution (e.g., from a batch type chemical reactor) into a carrier-wash solution, the detector of choice being operated in an impulse response mode. Thus a chemically modulated signal, continuously recorded as a series of peaks (injected material on a constant background of carrier-wash), offers a positive control of both the detector and of the process performance, detector fouling being prevented by means of a periodic wash sequence. Occasional alternate injections of standard solutions allows further control of detector performance thus assuring that the magnitude of the recorded peaks truly reflects the concentration of the monitored species (and not deterioration of the response characteristiques of the detector). During a typical one minute measuring cycle FIA allows dilution of up to several thousands times [10], chemical derivatization, separation or other solution handling tasks so well established by FIA laboratory praxis. Chemical production [28, 29], biological waste water treatment [30, 31], and biotechnological processes [32, 33, 34] were so far reported to be controlled by FIA techniques, and two FIA based process control chemical analyzers are now marketed [5]. However, in contrast to FIA systems used under protected laboratory conditions or in pilot plant scale, industrial FIA equipment, being exposed to operation under often hostile environment, obviously has to fulfill more stringent requirements. Such system therefore demands robust, custom designed apparatus which can operate reliably for longer periods of time and which inherently are intelligible in design, simple to use and preferably does not require special skills to operate.

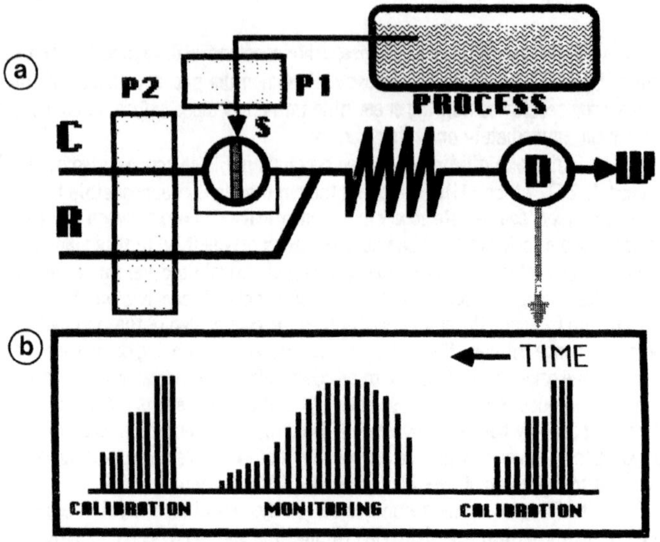

Figure 7. Continuous monitoring based on sample injection. The sample (S), aspirated from a process reactor (process) by pump P1 is by means of valve S intermittently injected into carrier stream C which downstream is merged with reagent solution R, propelled by pump P2. Prior to the monitoring sequence (monitoring) the system is calibrated by injecting a series of standards (calibration) which procedure is repeated after the monitoring period. Note that the baseline is reached between each injection thereby allowing control of the performance of the analyzer itself.

The recent analytical literature abunds with reports on the development of electrochemical, fiber optic, piezoelectric and other *sensors*. Novel detection principles are frequently announced, and attainment of improved sensor quality in terms of selectivity, sensitivity and lifetime are the goal of many established as well as newly formed research groups. The need for continuous monitoring of critical parameters in clinical chemistry, biotechnology, pharmaceutical, chemical and nuclear industry, chemical warfare or environmental control is the driving force behind the sensor research.

While the majority such research activities has novel aspects, sound theoretical basis and uses high technology to fabricate the sensing devices, the Achilles Heel of many chemical sensors is the lack of their time stability as reflected in deterioration of their response - up the the ultimate lack of it. Surprizingly, most of the research papers published to date do not contain the vital data on the performance of these sensors, such as speed of response, sensitivity and detection limit, as these parameters necessarily undergo changes during the lifetime of each device. Information on poisoning, or recovery, of sensors as well as on typical lifetimes at various operating conditions are even more scarce. Whether this is due to proprietory information or lack of data is difficult to judge.

Introduction of an automated, well reproducible rigorous testing procedure which would allow critical testing of chemical sensors should become a powerful tool for sensor development. There is increasing experimental evidence that FIA is suitable for sensor and material testing by means of the impulse-response technique [4].

Impulse-response FIA is based on repetitive action of a well defined zone of a selected chemical species (Reagent) on a target (T) situated in a continuous unsegmented carrier stream (C) consisting of an inert fluid (Fig.8). The target may be a sensor, the carrier stream may be an aqueous solution to which the sensor does not respond (or responds at the detection limit - thus establishing a continuous baseline). The injected species is the one to which the sensor is to respond. By continuously monitoring the sensor output, while repeatedly injecting, at a fixed frequency, the same species at the same concentration level in the form of a well defined zone, a series of peaks will be obtained, which would reflect the response of the sensor. Thus the series of peaks will show the response to either remain unchanged (Fig. 8b (A)) or to be deteriorating (Fig. 8b (B)) over a selected test period. Since the procedure is automated it is not prone to subjective judgement or manual operational errors. Indeed, instead of injecting repeatedly the same solution,

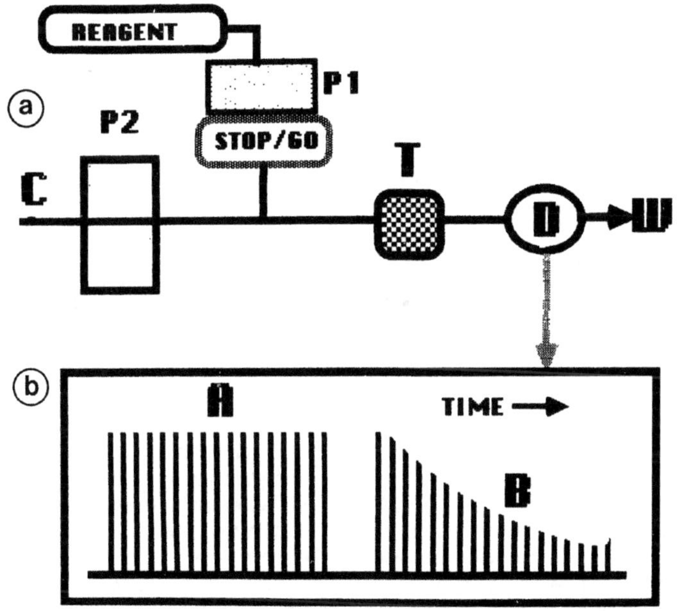

Figure 8. (a) Impulse response FIA system comprizing a target material T situated within carrier stream (C) propelled by pump P1. At preselected time intervals a metered volume of test liquid (reagent), aspirated by pump P1 by operating it for a fixed period of time (GO), is introduced into the carrier stream, the effect of the exposure being monitored by detector D. (b) In **A** is shown a run where the target is unaffected by the repeated bombardments of test solution, while **B** represents a situation where the target is gradually deteriorating due the repeated exposure.

a series of solutions, R1, R2, R3, of different concentrations, may be injected in preprogrammed sequences. Similarly, interfering or regenerative species, may be injected repeatedly or singly.

Target materials may range from inorganic materials to biologically active species. Thus surface corrosion may be investigated by repeatedly bombarding the target with a suitable corrosive agent and by measuring the released material by AAS or ICP. The rate of dissolution of coatings of surfaces, of pharmaceutical products, the stability of dyes against bleaching, of the degree of immobilization of enzymes, the rate of dissolution of solids, or, as recently suggested [35], for the study of drug-protein binding interaction using continuous monitoring of dynamic dialysis, all can be investigated by the FIA impulse-response technique with aid of suitable detectors. The lifetime and change of response characteristiques of sensors (fiber optic, electrochemical) can be tested and quantitatively evaluated as described above. Since target materials can be tested automatically with high reproducibility under prolonged periods, the FIA

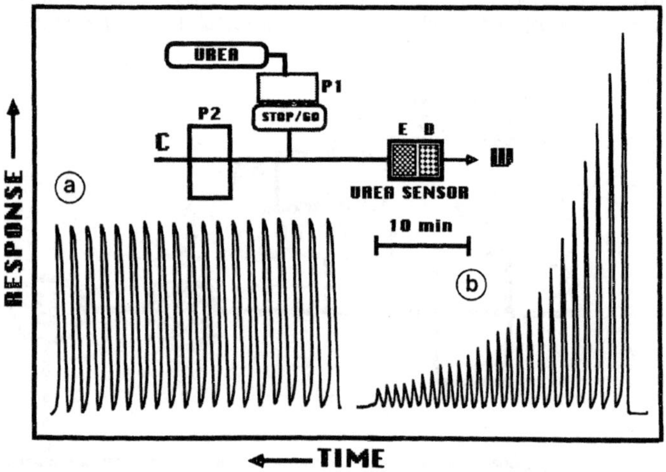

Figure 9. Impulse response FIA system used for investigating the performance of a urea optosensor comprizing a pad on which urease (E) and an acid-base indicator (D) are coimmobilized, the transmission og light to and from the flow cell being effected by means of optical fibers (not shown). First the indicator stability in terms of its covalent attachment to the supporting material was evalated by preparing a pad with acid-base indicator alone and subjecting it repeatedly and alternately to acidic and alkaline standard solutions. A series of readouts similar to those shown in (a) confirmed that no loss of immobilized dye is encountered. Two types of urease immobilization were then tested by repetitive impulses of the same urea standard (10 mM), operating the FIA system in the stopped-flow mode (stop time 60 sec): In (b) the indicator containing pad is sandwiched with an immunoaffinity membrane (Biodyne) to which urease is attached, yet the rapid decay in response as a function of time clearly indicates that the enzyme is gradually being washed out of the sensor. In (a) urease is coimmobilized by glutaraldehyde cross-linking on the cellulose indicator pad, the constant response signals demonstrating the stability of this sensor system [37].

impulse response technique might become, once its potential has become generally recognized and exploited, the most important modification of the flow injection method, perhaps even surpassing its applications as a tool for serial assays.

Besides examining a tendency of a series of response peaks, as illustrated in Fig. 8, a change in individual peak shape will be indicative of e.g., adsorption-desorption phenomena at the surface of a sensor. Since most chemical sensors utilize a chain of chemical reactions taking place in the sensing area, it is possible to examine each set of sensing reactions separately and thus localize the weakest link of the sensing process. Examples of such a systematic testing and development approach is the work on fibre optic pH sensors utilizing immobilized acid-base indicators, carried out by R. Singer at the at the Chemistry Department A in Denmark [36], and

The type of recordings obtained can be analysed to show how the lifetime and response that of T. Yerian at the Department of Chemistry University of Washington in USA [37] investigating the performance of an urea sensor, consisting of a sandwiched immobilized urease-covalently bound pH indicator, as studied by the FIA impulse response technique for different types of enzyme immobilization, the detection being effected by means of optosensing (Fig. 9).characteristiques of a pH sensor or a biosensor depend on the stability of: (a) The optical components (light source, detector and light losses in and the quality of the optical fibres), as well as on the physical changes occuring on the surface of the reflecting material in the flow cell; (b) The indicator stability in terms of its covalent attachment to the supporting material, and as affected by photobleaching; (c) The enzyme stability in terms of its physical or chemical attachment to the supporting material, and the activity loss as caused by thermal and chemical degradation during use; (d) The ability of the indicator-enzyme couple to respond to the analyte in a wide range of concentrations and to become regenerated/washed if a maximum level of analyte is repeatedly exceeded; and (e) The speed of response and the reversibility of the enzyme-indicator couple used.

Thus FIA has developed, during its first twelve years of existense, from a method for performing serial assay in analytical chemistry to becoming a versatile and even *diagnostic tool* in entirely new areas of solution handling.

Acknowledgement The author expresses his gratitude to the Danish Council for Scientific and Industrial Research for financial support of the current FIA research programme. Thanks are also due to the North Atlantic Treaty Organization (NATO) for facilitating participation in the Florence Meeting.

REFERENCES

1 J. Ruzicka and E. H. Hansen, *Anal. Chim. Acta*, **78** (1975) 145.
2 J. Ruzicka and E. H. Hansen, *Flow Injection Analysis*, Wiley-Interscience, New York, USA, (1981).
3 E. H. Hansen, *Int. Lab.*, **15**(8) (1985) 14.
4 J. Ruzicka and E. H. Hansen, *Anal. Chim. Acta*, **179** (1986) 1.
5 J. Ruzicka and E. H. Hansen, *Flow Injection Analysis 2*, Wiley-Interscience, New York, USA, (in press; to be publ. Fall 1987).
6 J. Ruzicka and E. H. Hansen, *Anal. Chim. Acta*, **180** (1986) 41.
7 Z. Fang, J. Ruzicka and E. H. Hansen, *Anal. Chim. Acta*, **164** (1984) 23.
8 J. Ruzicka and E. H. Hansen, *Anal. Chim. Acta*, **145** (1983) 1.

9 B. F. Reis, A. O Jacintho, J. Mortatti, F. J. Krug, E. A. G. Zagatto and H. Bergamin Fº, *Anal. Chim. Acta*, **123** (1981) 221.
10 M. B. Garn, M. Gisin, H. Gross, W. Schmidt and C. Thommen, *Anal. Chim. Acta*, (in press).
11 S. S. Olsen, J. Ruzicka and E. H. Hansen, *Anal. Chim. Acta*, **136** (1982) 101.
12 J. Ruzicka and E. H. Hansen, *Anal. Chim. Acta*, **106** (1979) 207.
13 J. Janata and J. Ruzicka, *Anal. Chim. Acta*, **139** (1982) 105.
14 Z. Fang, J. M. Harris, J. Ruzicka and E. H. Hansen, E. H., *Anal. Chem.*, **57** (1985) 1457.
15 S. Greenfield, *Ind. Res. Dev.*, **23** (1981) 140.
16 K. E. Lawrence G. W. Rice and V. A. Fassel, *Anal. Chem.*, **56** (1984) 289.
17 J. T. Dobbins Jr. and J. M. Martins, *Spectroscopy*, **1** (1986) 20.
18 J. Ruzicka, E. H. Hansen and H. Mosbæk, *Anal. Chim. Acta*, **92** (1977) 235.
19 A. U. Ramsing, J. Ruzicka and E. H. Hansen, *Anal. Chim. Acta*, **129** (1981) 1.
20 E. H. Hansen, J. Ruzicka, F. J. Krug and E. A. G. Zagatto, *Anal. Chim. Acta*, **148** (1983) 111.
21 J. Ruzicka and E. H. Hansen, *Anal. Chim. Acta*, **173** (1985) 3.
22 J. Ruzicka and E. H. Hansen, *Anal. Chim. Acta*, **161** (1984) 1.
23 A. Petersson, E. H. Hansen and J. Ruzicka, *Anal. Lett.*, **19** (1986) 649.
24 B. Olsson, B. Stålbom and G. Johansson, *Anal. Chim. Acta*, **179** (1986) 203.
25 Y. Hayashi, K. Zaitsu and Y. Ohkura, *Anal. Chim. Acta*, **186** (1986) 131.
26 B. A. Petersson, Z. Fang, J. Ruzicka and E. H. Hansen, *Anal. Chim. Acta*, **184** (1986) 165.
27 H. Wada, Y. Deguchi and G. Nagakawa, *Mikrochim. Acta*, **III** (1985) 393.
28 A. Schelter-Graf, H. L. Schmidt and H. Huck, *Anal. Chim. Acta*, **163** (1984) 299.
29 M. J. Whitaker, *Amer. Lab.*, **18** (1986) 154.
30 T. Korenaga and H. Ikatsu, *Bunseki Kagaku*, **31** (1982) 135.
31 K. M. Pedersen, P. Haagensen, M. Kümmel and H. Søeberg, *IFAC Proc. Ser.*, **10** (1986) 131.
32 C. F. Mandenius, B. Danielsson and B. Mattiasson, *Anal. Chim. Acta*, **163** (1984) 135.
33 A. Recktenwald, K. H. Kroner and M. R. Kula, *Enzyme Microb. Technol.*, **7** (1985) 146; ibid., **7** (1985) 607.
34 C. P. Parker, M. G. Gardell and D. DiBiasio, *Int. Biotech. Lab.*, **4** (1986) 33.
35 P. Macheras, M. Koupparis and C. Reppas, *Int. J. Pharm.*, **30** (1986) 123.
36 R. Singer, *Applications of Immobolized Indicators for Optosensing by Flow Injection Analysis*. (In Danish). M.Sc. Thesis, Tech. Univ. Denmark, (1986).
37 T. Yerian, G. D. Christian and J. Ruzicka, *Anal. Chem.*, (in press).

EXTENSION TO ENZYMATIC ELECTRODES OF A GENERAL PROCEDURE FOR THE CONSTRUCTION OF FLOW-THROUGH ISEs BASED ON THE USE OF CONDUCTIVE EPOXY AS SUPPORT OF THE SENSOR

Salvador Alegret
Department of Chemistry,Universitat Autonoma de Barcelona
E08193 Bellaterra,Spain

Adélio A.S.C.Machado*
Chemistry Department,Faculty of Science
P4000 Porto,Portugal

ABSTRACT.Work in progress to apply a previously developed general technique for the construction of ion-selective electrodes (ISEs) to the construction of enzymatic electrodes is presented. This technique,which consists in forming the sensor membrane on a conductive epoxy, has been found to be suitable for the preparation of crystalline membrane or PVC membrane (mobile carrier) electrodes and its most interesting advantage is that devices with unusual shape and miniaturised dimensions can be constructed (for instance,tubular and sandwich flow-through electrodes suitable for use in FIA). The present work is devoted to the preparation of flow-through enzymatic detectors of the potentiometric and amperometric types. Preliminary results for an urea potentiometric electrode and a glucose amperometric electrode are discussed.

1. INTRODUCTION

As a result of a collaborative project between our groups in the University of Oporto and the Autonomous University of Barcelona, we developed a general procedure for the construction of flow-through ion-selective electrodes (ISEs) with PVC or solid crystaline membranes applied to conductive epoxy supports (1-5). These sensors were designed for use in flow-injection analysis (FIA) and have been tested in FIA assemblies for sequential measurement of two species (3,6).

We have recently initiated some exploratory work with the purpose of extending this construction procedure to obtain enzymatic detectors. This note reports the first results of this

*) Author for correspondence.

G. G. Guilbault and M. Mascini (eds.),
Analytical Uses of Immobilized Biological Compounds for Detection, Medical and Industrial Uses, 309–317.

line of work.First,a brief review of the previuosly developed construction techniques for the preparation of classical and flow-through conductive epoxy ISEs is included.

2. MATERIALS AND PROCEDURES FOR ELECTRODE CONSTRUCTION

2.1.Electrodes of classical configuration

The basic feature of the general procedure developed in Oporto for assembling all-solid-state ISEs both with crystalline membranes (7-9) and with PVC membranes (10-12) (in this case,electrodes of the mobile-carrier type),is the use of a conductive epoxy base for application of the sensor membrane. This conductive epoxy base serves simultaneously as a physical support for the sensor and as an electrical conductor for internal contact.

With reference to practical details of the construction procedure,the bodies of the electrodes are prepared (7-9) from perspex (or glass) tubes of suitable length and diameter. A conductive epoxy layer is applied to one of the extremities of the tube,the internal electrical cable being fixed to it.The other extremity is sealed.

For application of a solid sensor (7-9), a conical cavity is drilled in the epoxy layer, after its hardening. This cavity is filled with a second layer of the same epoxy and,while it is still fresh,the finely powdered solid sensor is applied to its surface to re-cover the resin as completely as possible. After hardening,the surface of the electrode is polished,first over glass and then with polishing paper (in both cases,the surface is kept wet while being polished).Using this technique, electrodes with sensors made of Ag_2S or mixtures of Ag_2S and AgX (X = halide) (7,8,13), Ag_2S and MS (M = Cu,Cd, Pb) (7,14), and HgS and Hg_2X_2 (X = Cl,Br) (9,15),have been obtained whose response characteristics compare well with commercial electrodes.

Crystalline heterogenous membrane electrodes,whose membranes are constituted by crystalline sensors dispersed in non-conductive epoxy applied to a first layer of conductive epoxy, have also been prepared (16,17).

For electrodes with extracting systems dispersed in PVC membranes (Fig.1),a flat bottomed cavity for application of the plastic membrane is prepared at the extremity of the body where the conductive epoxy layer was applied. A solution of the sensor system (sensor system,PVC,and,in most cases,a plasticizer,in a suitable solvent,generally THF),is deposited in the cavity.After evaporation of the solvent, a PVC membrane with the sensor dispersed in it is formed,which shows good adherence to the epoxy. The application of the activating solution is repeated half a dozen of times in order to obtain an enough thick membrane. Electrodes of this type with several ionic or neutral extracting agents have been prepared (10-12,18,19). Their response characteristics are very similar to those of the

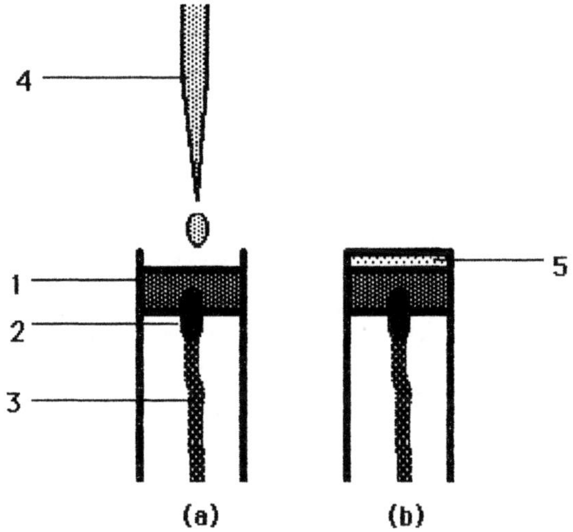

Fig.1.Preparation of the PVC membrane electrodes:(a) body with epoxy support and (b) electrode with PVC membrane.
1) conductive epoxy support 2) electric contact
3) shielded cable 4) solution of sensor and PVC in PVC
5) PVC membrane

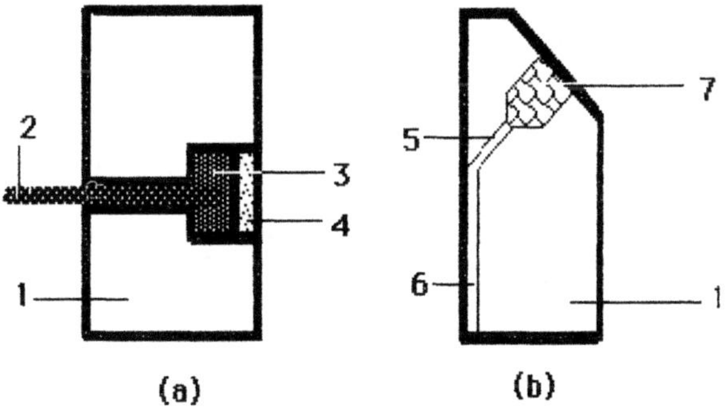

Fig 2.Basic design of sandwich flow-through cell:
(a) block with sensor chamber (b) cover,which are pressed together by four screws (not shown)
1) perspex block 2) shielded cable 3) conductive epoxy
4) PVC membrane 5) inlet channel 6) outlet channel
7) thread

classical Moody and Thomas' PVC membrane electrodes with internal solution and reference electrode (20),as shown by parallel evaluation of both types of assemblies activated for several species (p.ex.11,19). On the whole, PVC membrane electrodes with epoxy support show longer durability but slightly worse stability than electrodes with the same type of membrane but with internal solution and reference electrode.

As conductive epoxy materials,a commercial silver-filled resin (EPO-TEK 410,obtained from Epoxy Technology Inc) has been used in most cases,but a commercial copper-filled epoxy (EPO-TEK 410,obtained from the same manufacturer) as well as non--conductive epoxy (for instance,Araldite M with hardener HR) loaded with graphite powder,have also been tried with success.

For both cystalline membrane and PVC membrane electrodes, when the quality of response is no longer acceptable,the membrane can be easily removed and replaced by a new one,or the same body can be used for activation with another sensor.

In summary,the present construction technique is simple,general,economical,easy to implement and,specially with reference to PVC membrane electrodes,the procedure has proved to be very versatile as far as shape,configuration and dimensions are concerned (for instance,miniaturized electrodes of microconfiguration have been prepared [21]).

2.2.Flow-through electrodes

The constructional versatility of the present procedure led us to use it for preparing flow-through electrodes for application in FIA. Two basic designs have been developed with sandwich and tubular configurations.Both types of configuration have been used for constructing PVC membrane electrodes.The tubular design has been adopted for preparing crystalline heterogenous membrane electrodes.

With reference to sandwich electrodes,the basic configuration (Fig.2) consists of: (i) a perspex block where the PVC membrane is applied to the conductive epoxy in a cavity and (ii) a lid with an inlet channel provided with a screwing thread for connections and a semi-cylindrical channel which fits on the sensor membrane (2,5). More elaborated designs have been developed from this basic configuration (5). Designs with two sensor membranes in sequence have also been used. We have found that this type of sandwich electrode shows some advantages when compared with the cascade configuration (22),especially improved response stability. This results from improved mechanical stability,a close contact between the solution and the surface of the membrane and the area of solution path over the membrane is constant.

With respect to tubular electrodes with PVC membranes,the basic idea of the construction procedure was to have a channel drilled along a conductive epoxy cylinder,which is then fixed to a connecting cable and included in a perspex block where inlet/outlet channels are drilled (1). The conductive epoxy channel is then re-covered with a PVC membrane that is evaporated from a THF solution by the tecnique described before. However, in the present case, to warrant that the PVC membrane on the epoxy surface is continuous, the application of the membrane is more difficult.

After trying several variations in constructional design of this type of tubular electrode, devices in which a small/ /interchangable sensing element is placed between two holders proved of easier construction and more efficient use (4,5). Two tubular sensing elements can be used in sequence if an intermediate holder is placed between them. Moreover, to obtain sensing elements with crystalline membranes, the channel of the element is filled with a dispersion of solid sensor (fine powder) in non-conductive epoxy. After drying, a channel is drilled through this mixture (23).

The present procedure for construction of flow-through electrodes shows several advantages: it is simple and economical, general and versatile, and the tubular electrodes are specially suitable for use as on-line detectors in sequential determinations. This has been proven by detailed FIA studies of two problems: (i) determination of the contributions to water hardness of calcium and magesium, in which calcium was determined on-line by potentiometry and magnesium at the end-of-line position by atomic absorption spectrophotometry (3); (ii) determination of nitrate and calcium in water, by potentiometry. In this assay, calcium was measured first and a silver(I) solution was added before nitrate measurement to eliminate the interference of chloride in the nitrate sensor (7).

2.3. Immobilized enzyme electrodes

In exploratory work to extend the present technique to obtain enzymatic detectors, the sandwich type of assembly has been used. A potentiometic electrode for urea (24) and an amperometric sensor for glucose (25) have been tried.

For the preparation of a urea potentiometric detector (24), a modified sandwich cell with a graphite-loaded epoxy base was filled, by the procedure described before, with a PVC nonactin membrane (26) to construct an ammonium sensor. Urease was immobilised on a nylon mesh by a standard procedure (acrylamide gel) (27). This master membrane was kept in the refrigerator in 0.1 **M** TRIS solution (at pH=7, adjusted with hydrogen chloride). From this master membrane, a small rectangle was cut and placed on the PVC membrane. When the cell was closed with its lid, the nylon mesh piece was pressed against the ammonium sensing membrane. The device was included in a simple FIA manifold to test its response to urea.

To prepare an amperometric glucose detector (25), a sandwich cell was prepared with a base constructed by a mixture of graphite powder and non-conductive epoxy. After polishing, the surface of the carbon paste layer had to be flat, because any hollows would result in entrapment of oxygen bubbles formed by the anodic decomposition of hydrogen peroxide. This would provoke poor reproducibility of response.This device was activated with glucose oxidase by two procedures: (i) enzyme immobilized on bovine serum albumin placed over the carbon paste surface, and covered with a naked nylon mesh serving as a spacer for protection of the enzyme membrane and (ii) enzyme immobilized on a nylon mesh by the procedure described in detail by Moody,Shangera and Thomas (28), held against the carbon paste support.

The sandwich electrode was included in a FIA manifold (28), with a silver wire re-covered with silver(I) chloride and a platinum wire as the reference and auxiliary electrodes respectively,and calibrated against standard solutions using a Metrohm E-641 amperometric detector. The carbon paste electrode was calibrated before activation with hydrogen peroxide,and both immobilized enzyme electrodes against glucose.

3. RESULTS AND DISCUSSION

3.1.Potentiometric urea electrode (ref.24)

The nonactin PVC membrane electrode showed linear response to ammonium in the range 10^{-6}-10^{-1} **M** (at pH=7.5). The enzymatic sensor was calibrated with solutions in 10^{-2} **M** TRIS buffer with pH adjusted to 7 with hydrogen chloride (this same solution was used as carrier).Seven standards were used with urea concentrations in the range 10^{-5}-10^{-2} **M**. These calibrations were repeated during a fortnight (between the calibrations the electrodes were filled with 10^{-2} **M** TRIS buffer (pH=7) and stored in the refrigerator. After an initial period of stabilization,linear response with slopes of 40-50 mV/decade was obtained in the range $5x10^{-5}$-10^{-2} **M**. The shape of the signals produced in this case and in the case of the bare nonactin PVC membranes in response to ammonium ions were similar.

3.2.Amperometric glucose detectors (ref.25)

When the naked carbon paste electrode was calibrated with hydrogen peroxide standards (at +900 mV *vs.* the silver/silver(I) chloride electrode),it showed linear response in the range $5x10^{-5}$-$1x10^{-3}$ mol dm^{-3}. When calibrated with glucose in 0.1 **M** orthophosphate buffer,pH=6 (carrier: 0.1 **M** NaH_2PO_4 with NaOH),both enzymatic detectors exhibited the same linear range $1x10^{-4}$-$2x10^{-3}$ mol dm^{-3},but the nylon mesh electrode showed faster response (Table 1). The results are similar to those reported by Moody,Shangera and Thomas (28) in which a nylon mesh/platinum wire electrode was used in a modified Stelte micro-cell.

Table 1.Calibration of glucose amperometric detectors [a,b]

Electrode [c]	RLR mol dm^{-3}	RT s	WT s
BSA	$1x10^{-4}$-$2x10^{-3}$	60	50
NM	$1x10^{-4}$-$2x10^{-3}$	35	45

a) Ref.25.
b) At +900 mV *vs.* the Ag|AgCl electrode
RLR,range of linear response; RT,response time;
WT,wash time
c) BSA,bovine serum albumine; NM,nylon mesh

CONCLUSIONS

These exploratory experiments of activation with immobilized enzyme membranes of sandwich flow-through detectors with conductive epoxy supports, yielded acceptable results, although further studies for improvement of the preparation procedure and optimization of the working conditions are required.

ACKNOWLEDGEMENTS

To Dr.J.D.R.Thomas,of UWIST,Cardiff,UK,for many helpful suggestions and discussions.These and the exchange of researchers involved in the work between Cardiff and Oporto,were made possible by a Travel Grant from the Scientific Affairs Division of NATO (nr 069/84).

The Councils of Rectors of the Universities of Portugal and Spain are thanked for a Travel Grant under which the exchange of researchers between Barcelona and Oporto has been made possible.

One of us (AASCM) thanks INIC,Lisbon,for financial support (to CIQ(UP),Line 4).

REFERENCES

1. S.Alegret,J.Alonso,J.Bartroli,J.M.Paulis,J.L.F.C.Lima and A.A.S.C.Machado, 'Flow-through Tubular PVC Matrix Membrane

Electrode without Inner Reference Solution for Flow Injection Analysis', Anal. Chim. Acta, **164**, 147/52 (1984)
2. S.Alegret,J.Alonso,J.Bartroli,J.L.F.C.Lima,A.A.S.C.Machado and J.M.Paulís, 'Flow-through Sandwich PVC Matrix Membrane Electrode for Flow Injection Analysis', Anal.Lett., **18**, 2291/303 (1985)
3. J.Alonso,J.Bartroli,J.L.F.C.Lima and A.A.S.C.Machado, 'Sequential Flow-injection Determinations of Calcium and Magnesium in Waters', Anal.Chim.Acta, **179**, 503/8 (1986)
4. S.Alegret,J.Alonso,J.Bartroli,J.L.F.C.Lima and A.A.S.C.Machado, '"All Solid State" PVC Membrane Flow-through Potassium Electrode Based on Valynomicin', *in* J.L.Aucouturier,J.S.Cauhapé,M. Destriau, P.Hagenmuller,C.Lucat,F. Ménil,J.Portier and J. Salardenne (ed.),Proceedings of the Second International Meeting on Chemical Sensors, Bordeaux, 1986, p.751/4
5. S.Alegret,J.Alonso,J.Bartroli,J.L.F.C.Lima,A.A.S.C.Machado and J.M.Paulis, 'Construction of Equipment for Potentiometric Determinations in Flow Injection Analysis', Quim.Anal., accepted (1986)
6. J.Alonso,J.Bartroli,J.L.F.C.Lima and A.A.S.C.Machado, 'Sequential Flow-injection Determinations of Calcium and Nitrate in Waters with Flow-through Tubular Electrodes", in preparation
7. J.L.C.Lima and A.A.S.C.Machado, 'ISEs with Solid Sensor Coatings Applied to Electrically Conductive Epoxy', *in* J.Albaiges(ed.), Analytical Techniques in Environmental Chemistry,Vol.2, Pergamon Press, 1982, p.419/26
8. J.L.F.C.Lima and A.A.S.C.Machado, 'Construction and Evaluation of a Silver(I) Sulphide Ion-selective Electrode with Conductive Epoy Support", Rev.Port.Quím., **21**, 15/30 (1979)
9. J.L.F.C.Lima and A.A.S.C.Machado,'Preparation of a Chloride--selective Electrode Based on Mercury(I) Chloride - Mercury(II) Sulphide on an Electrically Conductive Epoxy Support', Analyst, **111**, 151/5 (1986)
10. J.L.F.C.Lima and A. A.S.C.Machado,'Procedure for the Construction of All-solid-state PVC Membrane Electrodes', Analyst, **111**, 799/802 (1986)
11. M.C.B.S.Montenegro,J.L.F.C. Lima and A.A.S.C.Machado, 'Mono-,Di- and Trichloroacetate Selective Electrodes with Sensors Dispersed in PVC Membrane Applied on Conductive Epoxy Support', Rev.Port.Quim., **26**, 197/211 (1984)
12. J.L.F.C.Lima and A.A.S.C.Machado,'Potassium Ion Selective Electrode with a PVC Membrane Applied to a Conductive Epoxy Support",Portugaliae Electrochim.Acta, **3**, 311/28 (1985)
13. J.L.F.C.Lima and A.A.S.C.Machado, 'Construction and Evaluation of Halide Ion-selective Electrodes with Conductive Epoxy Support', Rev.Port.Quím., **21**, 153/178 (1979)
14. J.L.F.C.Lima and A.A.S.C.Machado, 'Construction and Evaluation of a Copper(II) ISE with a Conductive Resin Support", Rev.Port Quim., **24**, 156/77 (1982)
15. J.L.F.C.Lima and A.A.S.C.Machado, 'Bromide Electrode Based on Hg_2Br_2/HgS Mixture Applied to Electrically Conductive Epoxy', Portugaliae Electrochim. Acta, **4**,33/50 (1986)

16. S.Alegret, J.L.F.C.Lima,A.A.S.C.Machado,E.Martínez-Fàbregas and J.M.Paulis, 'A New Procedure for the Construction of Mouldable Shape Membrane Ion Selective Electrodes with Crystalline Sensor Dispersed in Epoxy Resins', Quím. Anal., accepted (1986)
17. S.Alegret,A.Florido J.L.F.C.Lima and A.A.S.C.Machado, 'Chloride and Bromide Ion Selective Electrodes with Crystalline Sensor Dispersed in Epoxy Resin', Quim.Anal., accepted (1986)
18. J.L.F.C.Lima,A.A.S.C.Machado,A.Florido,S.Alegret and J.M.Paulís, 'Nitrate-selective Electrode with PVC Matrix Membrane Applied to Electrically Conductive Epoxy', Quim. Anal., **4**, 145 /58 (1985)
19. S.Alegret,A.Florido,J.L.F.C.Lima and A.A.S.C.Machado, 'Perchlorate Selective Electrode with PVC Matrix Membrane Applied to Electrically Conductive Epoxy', Quim.Anal., **5**, 36/48 (1986)
20. G.J.Moody and J.D.R.Thomas, 'Poly(Vinyl Chloride) Matrix Membrane Ion-selective Electrodes', *in* H.Freiser (ed.), Ion-selective Electrodes in Analytical Chemistry, Vol.1, Plenum Press, New York, 1978, p.287/309
21. S.A.H.Khalil,G.J.Moody, J.D.R.Thomas and J.L.F.C.Lima, 'Epoxy--based All-solid-state Poly(Vinyl Chloride) Matrix Membrane Calcium Ion-selective Microelectrodes', Analyst, **111**, 611/7 (1986)
22. J.Ruzicka,E.H.Hansen and E.A.Zaggato, 'Flow-Injection Analysis--VII-Use of Ion Selective Electrodes for Rapid analysis of Soil Extracts and Blood Serum.Determination of Potassium,Sodium and Nitrate', Anal.Chim.Acta, **88**, 1 (1977); **90**, 365 (1977)
23. S.Alegret,A.Florido J.L.F.C.Lima and A.A.S.C.Machado, 'Chloride and Bromide Tubular Ion Selective Electrodes with Crystalline Sensor Dispersed in Epoxy Resin', in preparation
24. S.Alegret and E.Martínez-Fàbregas, work in progress
25. A.A.S.C.Machado,G.J.Moody,L.M.B.C.A.Ribeiro,G.S.Sanghera and J.D.R.Thomas, work in progress
26. W.E.Morf and W.Simon, 'Ion-Selective Electrodes Based on Neutral Carriers', *in* H.Freiser (ed.), Ion-selective Electrodes in Analytical Chemistry, Vol.1, Plenum Press, New York, 1978, p.211/286
27. G.G.Guilbault and G.O.Neto, 'Immobilised Enzyme Electrodes', *in* J Woodward (ed.), 'Immobilised Cells and Enzymes, A Practical Approach', IRL Press,Oxford, 1985, p.55/74
28. G.J.Moody,G.S.Sanghera and J.D.R.Thomas, 'Amperometric Enzyme Electrode System for the Flow Injection Analysis of Glucose', Analyst, **111**, 605/9 (1986)

BIOSENSOR-BASED ANALYZERS : FROM DESIGN TO INSTRUMENT

P.R. Coulet
Laboratoire de Génie Enzymatique - Atelier de Biotechnologie
CNRS-Université Claude Bernard Lyon 1- 43 Blvd du 11 Novembre 1918
69622 Villeurbanne Cedex
France

ABSTRACT. Numerous papers appeared in the last decade in the field of biosensors but the gap between academic results and real industrial applications is still important. Reasons for such a discrepancy are detailed. Concerning the biosensor-based analyzers available on the market, it must be stressed that only oxidase-electrodes are really used and glucose is the most commonly determined analyte. In our group, different self-contained enzyme electrodes have been developed during the last decade and a microprocessor- based analyzer derived from this work has been marketed in France by Tacussel. Recently, "Biodyne Immunoaffinity" membranes supplied in a preactivated form by Pall were selected in our group for the preparation of enzymic discs bearing new oxidases. New analytes could be reliably and conveniently monitored. Improvements are still expected concerning the stability of the bioactive component and the immobilized enzyme-transducer interface.

INTRODUCTION

The concept of biosensor was proposed more than twenty years ago. Numerous papers and patents describing various biological compounds associated to several sophisticated types of transducers for the assay of a variety of analytes, were issued since then. However, only a few biosensor-based analyzers are commercially available. It appears that the most commonly used biosensor is the enzyme electrode and the transducer is the amperometric probe for O_2 or H_2O_2 detection associated to an immobilized oxidase. The most commonly detected analyte is glucose. When investigating the reasons for such a relatively slow development, it is clear that even with a quite simple association biocatalyst-transducer for the specific determination of a common analyte, the overall sensor design will mainly depend on the domain, i.e. medical, industrial, environmental, where it will be used. Requirements are then completely different when considering both the conditions of measurements, manual or automatic, in the laboratory or on a specific site, and the qualification of the personnel available for operating. Fundamental problems on storage and operational stability, catalyst poisoning, fiability of new transducers must also be solved.

G. G. Guilbault and M. Mascini (eds.),
Analytical Uses of Immobilized Biological Compounds for Detection, Medical and Industrial Uses, 319–327.

BIOSENSOR BASIC COMPONENTS

The concept of biosensor has been proposed and developed since the early sixties. Briefly, a biosensor consists of a

- A biological sensitive element

generally in an immobilized form able to recognize a target molecule among thousands of others.
The main Biological sensitive elements used for the design of biosensors are :

* Enzymes
* Antibodies
* Organelles
* Bacteria or whole cells
* Tissue slices

- A transducer

which will monitor modifications induced in the microenvironment of the biomolecule when in contact with its target-molecule and giving an electrical output. The transducers generally used for the design of biosensors are :

*Amperometric or potentiometric electrodes
*Field Effect Transistors (FET)
*Piezoelectric crystals
*Thermistors
*Optoelectronic systems

- A signal processing electronic system

more or less sophisticated with digital result display and, possibly, connections for recorder, printer, computer, actuators etc...

More details on the numerous systems decribed since then can be found in well documented reviews dealing with both theoretical and analytical aspects of biosensors (1-7).

ENZYME SENSOR-BASED ANALYZERS

Different self-contained biosensors have been described but obviously, the most commonly used transducer in commercialized biosensor-based analyzers is the classical amperometric probe for O_2 or H_2O_2 detection associated to an immobilized oxidase and the most commonly detected analyte is glucose. Lactate and alcohol can also be measured in several cases.
20 enzyme electrode -based instruments are proposed in different countries (Japan, USA, eastern and western Europe, USSR) but they are generally available only on the national market (8-9).

Basic requirements :

The design of the biosensor-based analyzer will mainly depend on its specific use in

* Medicine
* Industry
* Environment
* Defense

and even inside each sector the requirements may be totally different, due to the specific site where the system has to be used.

Medical applications

for example, different locations are possible :

* In the hospital
    - in the laboratory for routine determinations
    - in an emergency service
    - at patient bedside
* At the physician's office
* At patient's home

Even for the same analyte the requirements in these different cases will be totally different. For example :

- Will the device be operated by a technician or an unskilled person ?
- Is a very long stand-by without recalibration possible ?
- Is an automatic detection of errors possible ?
- Are simpler alternative methods available ? etc...

and the question then arises : is the "same" biosensor usable ?

In the fast growing field of biomedical engineering, improvements are expected in implanted sensor technology for in vivo monitoring but serious difficulties have still to be overcome, namely for :

* Biocompatibility
* Lifetime of the biosensor
* Drift and sensitivity changes
* The problem of recalibration in vivo
* Sterilization (autoclavability ?)

Industrial use

The analyzer must be robust but sophisticated to be used either in the control laboratory or on the production site ? For these purposes, the following characteristics are required

* Total fiability
* Alarm signal
* Easy to operate after simple training
* Wide dynamic range
* No pretreatment of the sample
* Online measurements possible

Environment and pollution control

Ideally the biosensor might fulfill the following requirements...

* Fully automatic system
* Trace monitoring or spike detection mode
* No drift during long periods
* Automatic replacement of poisoned biocatalyst
* Connection to a data processing network
* Periodic autotest

OUR CONTRIBUTION TO THE DEVELOPMENT OF BIOSENSOR-BASED INSTRUMENTATION

Several self-contained enzyme electrodes have been developed in our group during the last decade. The bioactive membrane was obtained from reconstituted collagen films after an acyl-azide activation procedure followed by a coupling step (10).

Briefly, discs of the desired size were cut out of the dry collagen membrane and immersed successively in different solutions. Available carboxyl groups from aspartate and glutamate were first esterified by treatment with a mixture of 0.2 M hydrochloric acid/methanol for 3-4 days at 20-22°C. After a fast washing in distilled water, the membranes were treated overnight in aqueous 1 % hydrazine solution. The acyl-azide derivative was formed immediately prior enzyme coupling by dipping the membrane into a mixture of 0.5 M $NaNO_2$ in 0.3 M HCl for 1-3 min.

After washing with a buffer solution, membranes free from activating reagents were ready for the enzyme coupling step which must be performed immediately.
In routine experiments, after a 2 hour coupling at 4°C, enzymic discs were rinsed in 1 M KCl to eliminate adsorbed enzyme molecules. Bioactive discs could then be stored either in buffer solutions or in dry state. Generally, the enzyme coupling solution could be reused for successive preparations of active discs.

Two different types of coupling, "random" or "asymmetric" could be performed. For the random process, activated membranes are directly immersed in the coupling solution and a surface covalent binding occurs spontaneously on both faces.

In the asymmetric mode, the coupling can be performed with different enzymes immobilized on each face. Such a process is important for the design of multienzyme electrodes. When the first enzyme is exposed to the reaction mixture, the product diffuses through the membrane and acts as substrate for the enzyme on the other face (11).

Self-contained enzyme electrodes

Determination of glucose

In a first approach, the design of a glucose electrode was achieved (12-13) . For this purpose, the random procedure for the immobilization of β-D-glucose oxidase on collagen was used.
In the presence of glucose oxidase, glucose is oxidized according to the reaction.

$$\beta\text{-D-glucose} + O_2 + H_2O \xrightarrow{\text{glucose oxidase}} \text{gluconic acid} + H_2O_2$$

Enzymatically generated hydrogen peroxide is detected by anodic oxidation at + 650 mV vs Ag /Ag Cl/Cl^- reference.

$$H_2O_2 \xrightarrow[\text{Ag / Ag Cl / Cl}^-]{+650\text{ mV}} O_2 + 2H^+ + 2e^-$$

When the electrode is plugged to a polarograph and its tip immersed in the suitable buffer in which glucose-containing samples are injected, the current output increases and a plateau is reached after a few minutes. The time necessary for reaching the plateau corresponds to the steady-state response time. The variation of current is proportional to glucose concentration.

For further assays, the samples can be added successively after obtaining each plateau provided the final concentration in the reaction vessel is kept below ca 10^{-3} M.

Characteristics of such an electrode were the following : detection limit 10^{-8} M; calibration range 10^{-7} M - 2.10^{-3} M; sensitivity 1-3 $mA.M^{-1}$; up to 3000 assays could be performed with the same membrane depending on the conditions of use; The operational stability was good over a 6 month period.
A glucose electrode "Gluc1" derived from the above mentioned work is now supplied by Solea-Tacussel, France.

Extension to other analytes

Other enzyme electrodes have been developed with single-enzyme systems like galactose oxidase, cholesterol oxidase, alcohol oxidase or with bienzyme systems like glucoamylase/glucose oxidase, invertase/glucose oxidase (14). Good performances comparing well with data reported in the literature by other authors were obtained in our laboratory but in our opinion, improvements are still necessary for an industrial use of such bioactive probes.

Measurement of enzyme activities

The rapid determination of activities of soluble or immobilized enzymes based on the electrochemical detection of hydrogen peroxide can be easily performed using the same probe equipped with a non-enzymic membrane (15).

Dry storage

One of the main drawbacks to be overcome for a practical use of biosensors in general is obviously the poor stability of the bioactive component. Assays have been performed in our group with the enzyme electrode stored dried in the air for long periods. After reimmersion of the electrode in the suitable buffer, measurements of concentrations as low as 2.5 10^{-4} M glucose were still possible with a glucose oxidase monoenzyme electrode after 160 days of dry storage (16)

Glucose analyzer

A microprocessor-based instrument called "Glucoprocesseur"® able to measure glucose in samples of various origins for both clinical analysis and food industry applications, is now marketed by Solea-Tacussel, Villeurbanne, France . The basic enzymic probes described above were used in the design of this instrument but improvements were made for shortening the response-time and removing the output current due the presence of interfering substances in complex mixtures.

Shortened response-time

The steady-state response-time i.e. the time necessary for the current to reach a plateau with the self-contained monoenzyme electrode was about 3 - 5 minutes . This time was considered as too long for practical applications and the height of the peak of the first derivative i.e. the dynamic response, was monitored (13) .This height was found proportional to hydrogen peroxide concentration over the same concentration range and used in the development of the "Glucoprocesseur".

Removal of interferences

Two types of interferences may occur when using enzymic probes : interferences due to the lack of specificity of the enzyme itself and interferences due to the presence of electroactive species in the sample. Practically, the first type is not encountered with glucose oxidase but various contaminants present with the analyte to be measured may be oxidized at the poised potential.

In our case, a high selectivity could be obtained with a differential system allowing the continuous subtraction of the interfering current due to electroactive species.This was first carried out with a second electrode covered with a non-enzymic membrane and a differential current amplifier (13).

With the "Glucoprocesseur", a special measurement cell including the two-electrode differential system has been designed and with appropriate electronics, the detection and signal subtraction of interfering species is automatically performed.

This instrument is very easy to operate even by unskilled personnel. As a matter of fact, after adjusting the balance between the two electrodes by injecting an hydrogen peroxide aliquot, and after a one point-calibration, numerous assays can be performed.

10 to 100 µl of the glucose-containing sample, depending on the concentration, are injected into the 3 ml measurement cell with a glass capillary micropipette. The result is displayed 30 seconds only after injections which can be performed at 90 second intervals. Results can be expressed either in mmol .l^{-1}, g. l^{-1} or mg.dl^{-1} . Washing and cell filling are automatic. The washing time depends on the concentration determined and manual extra washing can be obtained. A very reliable thermostating system (30°

or 37° ± 0.02°C) enhances the reproducibility of measurements and up to 3000 glucose assays and even more could be performed with the same collagen disc. However, the membrane lifetime may depend on the conditions of use. With samples like soft drinks, honey, dry grapes or redcurrant jelly coefficients of variation lower than 2 % could be obtained.

IMPROVEMENTS AND EXTENSION TO NEW ANALYTES

The developments under investigation in our group concern the enzyme support itself, new sources of enzymes and the development of multipurpose systems.

Ready-to-use "Biodyne ™ Immuno Affinity" membranes

Recently, a very fast and simple immobilization procedure for preparing enzymic discs has been adapted in our group to the design of enzyme electrodes using novel polyamide membranes provided in a preactivated form for immunodiagnostic use by Pall, Glen Cove, U.S.A. (17).

This new approach allows anyone who needs to prepare his own enzymic support to do so easily and reliably.

With the simplest method, enzyme immobilization can be obtained by simply wetting the membrane with 10 µl of concentrated enzyme solution (for instance, glucose oxidase 25 U in phosphate buffer, pH 7) applied on one or both sides of the disk, the reaction being achieved within 1 minute.

Biodyne ™ Immuno Affinity membranes (BIA membranes) were first used for glucose oxidase immobilization as model enzyme. Collagen membranes previously used with either the self-contained glucose electrode or the glucose analyzer were conveniently replaced by such membranes. The determination of glucose in foodstuffs has been performed with the "Glucoprocesseur" equipped with glucose oxidase-BIA membranes.For assays carried out on honey and jam the coefficients of variation obtained with series of 10 assays were comprised between 0.7 and 1.7 %.

Novel oxidases for use with the analyzer

At this time, the extension of the method to other analytes using new oxidases is investigated in order to propose a multiparametric apparatus. However, if we except glucose oxidase, only a few of the commercially available enzymes exhibit a stability really compatible with routine use under industrial conditions.

Recently, a novel oxidase specific for L-lactate from *Pediococcus species* was immobilized on these new membranes . A very sensitive L-lactate self-contained sensor with a detection limit of 1.25 10^{-7} M was obtained (18) .When the"Glucoprocesseur" was equipped with such membranes, the concentration of L-lactate could be determined in several dairy products including cream cheese, plain yoghourt, whey, red or white wine etc.. For undiluted samples, the working concentration range was 10^{-4} - 10^{-2} M .

A novel enzyme membrane electrode for oxalate determination in foodstuffs using the same pre-activated membranes was also reported recently (19). With the analyzer, oxalate could be determined in the range 5. 10^{-3} - 10^{-1} M in the sample. A high correlation was found with the conventional spectrophotometric method.

CONCLUSION

The constant search for simple and reagentless methods for the specific determination of various metabolites is still an important goal for numerous laboratories working in the attractive field of biosensors. Unfortunately, despite the fact that numerous systems have been described in the literature, only a few biosensor - based instruments are commercially available. The most commonly used biosensor is the enzyme glucose electrode using a conventional probe for O_2 or H_2O_2 detection associated to an immobilized glucose oxidase. If the design of associated electronics and software for signal processing appear less critical, immobilization techniques, stability of the bioactive component and improvements of signal transduction at the enzyme - transducer interface still appears as the major bottlenecks for such systems.

"Protein engineering" however, although still in its infancy, is expected to provide us in the near future with highly stable tailor-made enzymes and this may lead to a real take-off in the biosensor area.

REFERENCES

1. **G. G. Guilbault,** in H. H. Weetall (Ed.), *Immobilized enzymes, antigens, antibodies and peptides* , Marcel Dekker Inc., New York (1975), pp. 293-417.

2. **P. W. Carr & L. D. Bowers,** *Immobilized enzymes in analytical and clinical chemistry* , John Wiley & sons, New York (1980), pp. 1-460.

3. **B. Danielsson, B. Mattiasson & K. Mosbach,** ' Enzyme thermistor devices and their analytical applications ', in *Applied Biochemistry and Bioengineering*, Vol. 3, Academic Press, New York (1981) pp. 97-143.

4. **S. Suzuki & I. Karube,** ' Bioelectrochemical sensors based on immobilized enzymes, whole cells and proteins ', in *Applied Biochemistry and Bioengineering*, Vol. 3, Academic Press, New York (1981) pp 145-174.

5. **D. A. Gough & J. K. Leypoldt,** ' Theoretical aspects of enzyme electrode design ', in *Applied Biochemistry and Bioengineering*, Vol. 3, Academic Press, New York (1981) pp.175-206.

6. **G. G. Guilbault,** *Analytical uses of immobilized enzymes*, Marcel Dekker Inc., New York (1984) pp. 1-453.

7. **G. G. Guilbault & J.-M. Kauffmann,** ' Enzyme-based electrodes as analytical tools ' *Biotechnol. Appl. Biochem.*, **9** (1987) 95-113.

8. **R. D. Schmid,** ' Biotechnology in Japan 1985 ' *Appl. Microbiol. Biotechnol.*, **24** (1986) 355-365.

9. **V. M. Owen,**
' Commercial aspects of use of immobilized biological compounds ',
This book.

10. **P. R. Coulet, J. H. Julliard & D. C. Gautheron,**
' A mild method of general use for covalent coupling of enzymes to chemically activated collagen films '
Biotechnol. Bioeng., **16** (1974) 1055-1068.

11. **P. R. Coulet & C. Bertrand,**
' Asymmetrical coupling of enzymatic systems on collagen membranes. Application to multienzyme electrodes '
Anal. Lett., **12** (1979) 581-587.

12. **D. R. Thevenot, P. R. Coulet, R. Sternberg & D. C. Gautheron,**
' A highly sensitive glucose elctrode using glucose oxidase collagen film '
Bioelectrochem. Bioeng., **5** (1978) 548-553.

13. **D. R. Thevenot, R. Sternberg, P. R. Coulet, J. Laurent & D.C. Gautheron,**
' Enzyme collagen membrane for electrochemical determination of glucose '
Anal. Chem., **51** (1979) 96-100.

14. **C. Bertrand, P. R. Coulet & D. C. Gautheron,**
' Multipurpose electrode with different enzyme systems bound to collagen films '
Anal. Chim. Acta, **126** (1981) 23-34.

15. **L. J.Blum, C. Bertrand & P. R. Coulet,**
' Amperometric sensor and immobilized enzyme electrode for the determination of enzymatic activities '
Anal. Lett., **16** (1983) 525-540.

16. **G. Bardeletti & P. R. Coulet,**
' Air-dried enzyme electrode '
Anal. Chem., **56** (1984) 591-593.

17. **C. H. Assolant-Vinet & P. R. Coulet,**
' New immobilized enzyme membranes for tailor-made biosensors '
Anal. Lett., **19** (1986) 875-885.

18. **G. Bardeletti, F. Séchaud & P. R. Coulet,**
' A reliable L-lactate electrode with a naw membrane for enzyme immobilization for amperometric assay of lactate '
Anal. Chim. Acta, **187** (1986) 47-54.

19. **C.H. Assolant-Vinet, G. Bardeletti & P. R. Coulet,**
' A novel enzyme membrane electrode for oxalate determination in fodstuffs '
Anal. Lett., **20** (1987) 513-527.

COMMERCIAL ASPECTS OF THE USE OF IMMOBILIZED COMPOUNDS

Valerie M Owen
Biotechnology Centre
Cranfield Institute of Technology
Cranfield
Beds MK43 OAL
UK

ABSTRACT

The immobilized biological compounds which have found the largest markets in industry, detection and analysis are enzymes and antibodies. The total world market for enzymes is in excess of $500M. About one quarter of this is for the enzymes, frequently immobilized, used by the food industry in the production of fructose and glucose from starch. Detection and analytical markets using biological compounds are dominated by the medical sector. The world market for diagnostics is estimated at $4 billion with an increasing proportion of reagents containing either enzymes or antibodies. The world immunoassay market is in excess of $1 billion and rising. Other analytical applications of biological compounds are in defence, veterinary medicine, food industry and environmental monitoring.

Immobilized enzymes have been used in solid phase chemistries for some 30 years and Technicon have used immobilized enzymes in flow-through analytical systems. However, biosensors will be a major growth area for the use of immobilized biological compounds. The largest single market is in the defence industry but the majority of commercially available biosensors are designed for the clinical analysis of blood glucose using the enzyme glucose oxidase. The market for enzymes in biosensors will remain very small with the existing laboratory-based biosensors which use multi-assay membranes but the market for enzymes will expand with the introduction of disposible, one-shot, biological components for use in biosensors designed for operation outside the laboratory.

1. THE BIOLOGICALS

The biological compounds, which have been immobilized for use in existing, commercially available devices and processes, include enzymes, antibodies, whole cells and whole organisms, but it is anticipated that this list will soon be extended to include nucleic acids and cell receptors. The current US market for immobilized biocatalyst technology products has been put at nearly $6 billion in a market report from Business Communications Co Ltd (1) and they expect the most dramatic rise in the immobilized biocatalyst market to be in the medical and diagnostic sectors.

G. G. Guilbault and M. Mascini (eds.),
Analytical Uses of Immobilized Biological Compounds for Detection, Medical and Industrial Uses, 329–339.

1.1 Enzymes

Enzymes are the most important group of immobilized biologicals. The current world market for enzymes is put at around $500 million, but of this nearly half are used by the food and beverage industries in cheese making, brewing, bakery, confectionary and most importantly in the conversion of starch to glucose and high fructose syrups. The enzymes used for this conversion account for approximately one quarter of the world consumption of enzymes. These enzymes are largely used in immobilized forms and therefore, annually, about $100 million worth of enzymes are immobilized for use in the production of glucose and high fructose syrups. This market is particularly strong in the USA. However, growth is expected to be slow in this sector. There will be growth in the food sector for new enzymes such as lipases for the production of short to medium length fatty acids which are potent flavour compounds.

The other market for bulk commodity enzymes is in laundry detergents and this proably represents about one third of the total enzyme market.

The enzymes used in analysis form a small proportion of the total market but in a recent market reports by Business Communications Co Ltd (2) on the US market and by Frost and Sullivan (3) on the market for industrial enzymes in Western Europe it was expected that small-quantity, high-value, diagnostic enzymes will show the greatest increase in demand compared to the other enzyme-using sectors.

1.2 Antibodies

Antibodies have been used since the 1960s in immunoassay, particularly in the analysis of biological samples. Usually the antibody is immobilized on the surface of a test tube, microtitre plate or on particles such as latex or blood cells. The total world market for immunoassay is currently estimated to be about $1 billion and is expected to rise at a faster rate than the rest of clinical diagnostic analysis.

1.3 Whole cells and microorganisms

Immobilized whole cells and microorganisms are finding applications in the pharmaceutical and brewing industries. However, difficulties are still being experienced in using immobilized yeast in the brewing industry and this has limited the spread of this technology through the world's industry. Immobilized microorganisms are beginning the appear in analytical devices and biosensors, but the applications are still limited and the markets achieved have been small.

1.4 Nucleic acids

There is a demand for quick, easy analysis of genetic material using single strands of characterised DNA or DNA probes(4). Frost and Sullivan in a report on 'DNA probes in medicine' (5) considered that the market for DNA probes, mainly for the diagnosis of infectious disease and cancer applications, could reach $190 million by 1990.

At present the 'unknown' nucleic acid is immobilized and, then, probed with the 'known' DNA. However it might be less time consuming and easier to perform an analysis if an analytical system could be developed in which the 'known' DNA were immobilized and the 'unknown', sample DNA were added. This would make the analysis more analogous to the methods used in immunoassay.

1.5 Cell Receptors

The interest in cell receptors is due to the fact that they form the natural 'biosensors' of the living organism. The inherent properties of cell receptors may lead to their use as the biological components of sensitive, reuseable biosensors, but before this can happen more work is required to isolate, stabilise and characterise these molecules.

2. ANALYTICAL APPLICATIONS

The analytical applications which are using or are expected to use immobilized biologicals in the foreseeable future include healthcare, veterinary medicine, the food industry, environmental monitoring, and the defence and security industries.

2.1 Healthcare

This sector is receiving a great deal of attention. The world diagnostics market is estimated to be about $4 billion and is expected to increase. Similtaneously there is a movement to simpler methods of testing as clinical analysis moves out of the hospital laboratory into the physician's office. This trend is particularly seen in the USA. In a recent study commissioned by the Office of the Assistant Secretary for Health Planning and Evaluation of the US Department of Health and Human Services various reports were quoted showing the value of testing performed in the physician's office (6). The testing service performed in this setting was worth $5 billion and this was expected to grow at the rate of 15% per year. In Europe the trend to testing in the physician's office is seen to different extents in different countries. In West Germany the physician's office laboratory is a common feature, while in the UK and Scandinavia clinical testing is very centralised in the hospital laboratory.

2.2 Veterinary Medicine

There is a demand for testing farm animals to assess the status of the reproductive cycle. The farmer wants quick, easy testing 'cowside' and 'sowside'. If possible, the farmer wants to use samples which are easy to obtain such as milk. The economic importance of testing farm animals should not be overlooked as the cost of a 'lost' reproductive cycle may as high as $75 per sow to the pig farmer.

The demand for testing pets and small animals for a range of conditions and diseases is presently being met using the same tests as are used for humans. Frequently dipstick methods are used by the veterinary surgeon.

2.3 Food Industry

The food industry has a requirement for analysis to show the biochemical composition of food, its freshness and any contamination or adulteration (7). At present samples are taken and treated to produce a liquid, slurry or syrup for analysis, but there would be advantages in using probes which could be inserted into the food directly thus avoiding the preparation before analysis. At Cranfield Institute of Technology work has been carried out in collaboration with the Leatherhead Food Research Association to develop a probe which measures the glucose profile into a piece of meat (8). This indicated the level of bacterial contamination on the surface of the meat. The device is designed to work under the conditions and temperature at which meat is stored.

2.4 Environmental monitoring

Environmental monitoring is being increased in response to regulations and legislation. The European Economic Community Directives on enviromental standards, issued by the Council of the European Communities, specify the quality of water for different applications. The Directives are due to become fully operational this year, 1987, the European Year of the Environment, and the countries within the EEC are expected to enact legislation in line with the Directives. Frequently legislation specifies the appropriate methods of analysis of environmental samples. Often these are chemical methods of analysis involving spectrophotometry and chromatography. Methods of analysis involving the use of immobilized biologicals will have to demonstrate that they are appropriate for environmental monitoring and will have to become accepted by regulatory bodies before there is widespread application in environmental monitoring.

2.5 Defence and security industries

This sector is probably the largest market for biosensors at present. The UK Thorn EMI NAIAD nerve gas detector has found a multimillion dollar market, as has the Dutch Acal from 01 Delft, which performs a similar task using a biochemical method of analysis. However, the Dutch instrument does not contain an immobilized biological compound. Both these devices are used to detect chemical warfare agents. There is a demand to detect trace vapours from explosives and illicit drugs. Detection at present is by gas chromatography sometimes in combination with mass spectrometry. Antibodies have been raised to explosives and used in their detection (9). Another paper in this publication describes the use of proteins in the detection of atmospheric pollutants (Protein coated piezoelectric crystal detectors by G G Guilbault, J Ngeh, P Foley and J Jordan). It remains to be seen if inexpensive devices will become commercially available based upon the use of this technology.

There will be an increasing demand for the use of immobilized biologicals to detect agents of biochemical and microbiological warfare.

3. ANALYTICAL TECHNOLOGIES

The reasons for using immobilized biological components in analysis can be divided into three catagories:-

- as an integral part of the analytical system,

- to reduce to cost of analysis by reusing the biological component,

- and as part of an 'easy to use' analytical device in which the user does not have to prepare the reagent.

The commercially available devices incorporating immobilized biologicals have been divided into four groups for ease of consideration in this paper.

3.1 Flow-through chemistry with enzyme coils

This group included the Technicon developments (10) in which an enzyme was immobilized on the inside of a coil which was included as part of the flow-through Technicon analysers. Problems were found with the stability of the enzyme when it was washed with the chemicals normally used to rinse and clean the analyser. However, this technology was an attempt to incorporate a reusable biological component into existing technology to reduce the cost of the reagent used in the analysis.

3.2 Immunoassay with immobilized antibodies

In immunoassay the antibody or antigen may be immobilized on particles or cells which then agglutinate to indicate the analytical result. Alternately the antibody is immobilized on the sides of a well or test tube in order to help separate the bound and free antigen. In both cases the biological component of the analysis is not reused but the immobilization is an integral part of the analytical system.

3.3 Solid-phase, dipstick chemistry

Solid-phase, dipstick chemistry is presently the main technology used in testing outside the laboratory. It is used on the ward, in the physician's office, at home and by the veterinary surgeon. The tests are simple to perform but the immobilized biological component is used in a one-off, throw-away mode. Not all the tests available are based upon enzyme or biologically based analysis, however, the manufacturers are extending their test range to include immunoassays for drug assays and enzymatic reactions for the assay of creatinine.

The market for dipstick chemistry is large. Figures from the USA suggest that 100 million urinalyses are performed each year and that the annual market for glucose testing could be as high as 600 million tests.

The dipstick chemistries rely upon personal estimation of a colour development or upon insertion of the stick into a reflectometer for a reading. Human error or ´tampering´ with these processes can lead to incorrect analytical results, however the ease of use has ensured widespread application of this method of testing.

A similar ´easy to use´ approach is the use test tubes or cuvettes, inside which, reagents and enzymes are immobilized. The sample is added for a single shot analysis and the result read in a colorimeter or spectrophotometer. This sort of test is being developed for the food industry.

3.4 Biosensors

Biosensors are analytical devices in which a biological component is used in intimate contact with a transducer. The biosensors which have been developed commercially, to date, consist of a reusable biological component used with a transducer, usually an electrode. They have been an approach to reducing analytical costs. However, future biosensors may have throw-away biological components which will be used in conjunction with a reusable transducer/electronic component. One of this new breed is currently being launched in the USA, by Genetics International, for the estimation of blood glucose.

The technology of existing commercially available biosensors is extremely similar. Table I lists commercially available biosensors for the clinical analysis of glucose and indicated the technology employed in the instruments. Without exception they use the enzyme glucose oxidase in conjunction with either an oxygen electrode or a hydrogen peroxide electrode. Other clinically useful biosensors are given in Table II. The oxidase enzymes, particularly lactate oxidase, are well represented in this group.

Table III shows biosensors designed for use in the food and fermentation industries and again oxidase enzymes and, particularly, glucose oxidase dominate the technology. However, in this group there are some immobilized microbial, yeast cells.

There are some biosensors for applications outside the healthcare, and food and fermentation industries. These, listed in Table IV, use a range of immobilized biologicals including yeast cells and enzymes, which are not necessarily oxidases.

Table I. BIOSENSORS FOR CLINICAL GLUCOSE ASSAY

MANUFACTURER	INSTRUMENT	TECHNOLOGY
Analytical Instrument Corporation, Japan	Glucoroder-E	Immobilized glucose oxidase used with an oxygen electrode
Fuji Electric Co., Japan	Gluco-20 & Amylase-Glucose Analyser	Immobilized glucose oxidase used with a hydrogen peroxide electrode
Gambro AB, Sweden	Glucose Monitor System	Blood dialysate passed through an immobilised glucose oxidase reactor and oxygen consumption measured using oxygen electrodes
(developed by Kulys in Vilnius, USSR)	Enzalyst-G	Immobilized glucose oxidase used with a hydrogen peroxide electrode
Kyoto Daiichi Kagaku, Japan	Glucose Auto and Stat GA-1120	Immobilized glucose oxidase used with a hydrogen peroxide electrode
Omron Toyoba, Japan	Diagluca	Immobilized glucose oxidase used with a hydrogen peroxide electrode
Radelkis Electrochemical Instruments, Hungary	Electrode set	Immobilized glucose oxidase used with an oxygen electrode
Solea-Tacussel France	Glucoprocesseur	Immobilized glucose oxidase used with a hydrogen peroxide electrode
VEB-MLW Prufgerate-Medingen, GDR	Analyser ECA 20	Immobilized glucose oxidase used with a hydrogen peroxide electrode
Yellow Springs Instrument Co., USA	Glucose Analyser Model 23A & 23M	Immobilized glucose oxidase used with a hydrogen peroxide electrode

Table II. BIOSENSORS FOR CLINICAL ASSAY OTHER THAN GLUCOSE

MANUFACTURER	INSTRUMENT	TECHNOLOGY
Yellow Springs Instrument Co., USA	Lactate Analyser Model 23L	Immobilized lactate oxidase used with a hydrogen peroxide electrode
Omron Toyoba Japan	HER-100	Immobilized lactate oxidase used with a hydrogen peroxide electrode
Wolverine Medical USA	Lactate-Stat Analyser	Lactate oxidized to pyruvate in the presence of the enzyme cytochrome b_2 and the artificial electron acceptor hexacyanoferrate-(III) used with a platinum electrode
Fuji Electric Co., Japan	UA-300	Immobilized uricase used with a hydrogen peroxide electrode

Table III. BIOSENSORS FOR USE IN THE FOOD AND FERMENTATION INDUSTRIES

MANUFACTURER	INSTRUMENT	TECHNOLOGY
Analytical Instrument Corporation, Japan	Enzyme Electrode Analysers AD-300, AS-200, M-100	Immobilized oxidase enzymes used with an electrode
Oriental Electric Co., Ltd., Japan	ST-1	Immobilized glucose oxidase used with an oxygen electrode
Toa Electronics Ltd., Japan	Online Glucose Anal. Model FGA-1	Immobilized glucose oxidase used with a hydrogen peroxide electrode
Yellow Springs Instrument Co., USA	YSI Model 27 Industrial Anal.	Immobilized oxidase enzymes used with a hydrogen peroxide electrode
Denki Kagaku Keiku, Ltd., Japan	Biosensor Kit (BSK-1)	*Trichosporon brassicae* yeast membrane used with an oxygen electrode

Table IV. BIOSENSORS WITH APPLICATIONS OUTSIDE THE HEALTHCARE, FOOD AND FERMENTATION SECTORS

MANUFACTURER	INSTRUMENT	TECHNOLOGY
Nissin Electric Co., Ltd., Japan	BOD Quick Measuring Apparatus	Immobilised yeast cells used with an oxygen electrode
Thorn EMI Simtec Ltd., UK	NAIAD Automatic Chemical Warfare Agent Detector	Activity of the enzyme choline esterase continually monitored using a graphite electrode
Provesta Corp., USA	Multipurpose Enzyme Sensor Models 98 & 90	Enzyme supplied by user or manufacturer is trapped on the end of an electrode
Universal Sensors, USA	Enzyme Electrodes	Immobilized enzyme membranes with oxygen, ammonia or carbon dioxide electrodes
Chemical Center, University of Lund, Sweden	Enzyme Thermistor	Column containing immobilized enzyme used with a thermistor to detect heat production as the result of enzyme activity.

Table V. MEMBRANE STABILITY IN COMMERCIALLY AVAILABLE BIOSENSORS

MANUFACTURER	INSTRUMENT	STABILITY
Analytical Instrument Corporation, Japan	Glucoroder-E	6,000 assays
Kyoto Daiichi Kagaku, Japan	Glucose Auto and Stat GA-1120	1,000 assays
VEB-MLW Prufgerate-Medingen, GDR	Analyser ECA 20	1,000 assays
Yellow Springs Instrument Co., USA	YSI Glucose Analyser Model 23A and 23M	500 assays
Solea-Tacussel France	Glucoprocesseur	2000 assays

4. CONCLUSIONS

Immobilized glucose oxidase has a dominant position in the immobilized biological compounds used in commercially available biosensors. But does this mean a significant growth in the consumption of glucose oxidase and eventually of other oxidase enzymes?

The stability of the enzyme membranes allows for as many as 6000 assays to be performed using a single membrane as can be seen in Table V.

The workload for clinical glucose assays in Europe is approximately 300 million assays per year. On the assumption that a single glucose oxidase membrane will perform on average 2000 assays European clinical glucose assays could be performed using 15,000 membranes per year, in theory. If a membrane contains about 1mg of glucose oxidase the annual requirement for glucose oxidase for membranes, in Europe, would be less than 200g. In the USA the workload for clinical glucose assay is estimated to be about 600 million which would mean that the enzyme requirement would be only about 400g. This level of enzyme consumption would not mean a bonanza for the enzyme supplier. However, the development of biosensors using one-shot, disposible, biological components will lead to an increase in the amount of enzyme consumed and may improve the prospects for the producer of specialist enzymes.

Clinical applications for biosensors are likely to remain the largest sector, other than defence applications, in the forseeable future. As the range of biosensors for use in the clinical laboratory increases and more tests are performed using the technology of immobilized biologicals the range of enzymes and other biologicals required will increase but the quantity of each enzyme or compound needed may be very small. The quantity of the biological component required will only increase as disposables are developed for use in the patient´s home or physician´s office.

REFERENCES

1. New immobilization technology: commercial opportunities for reagents, catalysts and biocatalysts. (C-066), Pub. Business Communications Co. Ltd., Stamford, USA, 1986.

2. Commercial and industrial enzymes. (C-067), Pub. Business Communications Co. Ltd., Stamford, USA, 1986.

3. Industrial enzyme technology and market development in Europe. Report No. E814, Pub. Frost and Sullivan, London, UK, 1986.

4. ´A novel optical system for the detection of labelled DNA.´ Downs, M. E. A., Fothergill, M. F., Warner, P. J., and Turner, A. P. F., (in preparation).

5. DNA probes in medicine. Report No. 1479, Pub. Frost and Sullivan, New York, USA, 1986.

6. ´Quality assurance in changing times: proposal for reform and research in the clinical laboratory field.´ Kenney, M. L., Clinical Chemistry, 33, 328-336, 1987.

7. ´Seeking sensor improvements.´ Kress-Rogers, E., Food Processing, pp 37-39, Sept. 1985.

8. ´Biosensors for the food industry.´ Kress-Rogers, E., and D´Costa, E. J., Analytical Proceedings, 23, 149-151, 1986.

9. ´A sensitive bioluminescent immunoassay for dinitrophenol and trinitrophenol.´ Wannlund, J., and DeLuca, M., Analytical Biochemistry, 122, 385-393, 1982.

10. ´Immobilized glucose oxidase used in the continuous-flow determination of serum glucose.´ Leon, L. P., Narayanan, S., Dellenbach, R., and Horvath, C., Clinical Chemistry, 22, 1017-1023, 1976.

Design Considerations of an Immobilized Enzyme Electrode for Measurement of Glucose in Whole Blood

Christopher C. Feistel, Charles F. Sternburg, Clarence D. Luther, Larry L. Gayleard
Liston Scientific Corp., 3800 Campus Drive, Newport Beach, CA 92660, USA.

Abstract. A new clinical glucose analyzer (ESKALAB) capable of measuring the glucose concentration of undiluted whole blood or serum is described. The analyzer employs glucose oxidase which is covalently crosslinked within a multilaminant membrane using glutaraldehyde and bovine serum albumin. The membrane is positioned over a platinum electrode in an amperometric circuit and the current produced by anodic oxidation of hydrogen peroxide from the enzyme reaction with glucose is measured. The linear range for the instrument is 1 to 25 mmoles glucose/ L with a mean precision of less than 2% for whole blood, plasma, and serum samples. A proprietary limiting membrane effectively blocks all electrochemical interferents from detection at the working electrode, permitting unimpaired measurement of hydrogen peroxide. Evaluations in undiluted whole blood and plasma showed an offset between these sample types. A software correction was added which successfully eliminated this bias. The accuracy of the ESKALAB results was confirmed using the GA2 and ABA 100.

1. Introduction

The incorporation of immobilized enzymes into chemical sensors is a technological advancement which has simplified the design and maintenance of many chemical analyzers. Multiple assay capability and enhanced enzyme stability have dramatically reduced the quantity of enzyme needed and storage requirements compared with free enzyme reagents. Manufacturers of medical instrumentation have capitalized on the advantages of using immobilized enzymes, but the full exploitation of this technology has been continuously hampered by constituents of the blood matrix. An instrument which can measure an analyte in whole blood eliminates cell separation and serum dilution procedures providing the user with a result in the shortest time and a minimum number of sample handling steps. The measurement of analytes in undiluted whole blood is a major challenge to the field of medical instrumentation.

An exhaustive research effort has been directed toward the development of immobilized enzyme sensors for use in whole blood. The glucose sensor is perhaps the best recognized example of these efforts. Most glucose sensors reported have been based on the immobilization of glucose oxidase directly on an electrode surface (1,2), in enzyme reactors (3,4,5), or within a membrane (6,7). Each of these glucose sensor types has had technical problems when measuring whole blood directly. Sensor linearity is lost at elevated glucose concentrations due to oxygen depletion (2) and substrate saturation of the enzyme (8,9). Amperometric sensors based on hydrogen peroxide detection are subject to errors caused by interferents (10). Blood cells tend to foul reactors and some membrane types. This results in sensors which have a short useful life relative to the active life of the immobilized enzymes. Blood hematocrit (11) and catalase activity present on erythrocytes (12) have also been cited as problems which make

G. G. Guilbault and M. Mascini (eds.),
Analytical Uses of Immobilized Biological Compounds for Detection, Medical and Industrial Uses, 341–358.

accuracy difficult to achieve. These technical problems have required current commercially available instruments to resort to sample dilution to achieve clinically acceptable results.

A new instrument which measures glucose concentration in undiluted whole blood has been developed by Liston Scientific Corporation (Newport Beach, CA 92660) for the ESKALAB ECS (SmithKline Diagnostics, Inc., Sunnyvale, CA 94086-4118, a SmithKline Beckman company). The design, operation and performance of this instrument are discussed.

2. Materials and Methods

2.1. Chemicals

Monobasic sodium phosphate, dibasic sodium phosphate, HEPES, TES, BES, citric acid, EDTA, glutaraldehyde, bovine serum albumin (BSA), and glucose oxidase (*A. niger*) were purchased from Sigma Chemical Company, Saint Louis, MO. 63178, sodium azide from Aldrich Chemical Co., Milwaukee, WI 53233, sodium chloride from J. T. Baker Chemical Co., Phillipsburg, NJ 07420, sodium hydroxide from Mallinckrodt, Inc., Paris, Kentucky 40631. All chemicals were reagent grade.

2.2. Reagents

0.05 molar phosphate buffered saline (PBS), pH 7.5 containing 0.002 molar EDTA and 0.01 molar sodium azide was used to prepare solutions for the linearity and interferent studies. The linear range of the instrument was tested using aqueous ß-D-glucose calibrants which were prepared in PBS to concentrations of 1, 5, 10, 20, 25, 30 mmol/ L. To test the effectiveness of the limiting membrane configuration and to test the interferent blocking capabilities of the final membrane laminate, 10 mmol/L solutions of acetaminophen, ascorbic acid, cysteine, glutathione, and uric acid were prepared in PBS.

Buffers used for the stability study were: 0.05 M TES, pH 6.5; 0.05 M sodium citrate, pH 6.6; 0.05 M BES, pH 7.0; 0.05 M HEPES, pH 7.2; 0.05 M Bis-Tris, pH 7.3; 0.05 M PBS, pH 7.5.

2.3. Membrane Construction

Several different membrane compositions were investigated. All membranes were multilaminate composite membranes consisting of three functionally and physically distinct regions.

The limiting membrane was constructed by impregnating a proprietary porous film with cellulose acetate (14). Its purpose was to block interferents known to be found in whole blood. Each membrane was tested for blockage of 0.003% hydrogen peroxide in phosphate buffered saline (PBS). The percent blockage of peroxide for the membrane was measured against the same concentration of peroxide tested with a bare platinum electrode.

The immobilized enzyme layer was formed by covalently crosslinking bovine serum albumin and glucose oxidase via glutaraldehyde. The method used was similar to that previously described (17) with the exception that the mixed solution was applied between the limiting membrane and the protective membrane.

Several different protective membranes were constructed for evaluation (14). The first protective membrane consists only of a single layer of 0.01 micron pore polycarbonate purchased from Nuclepore, Pleasanton, CA 94566. A modified version of this membrane was fabricated by adding another layer of this polycarbonate. Next a layer of catalase was immobilized between two polycarbonate membranes. Finally a BSA layer was immobilized between two polycarbonate membranes.

2.4. Instrumentation

Two instruments were selected to compare results with the ESKALAB. The ABA 100 (Abbott Laboratories, So. Pasadena, CA 91030) was operated using a Glucose [HK] Diagnostic Kit and a 300 mg/dL glucose calibrant purchased from Sigma Chemical Company and a 150 mg/ dL glucose calibrant purchased from Beckman Instruments. The assay is based on the reaction of free hexokinase with glucose in diluted serum or plasma measured by U.V. absorbance of NAD. The GA2 (Beckman Instruments, Brea, CA 92621) assays oxygen consumed by the reaction of free glucose oxidase with glucose in diluted serum or plasma. All reagents and calibrants used for the GA2 were purchased from Beckman Instruments.

The ESKALAB ECS is a small (dimensions, H*W*D = 30 * 22 * 34 cm) portable multichannel clinical chemical analyzer which consists of a main chassis and at least one analytical module. The main chassis can house up to three analytical modules with accompanying reagents. It contains the display and microprocessor as well. Each analytical module contains a sensor, a peristaltic pump, an amplifier circuit board, and a stepper motor (Sodium/Potassium module is commercially available. Glucose was in pilot production and not commercially available at the time this was written.). The relationship of these components is shown in Figure 1.

3. Theory and Operation

Glucose is measured using an amperometric glucose electrode. A platinum working electrode is used with a counter electrode and reference electrode at an applied voltage of 0.65 V (15). A current is generated at the working electrode in the presence of hydrogen peroxide generated from the reaction:

$$(1)\quad \text{Glucose} + O_2 + H_2O \xrightarrow[\text{Immobilized glucose oxidase}]{} \text{Gluconic Acid} + H_2O_2$$

The glucose oxidase is immobilized within a membrane described in a previous section.

Sample, calibrant, and buffer are transported to the enzyme electrode through a stainless

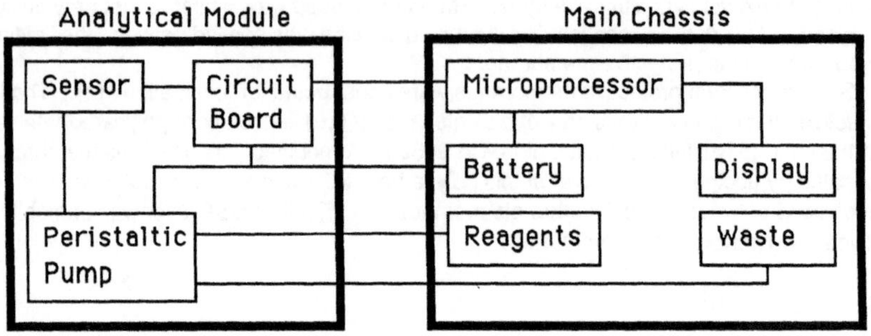

Figure 1. Block diagram of ESKALAB ECS instrument.

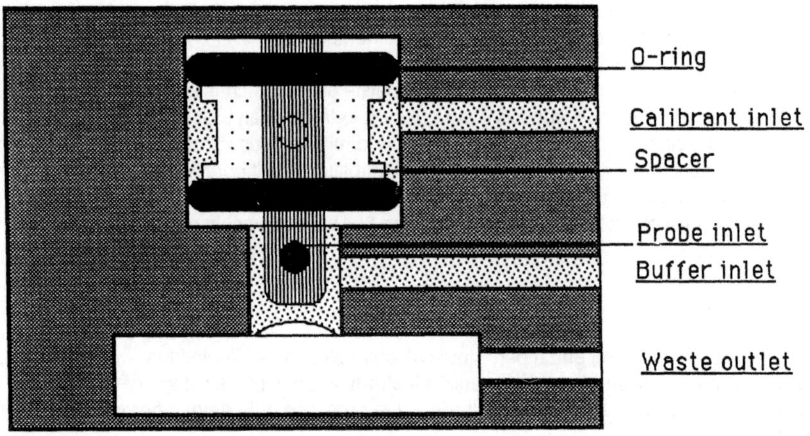

Figure 2. Diagram of wash cell construction. Probe inlet is raised into calibrant inlet chamber and calibrant is drawn to the enzyme electrode. The probe is shown positioned in buffer inlet chamber used for flushing and in standby mode. Probe is lowered into the sample through an opening in the bottom of the wash cell (not shown). Residual sample on the outside of the probe is drawn off through the waste outlet.

steel probe which acts as a passive heat exchanger to bring calibrant and sample to the same temperature. The probe is positioned to draw sample, buffer, or calibrant using a stepper motor. Buffer and calibrant are contained in a wash cell (see Figure 2). The wash cell consists of two chambers segmented using o-rings. The probe axis extends through the center of the o-rings and either calibrant, buffer, or sample is delivered to the enzyme electrode by raising or lowering the probe to the appropriate chamber and drawing the solution in using the peristaltic pump.

A one point calibration is automatically performed during each measurement cycle. Samples are dispensed into a sample cup designed to fit a stage located on the analytical module. When a user requests a test the probe is raised into the calibrant chamber and 100μL of calibrant are drawn to the enzyme electrode. The probe is lowered into the buffer chamber, the signal processed, and buffer is drawn past the enzyme electrode washing calibrant from membrane and electrode. The probe is next lowered into the sample cup and 100μL of sample are drawn to the enzyme electrode. The probe is raised into the buffer chamber, the signal is processed, and buffer is drawn past the enzyme electrode washing the sample from electrode and membrane. The result is calculated and displayed to the user.

4. Results

4.1. Interferent studies

It was found that a limiting membrane that effectively blocked 70% of the signal from 0.003% hydrogen peroxide was also effective in blocking 10 mM solutions of the known electrochemical interferents; acetaminophen, ascorbic acid, cysteine, glutathione, and uric acid. Cyclic voltametry of pooled serum and PBS gave no indication of the presence of native interferents at 0.65 V.

Glucose oxidase immobilized in BSA appeared relatively insensitive to buffer composition. When immersed in different buffers at 55 °C for 16 days only membranes stored in Bis-Tris buffer solution showed a significant decline in activity (Figure 3). It was also determined that 600 to 1000 whole blood tests could easily be performed using the same membrane stored in PBS.

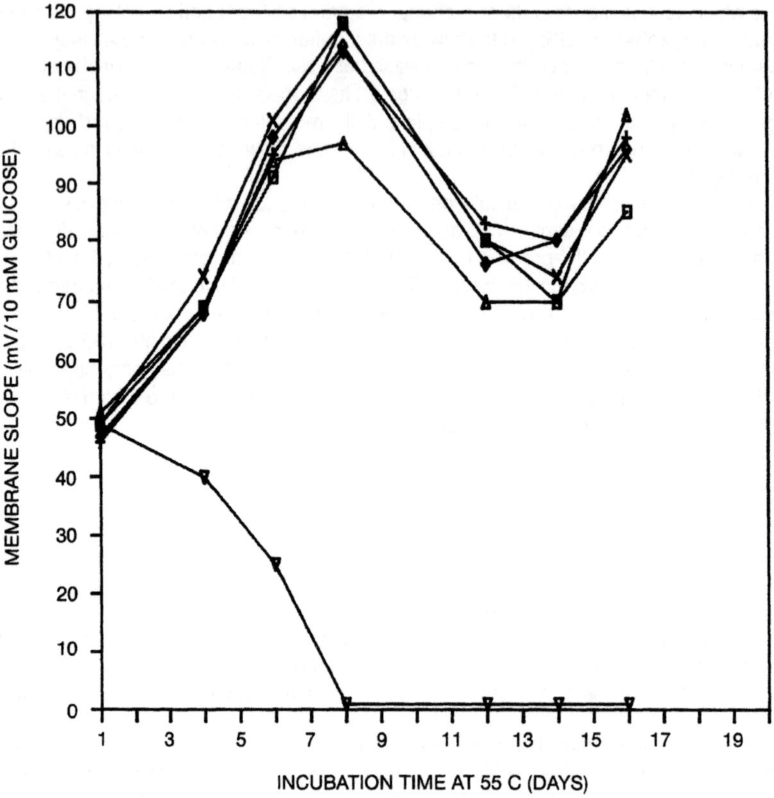

Figure 3. Effect of different buffers on glucose oxidase membrane stability: (▽) - .05 M Bis-Tris, pH 7.3; (□) - .05 M Citric Acid, pH 6.6; (△) - .05 M BES, pH 7.0; (◇) - .05 M HEPES, pH 7.2; (x) - .05 M TES, pH 6.5; (+) - .05 M PBS, pH 7.5.

4.2. Instrument Linear Range

The linear range of the system using a single ply polycarbonate membrane configuration was inadequate for clinical applications. The mean response to aqueous glucose standards is shown ir Figure 4. The recovery at 20 mmol glucose/ L was 94% while at 30 mmol glucose/ L it was only 84%. The response to glucose became nonlinear at approximately 15 mmol glucose/ L.

A second layer of polycarbonate was added to the membrane to further impede glucose diffusion to the enzyme layer. The data for these membrane indicate a significant improvement in linearity was achieved. The mean recovery at 20 mmol glucose/ L increased from 94% to 99% while the mean recovery at 30 mmol glucose/ L increased from 84% to 94%. An investigation was conducted to determine whether adding oxygen to the sample would further extend the linear response of this membrane. The linear response of membranes using two protective layers of polycarbonate was tested in oxygen saturated solutions. There was a 3.3% increase in recovery at 30 mmol/ L glucose when oxygen was added to this solution.

The possibility of regenerating oxygen from excess hydrogen peroxide was investigated by immobilizing catalase between the two layers of polycarbonate. This membrane configuration demonstrated a linear range similar to that obtained in oxygen saturated samples (Figure 5). Control membranes constructed using an immobilized BSA layer without catalase also showed linearity similar to that achieved in oxygen saturated solutions (Figure 5).

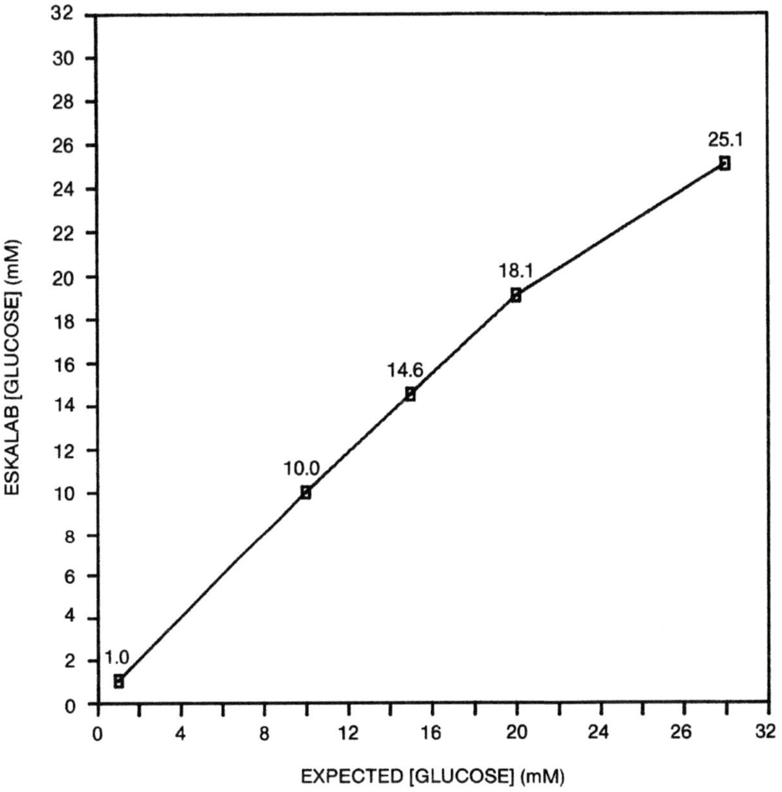

Figure 4. Mean linearity of eleven glucose oxidase membranes constructed using a single polycarbonate layer.

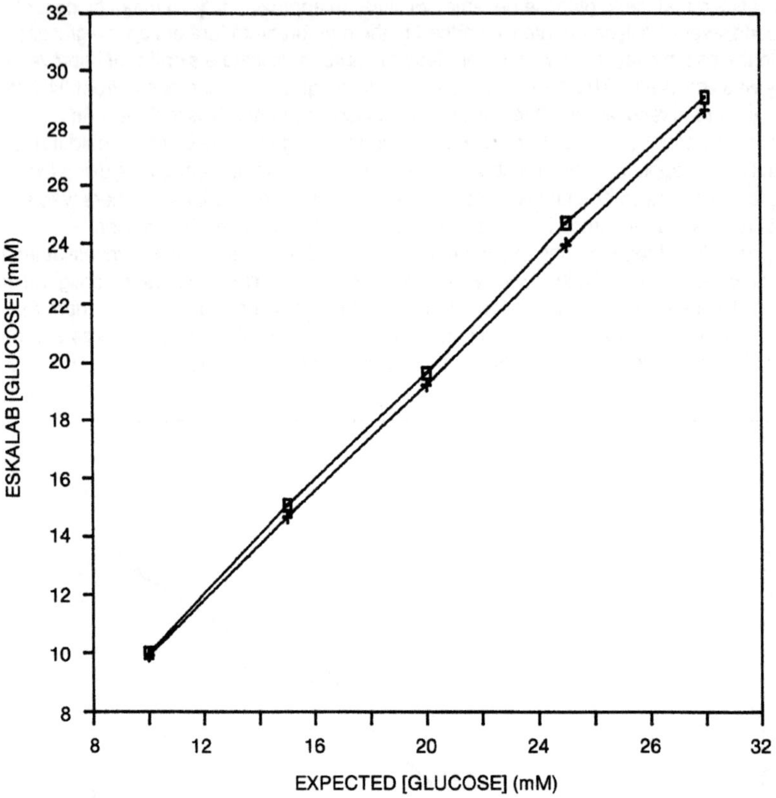

Figure 5. (□) — Mean linearity of four glucose oxidase membranes constructed with an immobilized catalase layer between two polycarbonate membranes; (+) — Mean linearity of four glucose oxidase membranes constructed with an immobilized BSA layer between two polycarbonate membranes.

4.3. Sample handling

Glycolysis and respiration of blood cells can reduce the glucose concentration in whole blood samples. Chemical agents such as sodium fluoride and potassium oxalate may be added to whole blood samples to arrest respiration and glycolysis. The use of whole blood samples which do not contain these agents was investigated. It was found that whole blood samples treated using sodium heparin anticoagulant could be maintained at room temperature for one hour without a significant decrease in glucose concentration compared with whole blood containing sodium flouride (Figure 6).

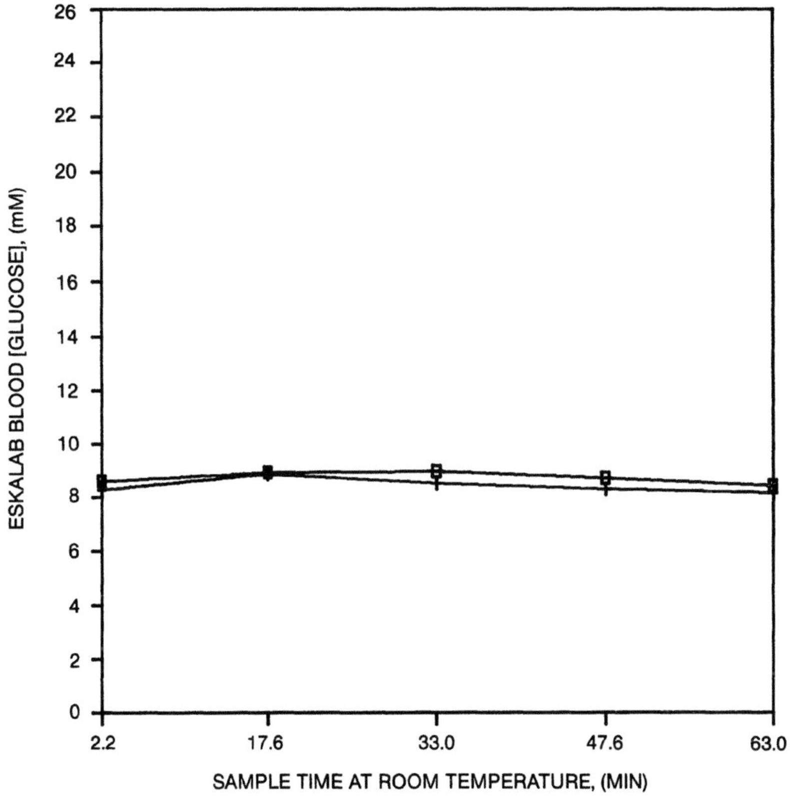

Figure 6. Effect of sample time at room temperature on ESKALAB glucose value in whole blood preserved with either sodium fluoride or sodium heparin: (□) — whole blood preserved with sodium fluoride; (+) — whole blood preserved with sodium heparin.

A correlation study comparing sodium fluoride and sodium heparin effects on results was performed using plasma samples. The results shown in Figure 7 indicate that the use of either anticoagulant does not effect the measurement.

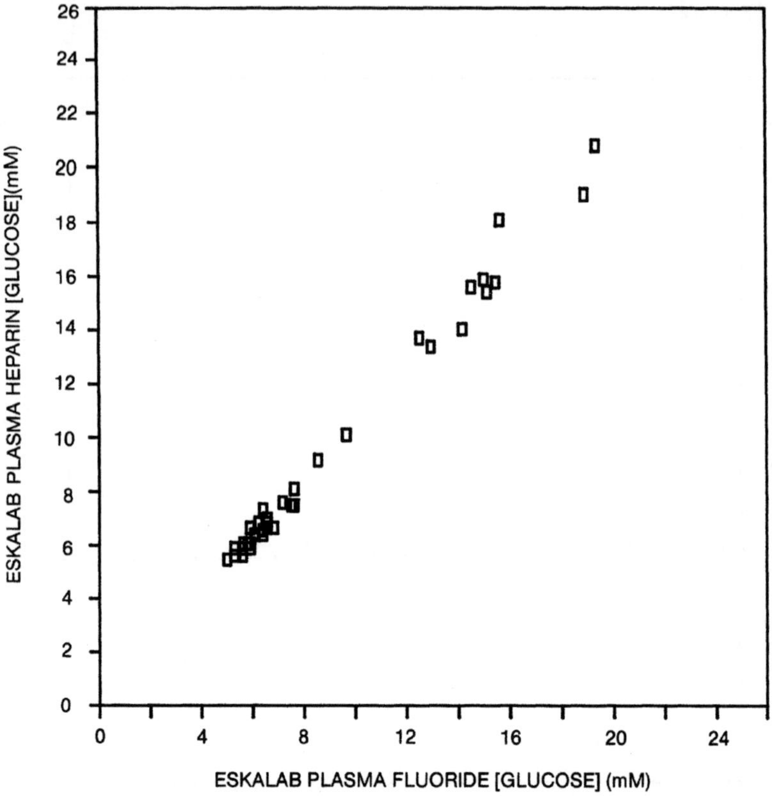

Figure 7. Correlation between ESKALAB glucose values in plasma preserved with sodium heparin and plasma preserved with sodium fluoride. y = 1.08 x .2 (r = .9938, n = 38).

4.4. Precision in Whole Blood and Serum

The precision of the system measuring whole blood, serum, or plasma was the same. The results of within run precision studies are shown in Table I. Within run precision was typically less than 2% for samples within the normal to high abnormal range. Low abnormal sample precision was less than 3%. Between day and within day precision studies were not performed.

TABLE I

Sample type	Mean ± S.D. (mmol glucose/ L)	N	% C.V.
Plasma	5.1 ± 0.03	10	0.6
Plasma	19 ± 0.25	10	1.3
Plasma	31 ± 0.53	10	1.7
Whole blood	2.0 ± 0.06	20	3.0
Whole blood	5.7 ± 0.09	20	1.6
Whole blood	23 ± 0.30	10	1.3

4.5. Instrument comparisons

ESKALAB measurements of glucose in undiluted serum were compared with the Beckman GA2 and the ABA 100 measurements. The results of these comparisons are presented in Figures 8, 9, 10. Linear regression analysis of these comparisons indicate that there is no significant difference between these instruments.

Early evaluations of ESKALAB performance in whole blood revealed a proportional bias between whole blood and plasma results. Results obtained from whole blood samples averaged 10% below plasma results. When 6 micron latex particles were suspended in glucose solutions a 5.5% reduction in signal was noted for each million particles per microliter. Blood cells were added to plasma to determine the response of the system to hematocrit (Figure 11). These data indicated that the effect of hematocrit decreased as the hematocrit increased. The effect of hematocrit between 0.30 and 0.60 was only 2.6%.

A correction was added to the instrument software to account for the difference between whole blood and plasma. A comparison of results for ESKALAB whole blood and plasma measurements is shown in Figure 12. Linear regression analysis of this comparison demonstrates that the bias between whole blood and plasma has been corrected.

5. Discussion

Although others have reported the potential presence of native interfering substances (10),

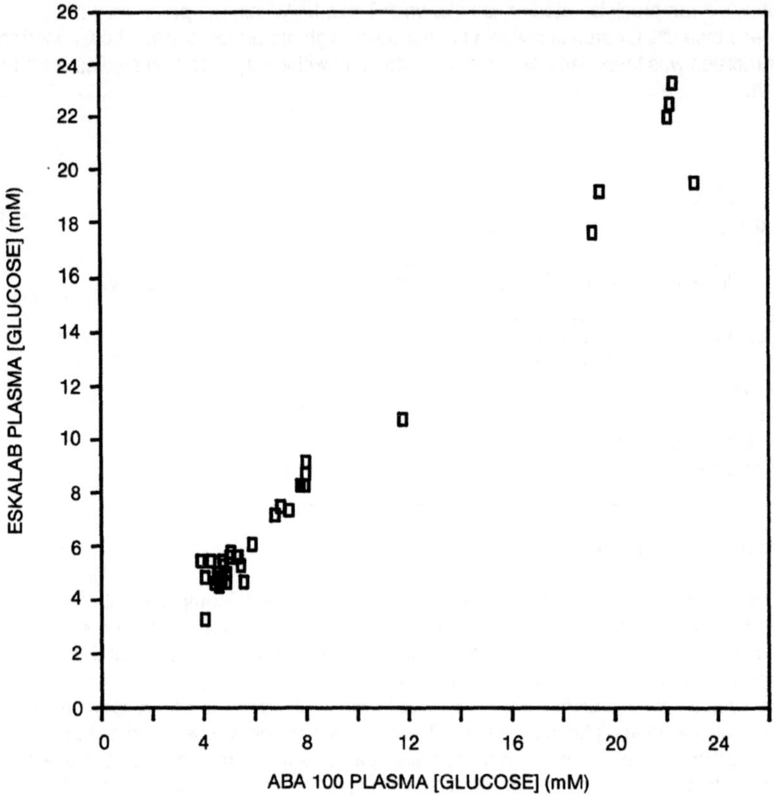

Figure 8. Correlation between ESKALAB plasma glucose and ABA 100 plasma glucose. y = .94 x .66 (r = .9906, n = 35).

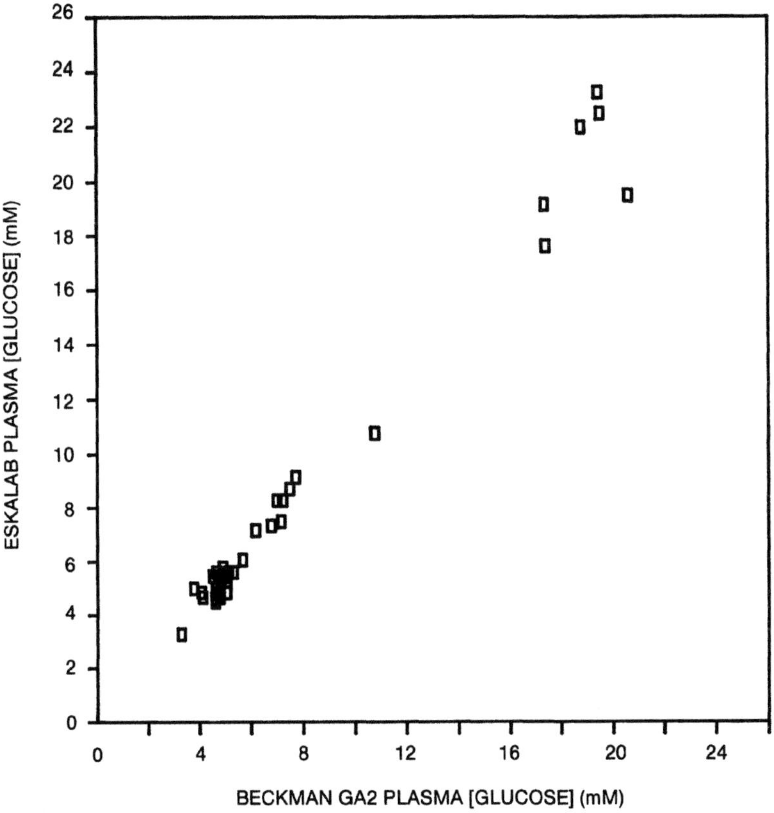

Figure 9. Correlation between ESKALAB plasma glucose and BECKMAN GA2 plasma glucose. $y = 1.09\,x + 2.9$ ($r = .9891$, $n = 35$).

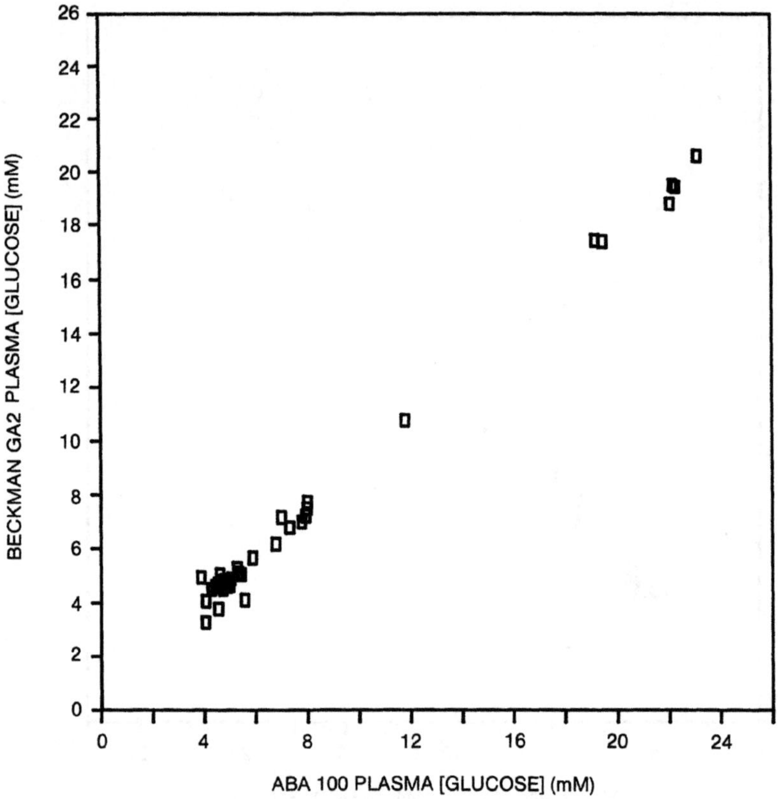

Figure 10. Correlation between BECKMAN GA2 plasma glucose and ABA 100 plasma glucose. y = .85 x + .61 (r = .9970, n = 35).

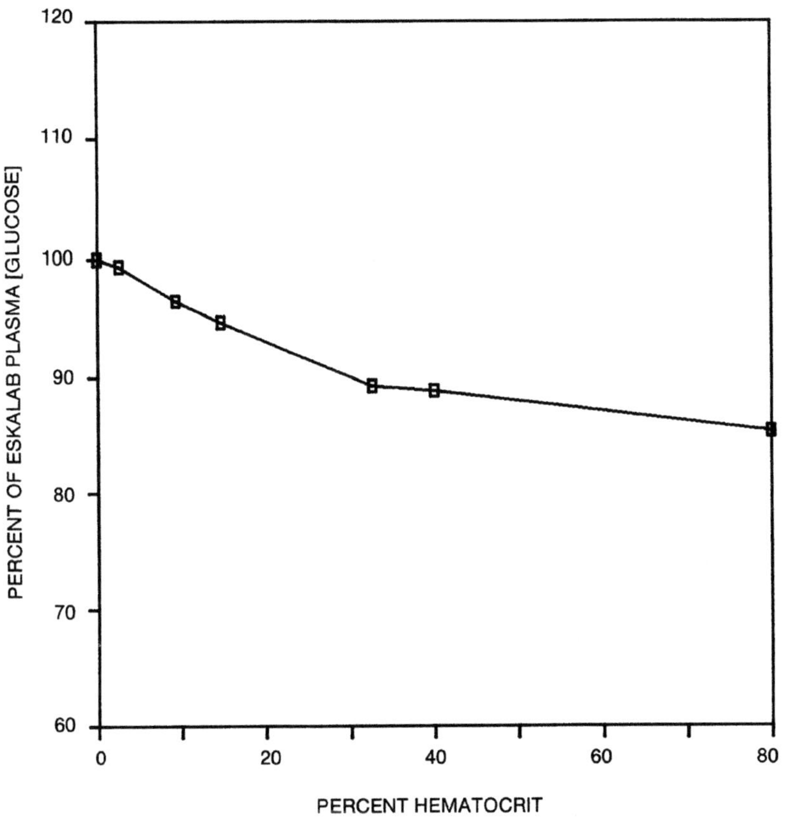

Figure 11. Effect of sample hematocrit on ESKALAB glucose value.

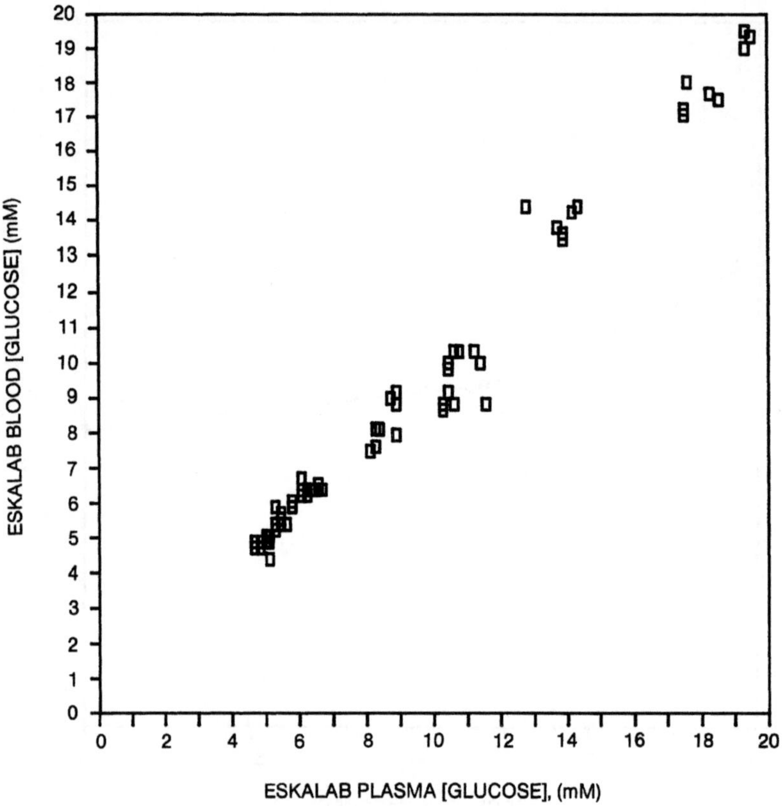

Figure 12. Correlation between ESKALAB whole blood glucose and ESKALAB plasma glucose. y = .96 x + .17 (r = .9906, N = 84).

none were detected in the cyclic voltametry of pooled serum reported here. It is possible that these substances were present at concentrations below the detection limit of this system.

The increase in activity which membranes exhibited from days 1 to 7 of the buffer comparison study has been reported by others (15). One explanation for this phenomenon is that glucose is able to penetrate closer to the working electrode surface as enzyme activity decreases (16). An increase in signal occurs because of the shorter distance hydrogen peroxide must diffuse to the working electrode. After enough enzyme activity is lost to allow glucose to penetrate the membrane completely the apparent signal of the enzyme electrode peaks and begins to decay. No experiments were conducted to test this hypothesis.

The nonlinearity of the glucose electrode using a single ply polycarbonate membrane was due to substrate saturation and oxygen depletion of the immobilized enzyme. This was demonstrated by the increase in linearity which occured after the rate of glucose diffusion was slowed by adding a second ply of polycarbonate. The observed increase in linearity which occured when oxygen saturated samples were measured using this membrane indicates that the available oxygen content of standing solutions was rate limiting at elevated glucose concentrations. All linearity data collected suggested that improvements could be made either by decreasing the apparent glucose concentration or increasing available oxygen.

Several investigators have proposed the use of mediator compounds with lower reduction potentials than oxygen to eliminate the oxygen dependency of glucose oxidase (2, 13). Ferrocene was the most applicable mediator for this system because of its low solubility and the possiblility of incorporating the mediator into the sensor construction. Unfortunately, electrodes constructed using the method described previously (2) proved to have a useful life which was too short and required considerably more maintenance compared with a platinum working electrode. Other investigators have reported supplying additional oxygen by adding hydrogen peroxide to the sample and then reacting it with immobilized catalase as shown in reaction 2:

$$H_2O_2 \xrightarrow[\text{Immobilized Catalase}]{} 1/2\ O_2 + H_2O \qquad (2)$$

Hydrogen peroxide can not be added to samples for this system. It was postulated that a substantial amount of hydrogen peroxide generated at the membrane was available because it diffused away from the surface of the working electrode and was not detected. The conversion of this hydrogen peroxide to oxygen was expected to provide additional oxygen without decreasing the hydrogen peroxide concentration apparent to the working electrode. When catalase was immobilized in the protective membrane an increase in linearity was achieved. However, it is possible that this linearity increase was not a result of recycling hydrogen peroxide. Control membranes constructed using an immobilized BSA layer without catalase also showed linearity similar to that achieved in oxygen saturated solutions. This indicates that the improved linear response may be a result of the immobilized catalase/ BSA layer forming an additional restriction to glucose diffusion as well as providing additional oxygen.

It is possible that the presence of blood cells at the membrane surface slows diffusion by partially occluding pores. This was demonstrated when 6 micron latex particles suspended in serum caused a reduction in the enzyme electrode response to glucose. A similar effect was noted when blood cells were added to plasma. The hematocrit effect was small within the hematocrit range compatible with human life. It was concluded that with the system corrected for a mean normal hematocrit, even in abnormal cases, would produce an error in accuracy of less than 3% due to the variations in sample to sample hematocrit. The comparison of whole blood and plasma results presented in Figure 12 confirmed this conclusion.

The accuracy of the ESKALAB in undiluted serum was equivalent to the GA2 and ABA 100. Comparison of the ABA 100 and GA2 results by linear regression analysis indicated an acceptable correlation coefficient, but a considerable proportional bias was noted. The ESKALAB results correlated well with either instrument, but less bias was noted when the ESKALAB was compared with the ABA 100.

The ESKALAB ECS will provide the user with accurate glucose results from undiluted whole blood, plasma, and serum. The ability to use sodium heparin anticoagulant permits the same sample to be used also for sodium and potassium measurements by the ESKALAB. There is no need to measure sample hematocrit because a correction for individual sample hematocrit is not necessary. Results can be obtained with the minimum sample handling possible.

6. References

(1) C. Bourdillon, J.P. Bourgeois, and D. Thomas, *J. Am. Chem. Soc.,* **102:12**, 4231 (1980).

(2) A. E. G. Cass, et. al., *Anal. Chem.,* **56**, 667 (1984)

(3) M. Mascini, M. Iannello and G. Palleschi, *Anal. Chim. Acta*, **146**, 135 (1983).

(4) R. Chirillo, G. Cainaro, B. Pavan, and A. Pin, *Clin. Chem.,* **25:10**, 1744 (1979).

(5) L. Gorton and L. Ogren, *Anal. Chim. Acta*, **130**, 45 (1981).

(6) D. R. Thevenot and R. Sternberg, *Anal. Chem.*, **51:1**, 96 (1979).

(7) D. P. Newman, *United States Patent no.* **4,073,713** (1978).

(8) G. G. Guilbault, *Enzymatic Methods of Analysis*, Pergamon Press, New York, NY, 1970.

(9) H. L. Pardue. In *Advances in Analytical Chemistry and Instrumentation*,**7**, C. N. Reilley and F.W. McLafferty, Eds., Wiley-Interscience, New York, NY, 1969.

(10) G. Palleschi, et. al., *Analytical Biochemisty,* **159,** 114 (1986).

(11) *YSI Model 24 Glucose Analyzer Instruction Manual.* Yellows Springs Insrument Co., Inc., Yellow Springs, Ohio.

(12) W. Dritschillo and M. K. Weibel, *Biomedical Medicine,* **9,** 32 (1974).

(13) L. Gorton and Khan M. Bhatti, *Anal. Chim. Acta,***105,** 43 (1979).

(14) U.S., European, and Japanese patents have been applied for.

(15) G. Guilbault and G. Lubrano, *Anal. Chim. Acta,* **64,** 439 (1973)

(16) G. Lubrano, unpublished communication.

DIFFUSION LIMITED ENZYME ELECTRODES

P. Vadgama
Department of Clinical Biochemistry
The Medical School
Newcastle University
Newcastle upon Tyne NE2 4HH
UK

ABSTRACT. Amperometric enzyme electrodes for glucose, lactate and urate based on oxidases have been studied. To increase the linear range of these devices and to improve biocompatibility, porous polymeric covering membranes treated with organosilanes were used. The membranes had relatively high permeability to O_2 co-substrate and enabled substrate assay at low background pO_2. The linear range of glucose electrodes was extended from 3mM to 30mM and the correlation for whole blood with a routine spectrophotometric method was y = 1.002x + 0.04, r = 0.998 (n = 44); for lactate the range was extended from 0.5mM to 18mM with a correlation of y = 0.9x + 0.11, r = 0.0966 (n = 46); for urate the range was extended from 0.5mM to 2.4mM with a correlation of 1.04x - 0.03, r = 0.98 (n = 40). Miniature needle-type devices were also constructed by embedding 25 - 125 µM Pt wires into stainless steel tubing (0.2 - 1.1mm o.d.) and dipcoating the requisite membrane laminates.

1. INTRODUCTION

1.1. Background

Enzyme electrodes [1] provide one of the most powerful illustrations of the benefits which can accrue from using immobilised enzymes for analysis. The close apposition of an immobilised enzyme layer to a detector has brought about a generation of electrochemical transducers capable of producing a direct signal in the presence of complex, organic analytes. The extension of the analytical repertoire of both potentiometric and amperometric electrodes in this way has renewed the interest in the use of these electrochemical devices for clinical measurement.

Hitherto, electrochemical techniques have made little impact in the hospital biochemistry laboratory, and measurement has been restricted to that of simple ions and gases. Most notably, practical assay has been made possible because of the ability of the enzyme to effect substrate conversion under very mild solution conditions and with a degree of selectivity that surpasses that of other analytical reagents.

G. G. Guilbault and M. Mascini (eds.),
Analytical Uses of Immobilized Biological Compounds for Detection, Medical and Industrial Uses, 359–377.

The commonly expressed concern about the high cost of enzymes as reagents in Clinical Biochemistry ceases to be a serious consideration here, since not only is the enzyme present as an immobilised, reusable component, but the small amounts required to construct an enzyme electrode with its relatively small sensing surface area makes it economically feasible to use even the most expensive and exotic of enzymes. This contrasts with the situation for enzyme column reactors, sometimes used in conjunction with chemical or thermal transducers, which require large quantities of the loading enzyme.

The key to successful practical deployment of enzyme electrodes lies in the initial, careful choice of the enzyme / sensor combination. If there are inadequacies at this stage in matching the two elements, then there is little chance of a practical system for clinical use emerging. This applies as much to sophisticated mediator - based systems as to the simple classical enzyme electrodes where the electrochemically active product of the enzymic reaction is usually detected:

$$\text{substrate} \xrightarrow{(a)} \underset{\text{(electrochemically active)}}{\text{product}} \xrightarrow{(b)} \text{electrode signal} \qquad (1)$$

(a) and (b) here constitute important control points for overall device construction, and are the focus of attention of most enzyme electrode studies. (a) represents the step for enzyme mediated substrate conversion and can be modulated according to the amount of enzyme loading, the solution variables that influence enzyme activity and to barrier membranes and materials used with the enzyme matrix to control substrate access to the enzyme molecule. The most frequently studied parameters, as in classical enzymology, are those which relate to the solution variables, and include pH, ionic strength, co-substrate concentration as well as the inclusion of enzyme activators and inhibitors. Sample stirring and viscosity can, of course, alter substrate delivery to the enzyme layer, but these are primarily problems associated with measurement in biological fluids rather than useful means of modifying electrode performance. Less commonly, covering polymeric membranes are used to modify the effective activity of an enzyme layer, and this is the approach to be described in this paper. Control point (b) is as yet much less amenable to manipulation, and involves factors such as alteration in configuration or the surface chemistry of the detector. Thus, for an amperometric device the working electrode material may be varied, a modified electrode may be used or the polarising voltage may be altered, eg by pulsing or resetting at a lower level to increase electrochemical selectivity. Here also a membrane may be used to effect control over the magnitude and selectivity of the final signal; the most important example of this is the use of a barrier membrane which has a low molecular weight cut-off ($\sim$200 M.W.) to screen out organic electroactive interferents in blood such as ascorbate [2] while allowing ready access to H_2O_2 generated in oxidase - catalysed reactions.

1.2. Operating Principles

Product release into bulk solution from an immobilised enzyme is mainly of interest to the process engineer. As regards enzyme electrodes, product concentrations achieved inside the enzyme layer are of greatest interest. In a given volume element of an enzyme layer, the concentration of product [P] is governed by Michaelis Menten kinetics and Ficks Law's of diffusion. The rate of change of product concentration is given by

$$\frac{\partial [P]}{\partial t} = Dp \frac{\partial^2 [P]}{\partial x^2} + \frac{k_2 [E][S]}{K_m + [S]} \qquad (2)$$

Where t is time, Dp is the diffusion coefficient for the product, x is the distance from the underlying sensor, k_2 is the second forward rate constant of the Michaelis Menten reaction scheme, K_m is the Michaelis constant and S is the substrate for the enzyme. Diffusion and reaction have opposing effects on the concentration of product (and substrate) and after a short lag phase, a steady-state concentration profile is achieved through the enzyme layer at which stage diffusion and reaction balance. An important analytical consequence of this is that despite the continued enzymic reaction, an enzyme electrode is able to achieve a steady-state response. For different substrate levels, different steady-state product profiles are attained, and the enzyme electrode thus operates as a substrate-dependent transducer capable of following substrate concentration on a continuous basis.

For classical amperometric devices used without mediators, product (or co-substrate) is electrochemically degraded at the underlying sensor, and so concentration profiles through the enzyme layer are distorted. Thus electrochemical degredation reduces the concentration of the detected species at the sensor surface, until at a sufficiently high polarising voltage surface concentration becomes zero and the flux of the electrochemically active species to that surface reaches a maximum. It is this rate of flux which determines the size of the electrochemical signal, given by the Cottrell equation :

$$i = nFAf \qquad (3)$$

where i is the current generated, n is the number of electrons transferred per molecule, F is the Farady constant, A is the area of the electrode and f is the flux of the electroactive species. For an amperometric enzyme electrode with high enzyme loading, some simplifying assumptions can be made [3], and the current output can be regarded as approximating to that of a non-enzymic sensor responding to an electroactive species, but with an intervening diffusion limiting covering membrane of thickness L :

$$i = \frac{nFAD_s}{L} [S]_b \qquad (4)$$

Where D_s is the substrate diffusion coefficient and $[S]_b$ is substrate concentration in bulk solution. Since i is proportional to $[S]_b$, the net outcome is that there remains a simple, linear relation between bulk substrate concentration and the electrode signal. If this were the case for the entire measurement range of enzyme electrodes then direct assay of biological fluids would become readily possible. However, the upper limit of linearity is restricted in practice by enzyme K_m such that at $[S]_b \gg K_m$ the electrode signal becomes zero order with respect to $[S]_b$ and the enzyme electrode is then no longer analytically useful. One mitigating factor is that the constraint of substrate diffusion to the enzyme membrane from bulk solution lowers the local, effective substrate concentration and raises apparent K_m. In general, however any major extension of the linear range necessitates either modification of the enzymic reaction [4] or, as outlined in this paper, a more drastic control of substrate/co-substrate diffusion.

1.3. Classical Oxidase-Based Enzyme Electrodes

Oxidase enzymes with their ability to consume O_2 and to generate H_2O_2 stoichiometrically have provided ideal catalysts for enzyme electrode studies using various Clark-type oxygen electrodes and H_2O_2 sensors. The gas-permeable membrane of the Clark electrode offers the advantage of conferring almost total selecticity, but the need then to correct for background pO_2 variation in blood has meant that a second, non-enzymic, O_2 sensor is required. The attendent problems of operating a differential amplifier circuit and compensating for variable drift and mis-match of dual electrodes do not bode well for practical use. However, twinned galvanic oxygen sensors were used by Layne et al [5] for blood glucose monitoring on an experimental basis employing glucose oxidase; the report well illustrates the practical difficulties of biochemical monitoring. One key problem was that of non-linear calibration, due almost certainly to the rate-limiting effect of oxygen co-substrate on the enzymic reaction. Updike et al [6] subsequently showed that useful linearity could be attained by placing a nylon membrane over the enzyme layer such that the (relative) amount of glucose reaching the enzyme layer was reduced; some difficulty was experienced, however, in obtaining reproducible membranes. Romette et al [7] eliminated background pO_2 effects on the initial slope of the glucose electrode response by employing gelatin - immobilised enzyme and carried out whole blood assay in the kinetic mode. Gelatin enhanced the membrane O_2 concentrations, but the steady-state response appears to remain vulnerable to sample pO_2 and also linearity is not significantly improved [8].

Selectivity approaching that of a Clark O_2 electrode was made possible at an H_2O_2 sensor with the development of cellulose acetate low molecular weight cut-off membranes [2], and practical blood measurements were then possible with a single electrode. The most potent illustration of the use of this electrode is its deployment in an artificial pancreas [9]. Here the output from an extracorporeal sensor is used to control the peripheral venous infusion of insulin (or dextrose) whereby blood glucose levels in patients with diabetes are stabilised. This particular sensor has an outer protective polycarbonate membrane to avoid fouling of the enzyme membrane by macromolecules, platelets, white cells and red cells in whole blood. However, despite such a modified surface, fouling causes electrode drift which is only partially reduced by employing a specially formulated diluent buffer having antiplatelet activity. Nevertheless, such an extracorporeal electrode provides a relatively accessible means of continuous patient monitoring. This is particularly so as the electrode does not then need to be sterilised or miniaturised and may be used under optimised solution and flow conditions. In order to eliminate the need for controlled sample presentation, to permit intravascular or tissue monitoring, Shichiri et al [10] devised a needle-type glucose electrode with a special low permeability outer polyurethane membrane. This miniature device showed considerable independence from external variables, and served to highlight the way in which enzyme electrodes might be used for invasive metabolite monitoring in future.

Numerous other oxidase-based electrodes have been reported; the ones most relevant to this paper and likely to find ready clinical application are electrodes for urate [11] and lactate [12]. The work described here demonstrates how such oxidase-based enzyme electrodes might be adapted for whole blood assay by control of substrate diffusion with appropriate covering membranes.

2. MATERIALS AND METHODS

2.1 Apparatus

A Radiometer (Copenhagen, Denmark) E 5046 PO_2 electrode was used for the extracorporeal and continuous flow studies. The electrode has a 20 µm platinum working and Ag/AgCl reference electrode, and was used in conjunction with a Radiometer PHM 72b Acid-Base Analyser with measurements recorded continuously on a strip chart recorder. For detection of O_2, the polarising voltage was -0.6V and for H_2O_2, +0.6V was used. When used in the continuous flow mode the electrode was housed in a Radiometer D616 measuring chamber and connected via transmission tubing to an Auto Analyser proportionating pump (Technicon Instrumentation Co, NY, USA. For continuous flow measurements using needle enzyme electrodes, a Perspex cylinder with a 0.5-1mm diameter flow channel was fabricated (Fig 1).

For discrete blood measurement a Rank oxygen electrode system (Rank Brothers, Bottisham, Cambridge, UK) which had a 2mm diameter platinum working electrode polarised at +0.65V against a silver reference was used.

2.2. Reagents

Buffer components, solvents and other standard reagents were obtained from BDH, Poole, Dorset and were of analytical reagent grade. Singly distilled water was used for making up solutions. Glucose oxidase (EC 1.1.3.4 from Aspergillus niger) was obtained respectively from Sigma, Poole Dorset (17 U/mg) and Boehringer, Mannheim (210 U/mg or 270 U/mg). Lactate oxidase (EC 1.1.32 from Pediococcus sp 12 U/mg) was obtained from Toyo Jozo Co, Tokyo. Lactate oxidase (EC 1.13.12.4, lactate 2 - mono-oxygenase from Mycobacterium smegmatis, 15 U/mg) and uricase (EC 1.7.3.3. from Aspergillus flavus 15 U/mg) were obtained from Boehringer, Mannheim. Cellulose acetate was obtained from Aldrich Chemical Co. (Milwaukee, WI, USA), polyethersulphone polymer was the gift of Dr G Scott, ICI Chemicals and Polymers Group (Runcorn, UK), polyurethane polymer precursors were the gift of Baxenden Chemical Co Ltd (Accrington, UK) and polyurethane polymers were the gift of B F Goodrich Chemical Co (Belgium).

2.3. Membranes and Electrode Materials

Oxygen - permeable polypropylene membranes of 37 μm thickness were either Radiometer Type D604 or Corning Scientific Instruments (Medfield, USA) cat no 477 367. Dialysis membranes (Cuprophan flat dialysing membrane 13.5 μm thickness) were obtained from Enka Glanzstoff AG (Wuppertal, FRG) Millipore ultrafiltration membranes were type HA, 0.22 μm pore size from Millipore Ltd (London, UK). Polycarbonate membranes (0.05 μm pore size) were from Nuclepore (Pleasanton, CA, USA).

Platinum wire of 25-125 μm thickness and stainless steel tubing of 0.2 - 1.1mm o.d. were purchased from Goodfellow Metals Ltd, Cambridge, UK.

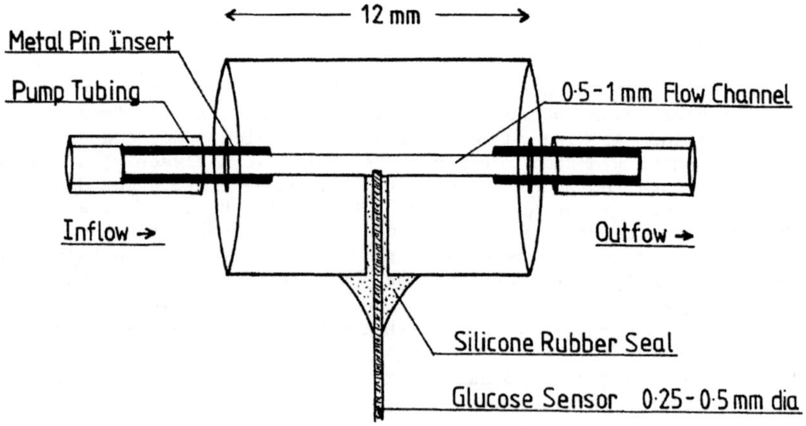

Fig 1 Miniaturised Perspex flow cell used with needle glucose sensor.

2.3 Membrane Preparation

Glucose oxidase, uricase and lactate oxidase, respectively, were cross-linked with albumin using glutaraldehyde according to a standard procedure [13]. Typically, in the case of 210 U/mg glucose oxidase 10 µl of a solution of enzyme (30 mg/ml) and albumin (200 mg/ml) were mixed with 5 µl glutaraldehyde (5% w/v) on a glass microscope slide and then applied either to the tip of an electrode, pre-mounted with membrane and nylon spacer, or applied to one side of a silanised polycarbonate membrane. Alternatively, in the case of lactate 2 - mono-oxygenase the enzyme was retained within a Millipore ultrafiltration disc by pre-soaking the disc in enzyme solution then allowing the disc to dry before mounting over an electrode and covering over with a Cuprophan membrane.

Silanised polycarbonate membranes were prepared by immersing membranes for 30 s in 2-5% v/v of methyltrichlorosilane or phenyltrichlorosilane solution made up in 1,1,1-trichloroethane. Membranes were then compressed between two glass slides and the solvent allowed to evaporate. Dimethyldichlorosilane and methyldichlorosilane solutions (2% v/v) were applied dropwise to polycarbonate or Cuprophan membranes placed flat on glass slides. Before use with any enzyme electrode, membranes were rinsed in a jet of water and allowed to dry at room temperature [14].

Cellulose acetate membranes were prepared by depositing 1.2% W/V cellulose acetate in acetone on a glass surface and allowing the solvent to evaporate; the formed membranes were left for at least one hour before use. Typically, membranes of 4-10 µm thickness were produced by applying 1.2 ml of cellulose acetate solution to a 5 x 5 cm surface. The membranes were used in conjunction with enzyme activated polycarbonate.

2.4. Needle Electrode Fabrication

Platinum wire was heat-sealed into glass capillaries and mounted within stainless steel tubes, which acted as the pseudoreference electrode; this furnished the basic H_2O_2 sensor. In an alternative design, Teflon spray-coated platinum was used as an insulator instead of glass (Fig 2). The needle tip was then sectioned and a smooth sensor surface ensured by diamond polishing.

Membrane laminates were deposited over the needle tips by a sequence of dip-coatings. First an inner perm-selective membrane was deposited using 4% polyethersulphone in dimethylsulphoxide, with solvent evaporated off in a vacuum oven at 40°C for 4 hours; next the enzyme was crosslinked over the resulting surface; finally an outer polyurethane layer was applied using a 6 - 10% solution of polyurethane in tetrahydrofuran. It also proved possible to mount prefabricated laminates of polycarbonate, enzyme and cellulose acetate over the needles.

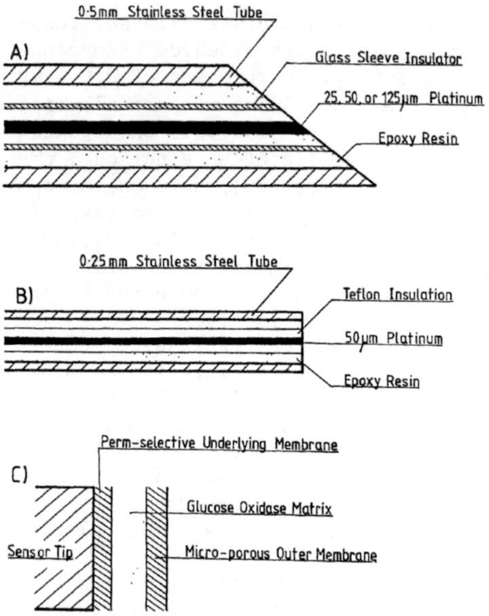

Fig 2. Design of needle enzyme electrodes using A) glass insulation. B) Teflon insulation. A dip-coated laminate is shown in C.

2.5. Extracorporeal Monitoring

To permit sampling of anticoagulated blood without systemic heparinisation, blood was withdrawn from a forearm vein by means of a double lumen cannula [15]. Sterile heparin in isotonic saline was infused into the cannula at a rate of 1.8 ml/h and a mixture of blood and heparin aspirated at 3 ml/h to give a net blood loss of 1.2 ml/h. The anticoagulated sample was then diluted in pH 7.4 phosphate buffer and air equilibrated in an AAI flow system to deliver a final dilution of 1 in 42.5. Two types of electrode flow cell were employed; initially an annular chamber was used where the entire electrode surface was exposed to sample (Fig 3A) and subsequently a narrow flow channel was created over the electrode (Fig 3B).

3. RESULTS

3.1. Extracorporeal Glucose Monitoring

Glucose electrodes housed in the annular flow chamber (Fig 3A) had response times of ~ 2.5 min at sample flow rates of 36 ml/l, and gave

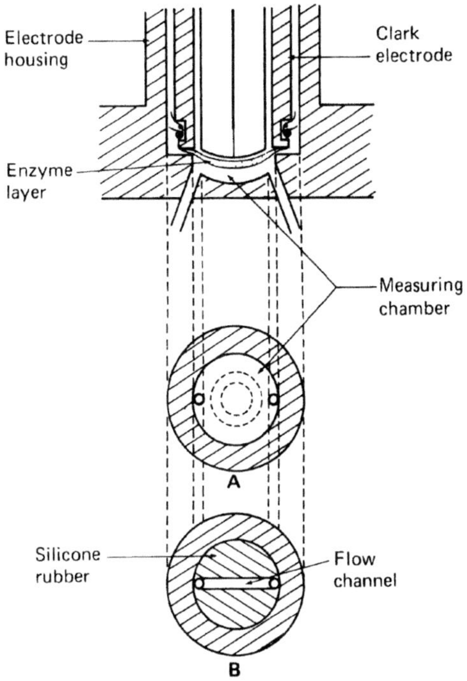

Fig 3. Electrode measuring chamber (A) unmodified (6mm diameter) (B) modified to give a 0.6mm diameter residual flow channel.

linear readings to 5mmol/l glucose. Relatively high enzyme loadings, obtained with thick (60 µm) enzyme layers, produced large glucose signals which showed only minor interferent effects (Table 1).
Bias due to this interference was found to be small in whole blood (duluted 1 in 50) when results were compared with a routine spectrophotometric assay; a correlation of $y = 0.94x + 0.13, r = 0.97$ ($n = 20$) was obtained. Use of the electrode for the extracorporeal monitoring of a fasted normal subject indicated that blood levels could be monitored reliably, and that response times were sufficiently short to allow short-term fluctuations in blood glucose levels to be detected (Fig 4). However, while stable readings were obtained in blood, any exposure thereafter to standards led to cumulative losses in sensitivity. The magnitude of the effect depended upon the duration of exposure to blood : typically a 1.2% loss was incurred after 3 min and 9% after 20 minutes. Examination of the external (Cuprophan) membranes of such electrodes by s.e.m. showed surface deposits of fibrinogen and platelets. Changes made to pH, ionic strength and the specific ion

TABLE 1

Electrochemical interference at a Clark Electrode with a blank membrane

Sample		[a]Interference
0.1 mmol/l ascorbate		4%
0.4 mmol/l urate		
Serum	(n = 6)	6 - 14% (mean = 9%)
Plasma	(n = 6)	4 - 12% (mean = 10%)
Whole blood	(n = 6)	4 - 10% (mean = 6%)

[a] Expressed as percentage of enzyme electrode response at 5 mmol/l glucose.

composition of buffers had no influence on the loss of sensitivity during the restandardisation. Similarly there was no effect on adding albumin, heparin dextran or platelet antagonists (viz acetyl salicylate, sodium salicylate, caffeine, adenosine) to buffers. Changes of the outer membrane similarly did not ameliorate the problem; loss in

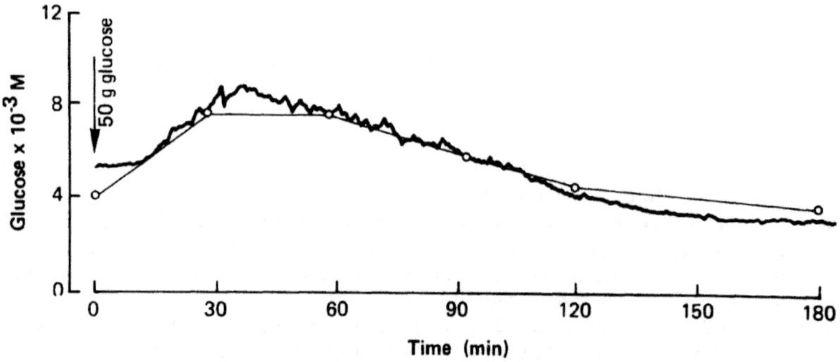

Fig 4. Glucose monitoring in the fasted subject after an oral glucose load. (o) blood assayed by a spectrophotometric method.

sensitivity was incurred when polycarbonate membranes were used (0.003 μm, 0.05 μm pore size) or when albumin crosslinked enzyme was exposed directly to blood. Substantial improvement in stability with blood was found, however, when dimethyldichlorosilane-treated Cuprophan was used; 20 minute exposure to blood led to only a 2% drop in sensitivity. Though loss in response was again a problem when such electrodes were used in the annular flow chamber (Fig 3A), reliable operation became possible on switching to the narrow channel flow cell (Fig 3B); a calibrating solution now had no effect on sensitivity at 50 min and only a 4% loss was observed at 160 min. The final flow system performed reliably during extracorporeal monitoring [14] and showed an acceptable correlation with the routine method during in vitro analysis (Fig 5).

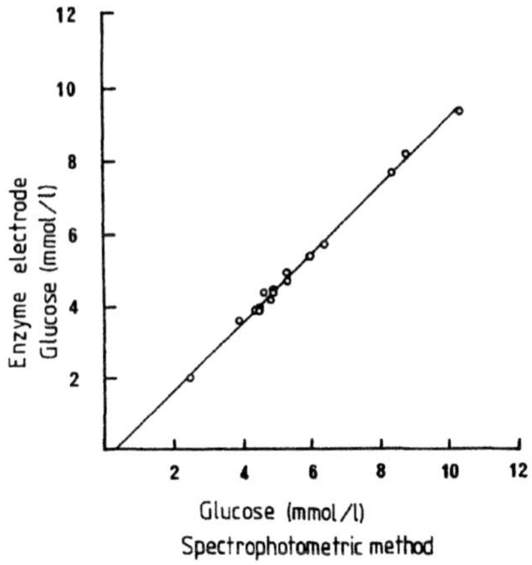

Fig 5. In vitro measurement of whole blood glucose using electrode with silane-treated membrane ($y = 0.94x - 0.21$, $r = 0.997$, $n = 16$).

3.2. Glucose 'drop' Sensor

Glucose enzyme electrodes made with organosilane-coated external polycarbonate membranes had extended linear ranges for glucose. The extent of this linearisation depended upon the particular organosilane used and with extensive treatment linearity increased substantially, though with a tendency towards slower responses (Table 2). Also, the electrode output was entirely independent of sample stirring or viscosity. In unstirred solution responses were stable for at least 1 hour. Inclusion of an underlying, permselective cellulose acetate membrane reduced interference from 0.1 mmol/l ascorbate to 1.9% of a 5 mmol/l glucose signal and from uric acid to $< 0.1\%$. The oxygen dependence of the system was low (Fig 6) and found to be acceptable for glucose assay

TABLE 2

Effect of organosilane treatment on glucose enzyme electrode with 0.05 µm pore size polycarbonate membrane.

Reagent	Maximum linear range (mmol/l)	Response times (t_{98})
Control	5	12 - 30 S
Methyltrichlorosilane	500	30 - 90 S
Phenyltrichlorosilane	500	30 - 90 S
Methyldichlorosilane	50	2 - 5 min
Dimethyldichlorosilane	70	2 - 5 min

of hypoxic blood. Also, sample volumes between 20 µl and 4 ml had no effect on electrode output and there was less than 10% change in the glucose signal over the pH range 6 - 8. Temperature dependence was $< 1.5\%\ ^{\circ}C^{-1}$ between 20 and 30°C; this was thought sufficiently small to allow assay without thermostatting. The electrode had the functional attributes of an electrochemical dip-stick and was therefore employed for assay of small volumes of undiluted whole blood. Initial blood exposure had no effect on electrode sensitivity, though small increments in response were observed after 5-10 measurements. The correlation of results on whole undiluted blood with an automated routine spectrophotometric assay using diluted blood was y = 1.002x + 0.24, r = 0.998 (n = 44).

3.3. Lactate Electrode Based on lactate 2-Mono-oxygenase (EC 1.13.12.4)

Measurement of lactate on the basis of H_2O_2 detection was not possible with the mono-oxygenase enzyme, but reliable calibration was obtained, at least in buffer, with the enzyme membrane mounted over an oxygen sensor. Reproducible calibration was achieved to 0.8 mmol/l lactate and the detection limit was 0.008 mmol/l; this allowed assay of air equilibrated blood samples at 1 in 40 dilution (clinical lactate range 0.3 - 10 mmol/l). The pH profile for the system showed a broad maximum around pH7.2, using 15U enzyme loading and 0.1 mol/l phosphate assay buffer. Even with the liquid membrane used, there was less than 2% change in sensitivity after 15 days of use in aqueous buffer. By contrast responses fell by 5-10% over 15 min after contact with whole blood. Without preconditioning of electrodes in whole blood, recoveries were low (56-83%, n = 11) but improved (83-102%, n = 9) with electrodes maintained in blood for several hours. A major drawback to the use of the electrode for routine blood analysis was the inhibition of the enzyme in the presence of oxalate as the anticoagulant; this precluded the use of standard fluoride/oxalate containing specimen tubes, and instead fluoride/heparin tubes had to be made up.

To avoid the need for a dual electrode arrangement to correct for blood pO_2 variation, a flow system was devised where air was introduced

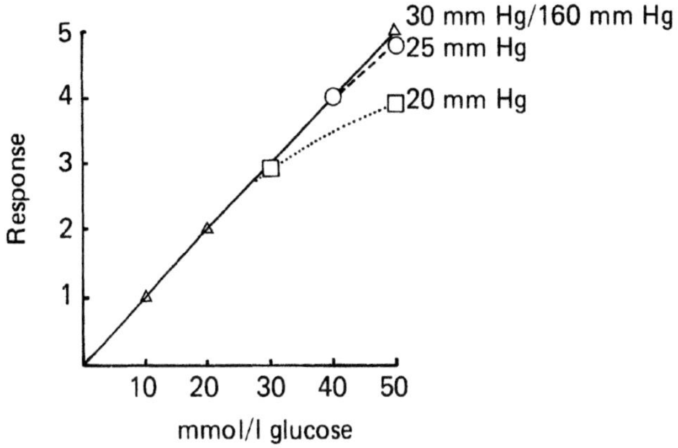

Fig 6. Effect of oxygen partial pressure changes on glucose enzyme electrode calibration. A silanised (methyltrichlorosilane) 0.05 µm pore size polycarbonate outer membrane was employed.

into the sample stream at different locations and equilibrated with sample by means of double mixing coils (Fig 7). A significant delay was inherent in this flow system (4 min from time of sampling) but the overall dynamic response allowed assay of 15-20 samples per hour *in vitro* and was acceptable for extracorporeal monitoring. Plasma lactate values obtained by enzyme electrode correlated with those by fluorimetric assay ($y = 0.978x + 0.085$, $r = 0.952$, $n = 40$) [17].

3.4. Lactate Sensor Based on Lactate Oxidase (Ec 1.1.3.2.)

Use of a peroxide generating lactate oxidase enabled lactate assay of blood along similar lines to that of the glucose 'drop' sensor (3.2). Methyltrichlorosilane treated 0.05 µm polycarbonate membranes enabled extension of linearity from 0.5 mmol/l to 18 mmol/l lactate giving response times of 1-3 min. Calibration curves up to 13 mmol/l lactate were unaffected by pO_2 changes between 16 and 160 mmHg, again indicating that electrodes were usable in hypoxic blood. A 22% increase in signal output was demonstrated on changing pH from 6 to 8 and the mean temperature dependence over 10-30°C was 1.8% $°C^{-1}$. Urate interference was negligible, though at 0.5 mmol/l lactate 0.2 mmol/l ascorbate produced a 10% increase in signal. Initial exposure to whole blood demonstrated no effect on response, and though there was a 20% fall in sensitivity after 3 samples, subsequent repeated use in blood caused no further reduction in current output. The correlation with the standard spectrophotometric assay was $y = 0.96x + 0.11$, $r = 0.996$ ($n = 46$) using undiluted, unstirred blood at an unthermostatted electrode. The within batch coefficient of variation was 1.1% at 2 mmol/l lactate, with a between batch coefficient of 2.8% [18].

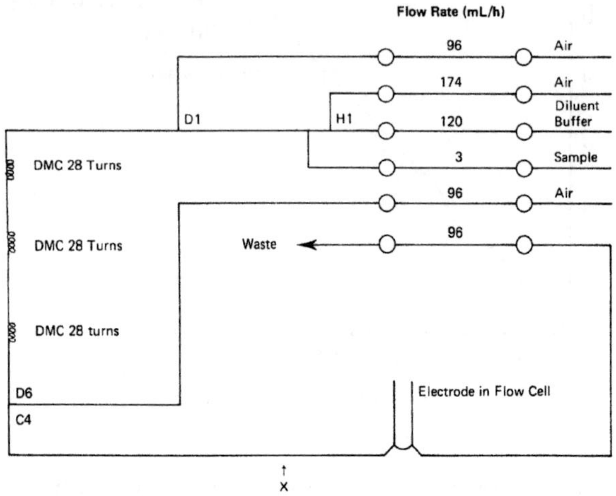

Fig 7. Extended flow system for blood lactate. DMC is a double mixing coil; C4 is a debubbler, D1, D6 and H1 are Technicon glass connectors. Standardisation is possible using calibrating solution introduced at "X".

3.5. Uric Acid Sensor

Spectrophotometric assay of uricase showed that the conditions used for glutaraldehyde crosslinking profoundly affected activity, and that at 12.5% glutaraldehyde activity was virtually abolished. Electrodes made with 1% glutaraldehyde were responsive to uric acid, and with dimethyldichlorosilane treated polycarbonate membranes their linearity was extended from 0.5 mmol/l uric acid to up to 2.4 mmol/l (reference range for uric acid, 0.15 - 0.65 mmol/l). Interference from ascorbic acid was equivalent to 0.04 mmol/l uric acid; no direct electrochemical signal to urate was observed at electrodes which had cellulose acetate membranes. A routine method for whole blood uric acid was not available, but measurements on undiluted serum gave a correlation with the routine, non-enzymic assay of $y = 1.0x - 0.03$, $r = 0.98$ ($n = 40$); there was a high degree of correlation between whole blood and plasma estimation by the electrode ($y = 1.02x - 0.11$, $r = 0.99$, $n = 25$) [19].

3.6. Needle Enzyme Electrodes

Following initial studies with silver plated stainless steel needles [10] it was found that potentials generated at steel itself varied within a sufficiently small range to provide for a reproducible H_2O_2 oxidation reaction at the platinum working electrode. The choice of working electrode surface area was a compromise between the largest current possible and the need for a large reference to working surface area ratio. A 0.25 mm needle with a 50 µm Pt wire provided a stable output even in the flow cell (Fig 1) where only 0.5mm was exposed to solution.

Dipcoated cellulose acetate [10] as a permselective membrane proved unsatisfactory here, as fragile, uneven membranes were formed. More robust membranes were formed using polyethersulphone, which also acted as an effective barrier against interferents. The combined interference with ascorbic acid (0.2 mmol/l), cysteine (0.1 mmol/l), glutathione (1.0 mmol/l) and uric acid (1.0 mmol/l) was $\leqslant$ 0.5 mmol/l glucose.

The activity of the enzyme layer used in a membrane laminate had little effect on signal size at enzyme loadings of 30 - 10000 U/ml glucose oxidase. Membrane thickness proved difficult to control with in situ crosslinking, and this had some bearing on response times, though in general these were complete within 3 min.

Membranes of polyhydroxyethylmethacrylate, cellulose acetate, silicone rubber, nylon and various silanes used externally either failed to give an extension of linearity, or virtually abolished the

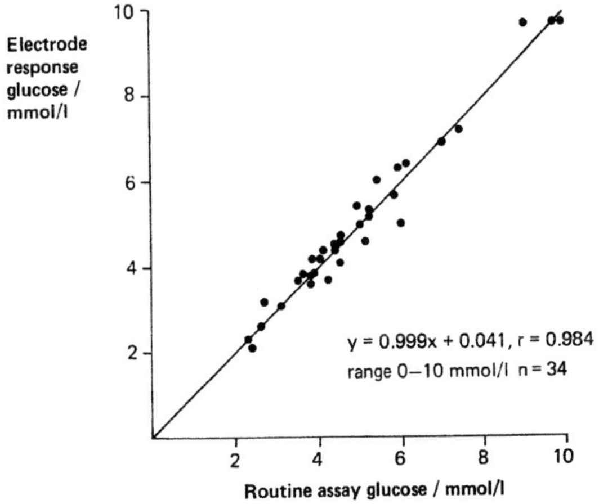

Fig 8. Needle glucose electrode correlation with routine spectrophotometric assay using undiluted blood.

glucose response. There was greater success with the various polyurethanes tried, and best results were obtained with a precursor polyurethane (Baxenden SC 762). Linearity then extended up to 70 mmol/l glucose, and readings became independent of sample stirring. With the covering polyurethane layers, the oxygen dependence of the needle electrodes was reduced sufficiently to allow in vitro assay of undiluted venous blood. The correlation with the routine method was $y = 0.954x + 0.202$, $r = 0.991$, ($n = 48$) over 0-20 mmol/l glucose and $y = 0.999x + 0.041$, $r = 0.984$ ($n = 34$) over 0-10 mmol/l glucose (Fig 8).

In some instances exposure to blood led to a loss in the signal level; this was particularly likely for electrodes with an uneven membrane surface. Dipcoated polyurethane membranes were found to have a microporous structure under the scanning electron microscope and pores of up to 2 µm diameter were observed [20]. Though homogeneous membranes can be produced, these are impermeable to glucose.

Use of needle glucose electrodes in miniature flow cells (Fig 1) enabled rapid assay of blood at 1 in 3 dilution using sample volumes down to 50 µl; whole blood correlation was $y = 0.97x + 0.024$, $r = 0.987$ ($n = 24$, glucose range 0-16 mmol/l).

4. DISCUSSION

4.1. General Comments

Substantial efforts have been made in the past to enhance the sensitivity of enzyme electrodes for metabolite analysis. The results of our studies indicate that sensitivity is adequate for blood glucose, lactate and uric acid, simply using commercially available enzymes. Rather than extending the detection limits, by far the most important challenge is that of modifying a reagent demonstrating saturation kinetics to provide for a non-saturating, linear signal. One approach is to alter the enzymic reaction such that (apparent) Km is raised. For oxidoseductases this is often achieved by using electron mediators [4]. We chose to reduce substrate entry into the enzyme layer and to extend linearity at the expense of sensitivity. The advantage of this latter approach is that no special demand is made on enzyme chemistry; a possible weakness lies in a tendency to increased electrode response times. However, 'acceptable' response times in the clinical context remain undefined entities for either in vivo or in vitro analysis; for the former some have found even 10 minutes to be tolerable [5]. The response times we have achieved are comparable to other simplified analytical systems. A further benefit accruing from a diffusionally restricted system may well be that the impact of any additional barrier, such as the cellular fouling layer which develops in biological fluids, is diminished.

4.2. Extracorporeal Monitoring

Continuous, extracorporeal monitoring [8,12] has provided a valuable means of following short-term metabolite fluctuations in critically ill patients, and has allowed the institution of therapy on a more efficient basis than hitherto possible with intermittent sampling and analysis. While there are practical disadvantages to using a cumbersome extracorporeal flow system, the fine control possible over assay conditions has enabled many electrodes to be used which are intrinsically unsatisfactory for bedside measurement. The one problem which has proved difficult to resolve is the loss in response arising from surface fouling of electrodes. The main initiators responsible for such loss are platelets and fibrinogen [9]. Our observation of a loss in sensitivity when electrodes are transferred from whole blood to standard [8] is likely to be due to the labile nature of this predominantly platelet fouling layer which remains responsive to a change in the solution environment. It is probable that the plasma matrix of blood helps stabilise platelets; thus, platelets suspended in simple aqueous solution show increased tendency to adhere to foreign surfaces [21], and subsequent surface behaviour with regard to spreading and degranulation may be triggered by such environmental change. In reality, the enzyme electrode comprises one additional outer membrane over which the analyst has little direct control : the biologically deposited layer of proteins and cellular elements.

The success of silane treatment in improving haemocompatibility is likely to have been the result of a retardation in surface coagulation; clotting time is known to be increased at silane-treated glass [21]. The precise effect on platelets is unclear, certainly platelet degranulation seems to be unaffected by silane [22]. Hydrodynamic shear forces are a potent trigger for platelets, and the increased stability of the glucose electrode in the narrow channel flow cell was most probably due to more laminar flow.

The flow system for lactate has demonstrated that a single O_2-based enzyme electrode can be employed for continuous measurement. Although hydrogen peroxide is an intermediate of the enzyme reaction [23], free quantities were not detectable [24]. The system devised here for lactate should be adaptable to other oxidases where free H_2O_2 is not detected. The drift observed in blood with this electrode was again the manifestation of an interfacial reaction with blood, and one which is likely to be refractory to refinements in electrode chemistry alone.

4.3. 'Drop' Sensors for *in vitro* Use

Organosilane-treated membranes for the *in vitro* glucose, lactate and urate electrodes had the combined advantage of improving haemocompatibility and extending linearity[25].In part, the extension of the linear range was due to the high permeability of organosilane to oxygen co-substrate; oxygen limitation for the reaction is likely at even modest glucose levels since O_2 is normally present at low concentrations in solution ($\sim$ 0.15 mmol/l). One further factor helping to maintain the pO_2 level in the enzyme membrane is the oxygen regenerated at the working electrode via electrochemical oxidation of hydrogen peroxide :

$$H_2O_2 \longrightarrow O_2 + 2H^+ + 2e^- \qquad (5)$$

Ability to measure undiluted specimens is an advantage in any clinical analytical system; thus, the electrodes described here could eventually be considered for use by hospital personnel untrained in laboratory techniques, or for home monitoring by the patient. In addition, errors from the redistribution of analyte between the red cell and plasma compartments on dilution can be avoided. Thus, lactate concentrations in red cell water do not equal those in plasma, and can distort measured,diluted plasma levels.

4.4. Needle Enzyme Electrodes

Needle electrodes are structurally simple and could be used for invasive monitoring with minimum trauma to the patient [10]. Their basic construction, furthermore, does not require sophisticated microfabrication techniques [20] and the requisite membranes can be coated without recourse to complicated techniques. They are therefore a highly acceptable type of electrode for fabrication in a clinical laboratory. The present study has validated in vitro use and effected a functional adaptation similar to that for 'drop' sensors (4.3). It remains to be seen whether needle electrodes will allow accurate measurement in the highly diffusionally restricted environment of tissue. Local distortion of tissue metabolite levels is likely due to substrate consumption by the electrode and it may, for example, be possible to only achieve trend monitoring, as is already the case for transcutaneous pO_2 / pCO_2 measurement in adults. The lack of agreement in reported tissue glucose concentrations using implanted electrodes is some indication of such complicating factors.

The intravascular compartment is analytically more satisfactory. However, the safety hazards posed are more acute, and intravascular probes, regardless of design, pose special risks of infection, vascular injury, clot formation and clot dissemination. More extensive appraisal of needle electrode design is therefore needed before such devices are used intravascularly, and the blood compatibility of the overall electrode structure needs to be improved rather than simply that of the sensor membrane. In any case intravascular, drift-free monitoring is likely to be possible for only short periods; turbulance created in the vicinity of even a highly 'biocompatible' probe will promote cumulative deposition of platelets leading to eventual surface clot formation. It is unlikely that even intramembrane incorporation of anticoagulant (eg heparin [10]) can provide sufficient protection to allow chronic vascular implantation. Additional safety concerns for long-term use include local toxic effects of metal ions from the reference electrode and of polymer / plasticiser materials from the membrane laminates; the antigenicity of the enzyme protein; maintenance of sterility. Functional aspects to be considered include movement artefact in the signal, signal drift and reliability of the power supply. The problems of invasive monitoring thus extend well beyond the realms of analytical chemistry. If current research into needle-type devices being carried out in various centres is to lead to sensors of medical relevance, active research into the biological effects of implantation is vital.

5. ACKNOWLEDGEMENTS

The author wishes to express his gratitude to MRC, SERC and ICI for financial support for this work.

6. REFERENCES

1. Vadgama P (1979) 'Enzyme electrodes' In: Ion Selective Electrode Methodology Vol 12 ed Covington AK, CRC Press, Florida.
2. Clarke WL and Santiago JV (1977) Artif. Organs, **1** 78.
3. Bowers LD and Carr PW (1976) Anal Chem **48** 545A.
4. Cass AEG, Davis G, Francis G, Hill HAO, Aston WJ, Higgins IJ, Plotkin EV, Scott LDL and Turner APF (1984) Anal Chem **56** 667.
5. Layne EC, Schultz RD, Thomas LJ, Slama G, Sayler BS and Besseman SP (1976) Diabetes **25** 81.
6. Updike SJ, Shults M and Ekman B (1982) Diabetes Care **5** 207.
7. Romette JL, Froment B and Thomas D (1976) Clin Chim Acta **95** 249.
8. Hanning I. MSc Thesis, Newcastle University, 1980).
9. Fogt EJ, Dodd IM, Jenning EM and Clemens AH (1978) Anal Chem **24** 1366.
10. Shichiri M, Kawamori R, Yamasaki Y, Hakui N and Abett (1982) Lancet **2** 1129.
11. Nanjo M and Guilbault GG (1979) Anal Chem **46** 1769.
12. Mascini M, Fortunati S, Moscone D, Palleschi G, Massi-Benedetti and Fabietti P (1985) Clin Chem **31** 451.
13. Guilbault GG, (1975) 'Enzyme electrode probes'. In: Immobilised Enzymes, Antigens, Antibodies and Peptides Vol 1 ed Weetall HW Marcel Dekker Inc, New York.
14. Mullen WH, Keedy FH, Churchouse SJ and Vadgama PM (1986) Anal Chim Acta **183** 59.
15. Noy GA, Buckle ALJ and Alberti KGMM (1978) Clin Chim Acta **89** 135.
16. Hanning I, Vadgama P, Covington AK and Alberti KGMM (1986) Anal Lett **19** 461.
17. Weaver MR, Vadgama P (1986) Clin Chim Acta **155** 295.
18. Mullen WH, Churchouse SJ, Keedy FH and Vadgama P (1986) Clin Chim Acta **157** 191.
19. Keedy FH MSc Thesis, Newcastle University, 1986.
20. Churchouse SJ, Battersby CM, Mullen WH and Vadgama P (1987) Biosensors (in press).
21. George GN (1972) Blood **40** 862.
22. Brash JL (1977) Ann. N.Y. Acad. Sci **283** 356.
23. Lockridge O, Massey V and Sullivan A (1972) J. Biol. Chem. **247** 8097.
24. Weaver MR MSc Thesis, Newcastle University, 1983.
25. Vadgama P U.K. Patent Application 8514176, 1985.

ENZYME ELECTRODES FOR SUGAR SUBSTITUTE ASPARTAME

G.G. Guilbault*, G.L. Lubrano*, J-M. Kauffmann** and G.J. Patriarche**
*Universal Sensors, P.O. Box 736, New Orleans, La. 70148
**Free University of Brussels, Pharmaceutical Institute, 1050 Brussels, Belgium

ABSTRACT. An aspartame selective electrode has been developed for the rapid determination of this artificial sweetener in various dry powder mixes. The aspartase enzyme has been immobilized over an ammonia gas sensing electrode. The potentiometric responses of a soluble and a chemically immobilized enzyme device have been evaluated. The latter offers greater stability, more than 7 days, with a linear dependence of the response from 1 mmol/l to 10 mmol/l aspartame. Successful determinations of aspartame in commercially available powder mixes have been obtained with no special sample pretreatment.

1. INTRODUCTION.

Aspartame is an artificial sweetener which possess about 180 times the sweetness of sugar. Structurally the molecule is a dipeptide which is the methyl ester of aspartyl-phenylalanine (Fig. 1). Since the early eighties aspartame has been introduced in several countries as a sweetener of choice for various food products: powdered table top artificial sweetener, diet drink, chewing gum, dairy products. The increased popularity of aspartame has prompted many laboratories to develop fast and accurate methods of analysis. Most of the proposed techniques are chromatographic ones (TLC, GC, LC, HPLC). They allow achievement of relatively good separation of aspartame from its surrounding matrix, and quantitation (1, and references cited therein). In an effort to produce simpler analytical methods, a recent work has been devoted to the development of a microbial electrode for aspartame measurements (2). The sensor operates amperometrically and measures the change in respiration rate (oxygen change) by the uptake of substrate. Successful results have been reported for aspartame within the range 0.07 - 0.6 mmol/l, the sensor being stable for 8 weeks. The selectivity of the method might suffer from the presence, in the sample, of other microbial nutrients.

G. G. Guilbault and M. Mascini (eds.),
Analytical Uses of Immobilized Biological Compounds for Detection, Medical and Industrial Uses, 379–388.

The purpose of this work is to develop an immobilized enzyme electrode specific for the assay of aspartame in food products with no special sample pre-treatment, or requirement for reagents (9).

ASPARTAME

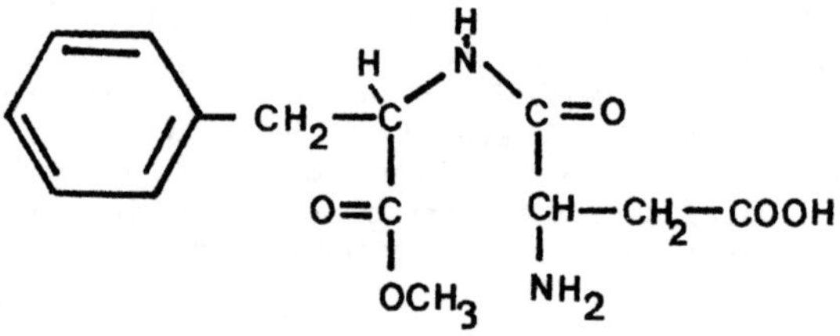

Fig. 1 - Schematic structure of aspartame

2. MATERIALS AND METHODS.

2.1. Reagents.

All chemicals used were of analytical reagent grade. Aspartame, L-aspartic acid and the enzyme L-aspartase E.C 4.3.1.1. from Hafnia alvei were obtained from Sigma Chemical Company. Stock solutions of aspartame and aspartate were prepared in 0.1 mol/l Tris-HCl buffer, pH 8.5 containing 2.5 mmol/l $MgCl_2$. This buffer has been shown to be optimum for the aspartate enzyme electrode (3). The aspartame and aspartate solutions were prepared fresh daily.

2.2. Apparatus.

The pH measurements have been made with a HANNA Instruments pH-meter (Limena-ITALY). The electrode used was an ammonia sensitive electrode set, Type 4000-4 (Radelkis, sold in the U.S. by Universal Sensors, P. O. Box 736, New Orleans, LA 70148). Potentials were monitored with a Corning pH-meter, Model 110, coupled to an Omniscribe recorder through a dc offset module (Schlumberger). The hydrophobic membrane utilized was a Celgrad polypropylene type 2400 (0.02 m). The membranes were sealed on the electrode jackets with rubber type adhesive and o-ring.

2.3. Procedure.

The "soluble" enzyme electrode was prepared by placing approximately 2.5 Units of lyophilized enzyme on the gas permeable membrane of the ammonia electrode. The moist dialysis membrane (m.w. cutoff: 50,000) was used to hold the enzsyme in place close to the gas selective electrode. The electrode was conditioned in buffer (tris 0.1 mol/1, pH 8.5 with $MgCl_2$, 2.5 mmol/1) for 20 hrs before using. Storage of the electrode at room temperature has been realized in 1.0 mmol/1 dithiothreitol solutions.

Enzyme immobilization was carried out by a covalent cross-linking method with glutaraldehyde. The lyophilized enzyme (2.5 Units) was deposited on the dialysis membrane, then 20 µl of phosphate buffer pH 8.5 was added in order to solubilize the enzyme. 2 µl of 6.25% glutaraldehyde is gently added and mixed for 30 sec. The resulting membrane was allowed to dry at room temperature for 2 hrs, and was mounted against the gas selective membrane electrode by taking special care to avoid bubble entrapment. The electrode was washed in a 0.1 mol.1 glycine solution (stirring 1 hr.) then equilibrated in 0.1 mol/1 Tris buffer, pH 8.5 (2.5 mmol/1 $MgCl_2$) for 20 hrs at room temperature before use. All experiments and storage were made at room temperature 25 ± 1^oC. Aspartame and aspartic acid stock solutions (10^{-1} mol/1) were prepared weekly and stored at 5^oC in buffer pH 8.5.

No BSA was used, due to the high protein content (32%, 1.30 U/mg solid) of L-aspartase.

3. RESULTS AND DISCUSSION.

Referring to the aspartame structure, several enzymic systems might be able to convert the molecule into electrode detectable species. In figure 2 are reported various possible ways of using commercially available enzymes for aspartame determination. The methylester can be cleaved by chymotrypsin to give methanol, detectable with a second enzyme, alcohol dehydrogenase. The dipeptide could be cleaved into its amino acid constituents using carboxypeptidase and the molecules subjected to further enzymic reactions. However, the requirement of aspartame splitting might be effected by the pH 8.5 used, thus eliminating some drawbacks encountered with bi-enzyme systems. Assuming that aspartame reacts the same way as aspartic acid and phenylalanine, we may have the general pathway scheme illustrated in figure 2.

Steps I, II, III are direct processes involving only one enzyme and might recognize aspartame. However, limitations are electrode stability as well as low electrode-enzymic activity. Step IV - Seems to be an interesting one due to the high activity of the enzymes involved and to a cycling process increasing the sensitivity of the method. Steps with CO_2 liberation are worthwhile since the enzymes are very stable and active. However, aspartame has to be detected in high CO_2 soft drink levels and it will need a differential operating device using a nonenzymic CO_2 electrode. In step IV, NH_3 can be detected potentiometrically or $NADH_2$ amperometrically using oxidative processes with or without a redox mediator.

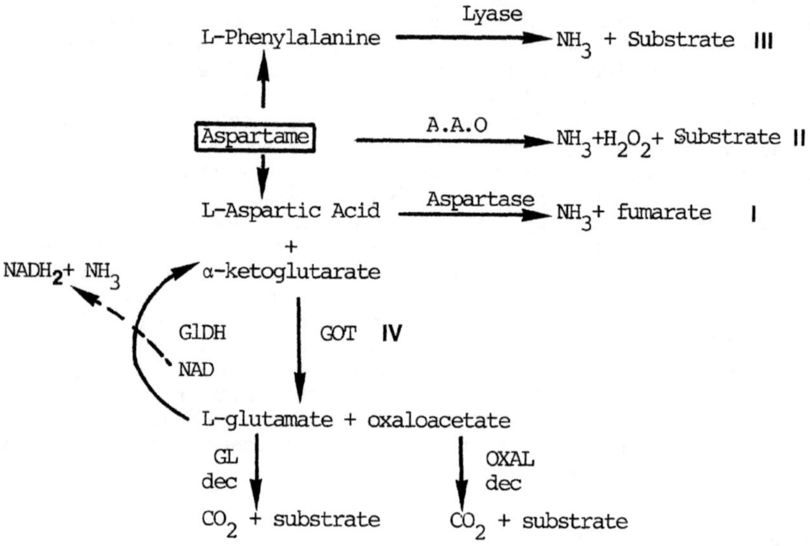

Fig. 2. Ways of using enzymes for aspartame study.

3.1 Aspartase electrode

The presence of an aspartate moiety in the aspartame structure led us to investigate the response of an aspartase immobilized electrode. The choice of step I in figure 2 relies upon the high selectivity achieved by the ammonia gas sensing electrode and by the enzyme aspartase itself (4,5). In reaction I liberation of ammonium ions could be monitored potentiometrically with an ammonium ion selective electrode as already reported for aspartate (6). However, potassium ions interfere, which constitutes a serious limitation wher operating in complex media. Thus, it was more judicious to take advantage of the gain in selectivity achieved with the gas sensing probe. The literature reports one study with an aspartase electrode (3). The soluble enzyme is maintained close to the ammonia electrode, the resulting device detects aspartate successfully within the range 0.3 - 7 mmol/l. The poor electrode lifetime (one day) could be improved to three days by a storage of the enzyme electrode in a 1 mmol/l dithiothreitol solution, a protective reagent for sulfhydryl groups. A bacterial electrode has also been proposed; the device performs similarly with a lifetime higher than 10 days (3).

3.1.1. Aspartase electrode response towards aspartate and aspartame.

Considering the stability of aspartase (max. at pH 7.5) and the Vmax optimal at pH 8.5 (7) and taking into account that the ammonia gas sensing response is higher at alkaline pH, a good compromise of operating pH would be pH 8.0. However, in order to have the best response and assuming that the enzyme immobilization will increase its stability, experiments were performed at pH 8.5.

3.1.1.a. Response of the "soluble" enzyme electrode

In figure 3 is illustrated the response towards aspartic acid (curve a) and aspartame (curve b), under identical experimental conditions, using the standard addition method. The response is linear for aspartic acid within the range of concentrations 0.4 - 6 mmol/l with a slope of 40 mV/decade, a response time of 9 min. and a recovery of base line around 20 min. In case of aspartame, the electrode response is more sluggish with a value of response time equal to 10 min, a slope of 40 mV/decade within a linear range of 1.5 to 8 mmol/l.

As already reported earlier for the aspartase electrode by Kobos and Rechnitz (3), the lifetime is poor; the electrode response decays rapidly after one day and is only slightly extended (3 days) by storing the electrode in protective media for sulfhydryl groups (3). Moreover, within the three days of use, the linear range of concentration becomes progressively narrower. Fig. 4 curve c shows the response after 3 days for aspartame.

3.3.3.b. Performance of the immobilized enzyme electrode.

It is well known that the immobilization of an enzyme tends to produce a diminution in enzyme activity, but a longer useful period of work. As expected, the electrode response is less pronounced to aspartame upon immobilization of aspartase, with a slope of 30 mV/decade but in a longer available range of concentration: 1 mmol/l - 10 mmol/l. The response time is also fairly slow, approximately 12 min. within a complete base line recovery after 28 min. However, in contrast to the "soluble" enzyme electrode, the stability is improved. The electrode responds for periods longer than a week with a slope of

30 mV/decade and a range of concentration comprised between 1 mmol/1-10 mmol/1 aspartame (fig. 5).

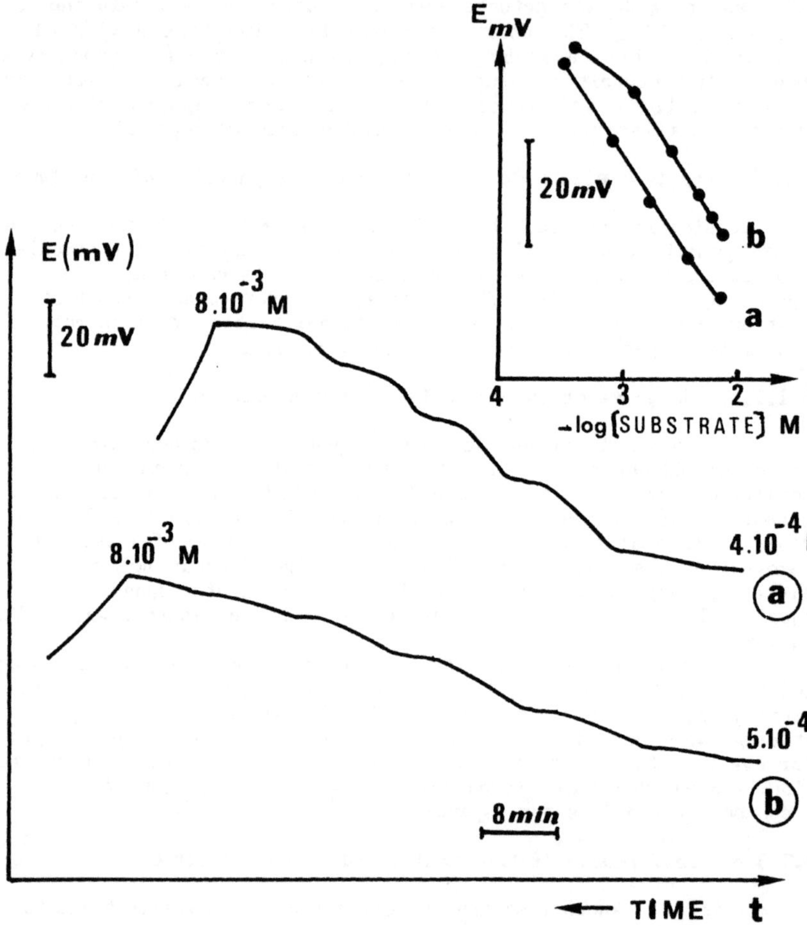

Fig. 3. Response curves of the "soluble" aspartase electrode towards: a, aspartic acid; b, aspartame

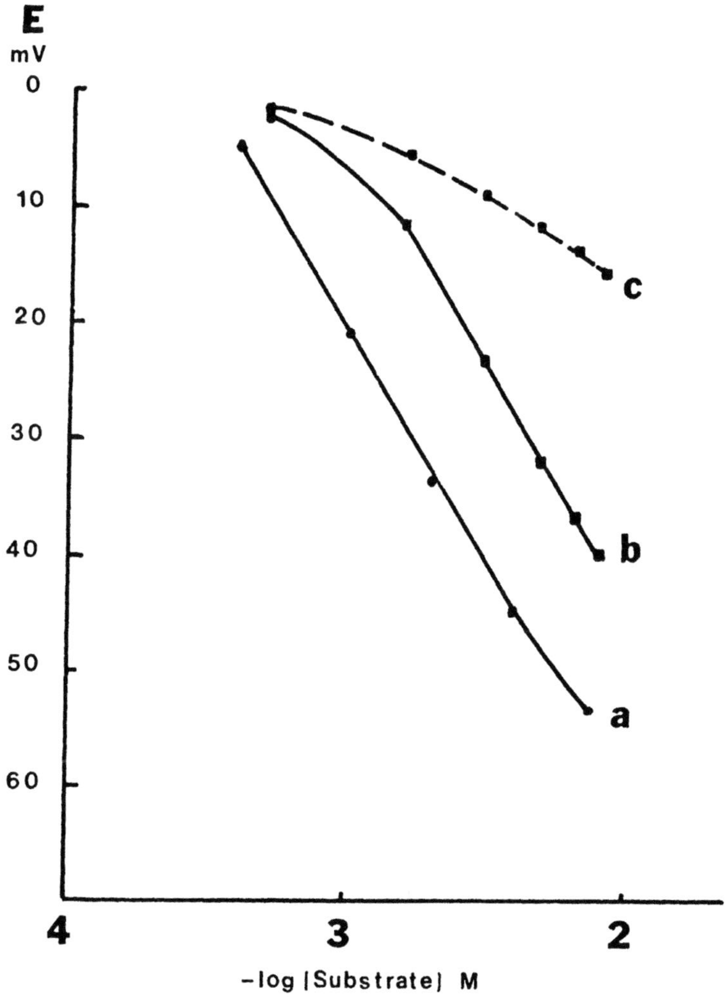

Fig. 4. Soluble electrode response after 3 days towards: a, aspartic acid first day; b, aspartame first day; c, aspartame third day (storage in dithiothreitol)

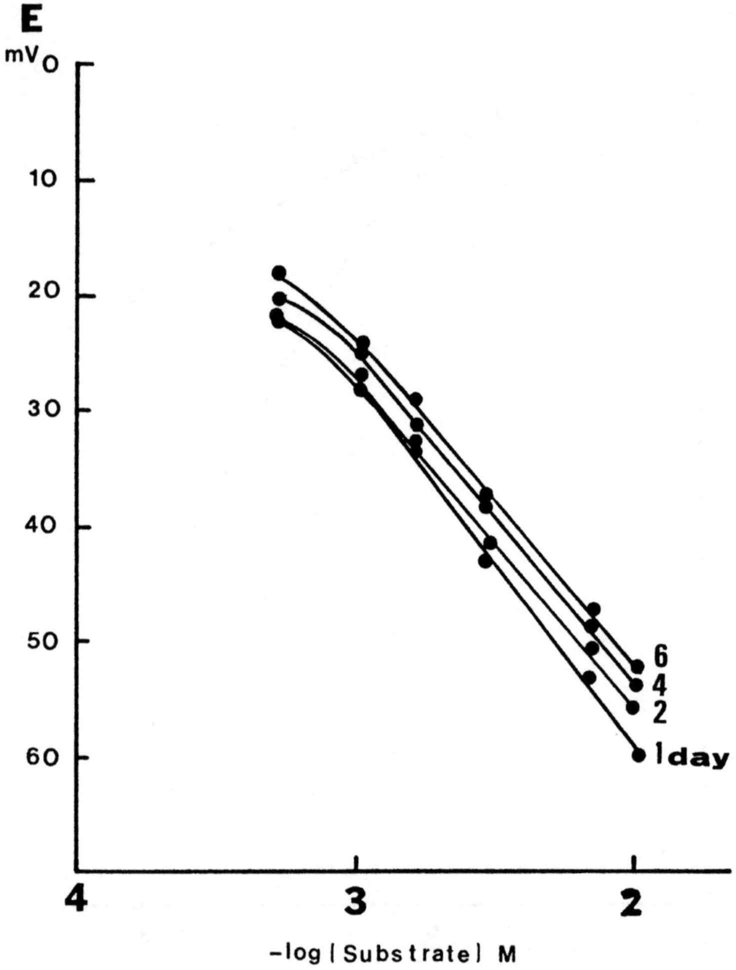

Fig. 5. Immobilized electrode response towards aspartame as a function of time.

3.1.1.c. Determination of aspartame in dry powdered mixes.

In table 1 are summarized some of the data obtained with the various samples analyzed. Measurements have been realized after direct dissolution of the substance in buffer pH 8.5. Filtration was not required, and the electrode response has been compared to the aspartame standard curve. Results were in close agreements with assessed values of aspartame (number of trials = 4, c.v. between 1 and 2%). Measurements realized amperometrically by oxidation of aspartame at +0.7 V vs Ag/AgCl KCl sat. using a glassy carbon electrode were less accurate due to the influence of the other constituents of the sample.

Table 1. Assay of aspartame in various products.

Sample	Assessed g%	Found NH_3*	G.C**
Equal	3.5	3.57	3.76
Canderel a	3.0	3.05	2.98
Canderel b	3.6	3.50	3.20

Equal = aspartame + dextrose
Canderel a = aspartame + maltodextrin
Canderel b = aspartame + lactose

* ammonia electrode
** glassy carbon electrode

4. CONCLUSION.

We have demonstrated the possibility of using an enzyme immobilized electrode for the determination of the dipeptide aspartame in complex media.

The presence of two amino acids in the aspartate molecule permits the construction of various enzyme electrodes, since the enzyme may recognize the amino acid as is the case with aspartase. Electrode rsponse is somewhat slower for aspartame than for aspartate probably due to the larger molecule size, lower diffusion coefficient and affinity of aspartame towards the enzyme. Further studies are in progress to establish the performances of the other enzyme systems reported in figure 2, and additional results will soon be published (8).

5. ACKNOWLEDGEMENT.

Thanks are expressed to the Fonds National de la Recherche Scientifique (Belgium) for help to one of us (G.J.P.), to the SPPS (Belgium Politic Research, ARC), contract n° 86/91-89 (JMK), and the U. S. Department of Agriculture (Grant 86-SBIR-8-0096) for financial support of this research at Universal Sensors.

6. REFERENCES.

1. H.J. Issaq, D. Weiss, C. Ridlon, S.D. Fox, G.M. Muschik, J. Liquid. Chromatogr. 9, 1791 (1986).
2. R. Renneberg, K. Riedel, F. Scheller, Appl. Microbiol., Biotechnol. 21, 180 (1985).
3. R.K. Kobos, G.A. Rechnitz, Anal. Letters, 10, 751 (1977).
4. G.G. Guilbault, Analytical Uses of Immobilized Enzymes, Marcel Dekker, N.Y. (1984).
5. G.G. Guilbault, J-M. Kauffmann, Biotechnol. Applied Biochem., 9, 95 (1987).
6. G.M. Davis, R.S. Haddad, W.R. Hussein, Anal. Letters, 20, 259 (1987).
7. V.R. Williams, D.J. Lartique, Methods Enzym. 13, 354 (1969).
8. G.G. Guilbault, G.J. Lubrano, J-M. Kauffmann, G.J. Patriarche, Anal. Chim. Acta, Submitted for publication, 1987.

AUTHOR INDEX

SUBJECT INDEX

D

E

F

P

R

S

T

U

V

W

Y

GPSR Compliance

The European Union's (EU) General Product Safety Regulation (GPSR) is a set of rules that requires consumer products to be safe and our obligations to ensure this.

If you have any concerns about our products, you can contact us on ProductSafety@springernature.com

In case Publisher is established outside the EU, the EU authorized representative is:

Springer Nature Customer Service Center GmbH
Europaplatz 3
69115 Heidelberg, Germany

Batch number: 10815974

Printed by Printforce, the Netherlands